BIOLOGICAL INHERITANCE

BIOLOGICAL INHERITANCE

AN INTRODUCTORY GENETICS TEXT

W.J.C. Roberts

The Book Guild Ltd.
Sussex, England

This book is sold subject to the condition that it shall not, by way of trade or otherwise, be lent, re-sold, hired out, photocopied or held in any retrieval system or otherwise circulated without the publisher's prior consent in any form of binding or cover other than that in which this is published and without a similar condition including this condition being imposed on the subsequent purchaser.

The Book Guild Ltd.
25 High Street,
Lewes, Sussex

First published 1994
W.J.C. Roberts 1994

Set in Baskerville

Typesetting by Wordset
Hassocks, Sussex

Printed in Great Britain by
Butler & Tanner Ltd.
Frome, Somerset

A catalogue record for this book is
available from the British Library

ISBN 0 86332 729 X

*To beneficial genes
in my immediate family;
and to beneficial genes
everywhere else.*

CONTENTS

Acknowledgements	15
Preface	17
Introduction	19

Chapter 1 – Classic Experimental Work in Genetics.   21

a) The Discovery of the Particulate Nature of Inheritance by Gregor Mendel.   23

b) The Contribution to the Modern Interpretation of Inheritance made by Charles Darwin and Alfred Wallace, following on from Jean-Baptiste Lamarck.   31

c) The Proof that DNA is the Material of Inheritance.   33

d) The Structure of DNA.   41

e) Genetic Mapping in Bacteria by:-
   (i) Interrupted Mating.   56
   (ii) Recombination Frequency.   67

f) The Proof of the 'One Gene – One Enzyme' Theory.   78

g) The Genetic Triplicate Code.   80

h) How The Genetic Code was Worked Out Using Frameshift Mutations.   86

Chapter 1 – QUESTIONS   90

Chapter 2 – The Transcription and Translation of the Triplicate DNA Codes into a Sequence of Amino-Acids in Polypeptides.   91

Polypeptides.	92
Naturally Occurring Amino-Acids.	93
Structure of Polypeptides.	96
Shape of Polypeptide Molecules.	98
Introduction to Polypeptide Synthesis.	102

Ribonucleic Acid (RNA)	102
Transcription	105
Ribosomal-RNA (rRNA)	109
Transfer-RNA (tRNA)	110
Ribosomes	111
The Function of Ribosomes	111
Genes	111
Coding for Amino-acids on the mRNA	112
Starting the Synthesis of a Polypeptide-Translation	113
Elongation of the Polypeptide Chain	116
Ending a Polypeptide Chain	119
Ribosome Movement Along mRNA	120
Summary of Polypeptide Synthesis in E. coli	122
Polypeptide Synthesis in Eukaryotic Cells	124
Ambiguity in Codon/AntiCodon Recognition – the Wobble Hypothesis	125
The Central Dogma	126
Recent Discoveries About Eukaryotic Species	127
Glycopolypeptides in Eukaryotes	128
Chapter 2 – QUESTIONS	130

Chapter 3 – Ribosome Synthesis in the Nucleoli of Eukaryotic Cells 131

Ulstrastructure of Nucleoli	132
The Synthesis of Ribosomes from rRNA and Polypeptides	135
Ribosome Synthesis Scheme	136
The Rate of Ribosome Production	138
The Relationship Between the Activity of the Nucleolus and Other Factors That Control Gene Expression	138
The Replacement of Missing Nucleolus Genes in Mutants During Cell Division	138
The Number of rRNA genes	139
Non-specific Recognition of mRNA by Ribosomes	139
The Nucleoli at Cell Division	139
Transport of Ribosomes	139

Human rRNA genes	142
Chapter 3 – QUESTIONS	142

Chapter 4 – Genes, Super-Genes, Alleles and Super Alleles, with Some Comments on How Genes are Controlled — 143

Genes	143
An Attempt to Define a Gene	148
Problems Related to the Definition of a Gene	149
Super-genes	154
Alleles	155
Super-alleles	156
The Control of Genes	157
Zinc Fingers in Gene-Controlling Polypeptides	160
Genes in the Chromosomes of Plants and Animals	160
Chapter 4 – QUESTIONS	165

Chapter 5 – Inheritance Determined by Chromosomal DNA — 166

Genetic Crosses Depending on Chromosomal Genes	166
The Study of Genetics Using Drosophila melanogaster	172
Drosophila Crosses	175
Crosses in Species Other Than Drosophila melanogaster	189
Chapter 5 – QUESTIONS	207

Chapter 6 – Extra-Chromosomal Inheritance — 209

Eukaryotic Extra-chromosomal DNA	209
Prokaryotic Extra-chromosomal DNA	209
A Pointer to Extra-chromosomal Inheritance	210
Mitochondrial/Chromosomal DNA Relationship	211
The Study of Mitochondrial Inheritance in Yeasts	211
Chloroplast/Chromosomal DNA Relationship	213
Inheritance Controlled by Endosymbiosis in Paramecium aurelia	214
Plasmids in Prokaryotic Cells	216

The Usefulness of Prokaryotic Plasmids	217
Chapter 6 – QUESTIONS	217

Chapter 7 – Some Examples of Inheritance Determined by Extra-Chromosomal DNA — 218

Extra-Chromosomal Inheritance in Prokaryotic Cells – Examples	219
Extra-Chromosomal Inheritance in Eukaryotes – Examples	222
Chapter 7 – QUESTIONS	228

Chapter 8 – Genetic Engineering — 229

Practical Uses for Genetic Engineering	231
Plasmid and Viral DNA Used for Genetic Engineering	234
The Properties of Plasmids That Make Them Suitable for Genetic Engineering	236
The Synthesis of a 'Desirable' Gene and the Method by Which It is Put Into a Bacterial Plasmid	236
Screening	242
Some Uncertainties about Genetic Engineering	243
Eukaryotic Genetic Engineering	244
Genetic Engineering and AIDS	258
Chapter 8 – QUESTIONS	263

Chapter 9 – The Behaviour of Chromosomes During Cell Divisions in Diploids – Mitosis and Meiosis — 265

Mitosis in Diploids	265
Meiosis in Diploids	277
Mitosis Photographs	289
Meiosis Photographs	295
Genetic Recombination Illustrated by Spores in Sordaria brevicollis	301
Comparisons Between Mitosis and Meiosis	310
Chapter 9 – QUESTIONS	311

Chapter 10 – Mutations — 312

Broad Categories of Mutation	313

Successful Mutations	314
Mutation Rate	315
The Breeding System and Mutation	316
The Types of Genes That Are Changed by Mutations	317
Reverse Mutations	318
Repair of Mutations	318
Natural Selection of Mutant Phenotypes	318
Naturally Occurring Mutations in Chromosomes	319
General Requirements of Mutation Viability	320
Categories of Viable Structural Changes in Chromosomes	322
Alterations to the Number of Chromosomes	330
Small Supernumary Chromosome Segments	334
Evolutionary Changes Caused by Mutations and Other Genetic Factors	334
Chapter 10 – Questions	336

Chapter 11 – Evolutionary Studies Comparing Polypeptides That Are Widely Spread Among Many Species — 337

Amino-acid Substitution	339
Probability of Amino-Acid Substitution	346
The Incorporation of Tautomeric Point Mutation Into Somatic Diploid Cells	348
Tautomeric Shift in Asexual Species	353
Mutations Derived from Gamete Mother Cells	353
Mutations Arising During Gamete Formation	354
Genetic Drift	356
The Rate of Amino-acid Substitutions to Make New Genes	356
Unit Evolutionary Time	357
Evolutionary Trees Deduced from Differences in Amino-acid Sequences in Widespread Polypeptides	357
Evolutionary Mutations Inferred by Comparing The Globins That Carry Oxygen in Different Species	364
Chapter 11 – Questions	368

Chapter 12 – Evolution — 369

Mutations	370

Recombinations and Best Sets of Alleles	370
So-called Invariable Genes	370
Changes to Allele Frequencies in Populations	371
The Creation of a New Invariable Genes	371
The Evolution of Races and Species	371
Genetic Drift	372
Classification	372
Population Genetics and Evolution	373
Races	373
Species	374
The Origins of Genetic Isolation	379
The Breeding Mechanism	390
The Cellular Requirements for Establishing a New Mutation	392
The Establishment of New Mutations in Populations of Sexual Species	393
Ecological Genetics	397
Some Ecological Studies of the Genetics of Natural Populations	398
Chapter 12 – QUESTIONS	424

Chapter 13 – Mobile Genetic Elements – Their Influences on Evolution and Their Potential Usefulness to Man — 425

Categories of Mobile Genetic Elements	426
Transposons	427
Mobile Inter-Cellular Genetic Elements	427
Base Sequence Comparisons of Viruses, Viroids and Host Cell Genomes – Some Evolutionary Implications	435
Uses to Which Viruses can be Put	439
Chapter 13 – QUESTIONS	443

Chapter 14 – The Modification of Gene Expression by Differentiation, Hormones and Ageing — 444

Differentiation	444
Genes that Control Development in Animal Embryos	454
The Control of Development in Drosophila melanogaster Embryos	458

Differentation in the Producction of Mammalian Antibodies Caused by Gene Mobility	464
'Mother Effect' Demonstrated by the Right Handed or Left Handed Coiling in Limnea pereger shells	469
The Modification of Gene Expression Caused by Hormones	471
Hormonal Control of Animal Gene Expression	473
Hormonal Control of Plant Gene Expression	476
The Molecular Biology of Ageing	480
Chapter 14 – QUESTIONS	485

Chapter 15 – Cancer 486

Metastases	486
The Genetic Basis of Cancer	487
Cancerous Mutations That Have Been Identified	488
Viruses that Cause Cancer	489
The Relationship Between a Human Gene That Causes Controlled Cell Division When in Blood Platelets and an Oncogene in Simian Sarcoma Virus	496
Human Cancers	497
Classic Experiments in Cancer Research	499
Chapter 15 – QUESTIONS	507

Chapter 16 – Human Genetics 508

Primate Chromosomes	508
The Hardy Weinberg Equations	515
The Identification of Human Chromosomes and Aberrations in Them	518
Genetic Human Defects	519
Genes for Human Sex Determination	533
The Selective Advantage or Disadvantage of the Heterozygote	548
The Association of Genetic Inheritance with Disorders That Do Not Seem to Be Directly Caused by the Genes	549
Linked Genes	550
The Geographical Distribution of Human Blood Groups	555
Genetic Engineering in Humans	555
Chapter 16 – QUESTIONS	557

Postscript to Chapter 16	558
Appendix 1 – The Organisation of Non-Dividing Cells	561
Table of Cellular Organelles	563
Appendix 1 – QUESTIONS	570
Appendix 2 – The Mathematics of Population Genetics	571
Subjective and Objective Decisions	572
General Criteria for Carrying Out Scientific Experiments	572
The Mathematical Analysis of Quantified Data	574
Confidence Levels	574
How to Decide What Sort of Statistics to Use – Discontinuous Variation and Continuous Variation	576
The Application of Statistics to 'Normally' Distributed Continuous Variation	585
Standard Error	596
Standard Error of the Difference	598
Appendix 2 – QUESTIONS	606
Bibliography	610
List of Photographs	630
Index	632

ACKNOWLEDGEMENTS

My thanks and appreciation go to several people and to a few institutions, either for their direct help in preparing this book or for allowing me access to the books and papers used as reading material.

The Bodleian Library, Oxford accepted me as a Reader and allowed me to use the science section of the library. A large part of the material used in this book was found in Oxford.

Mrs Hazel Coram very accurately typed many of my teaching notes from which this book gradually emerged.

Dauntsey's School provided me with a steady stream of bright students to whom the contents of this book were imparted.

Dr Audrey Davies painstakingly edited the manuscript and corrected it wherever errors of typing or of grammar occurred. Her comments on certain aspects of the subject matter and presentation were invaluable.

Professor J.G. Edwards, Professor of Genetics at the University of Oxford provided some specific references for chapter 10.

Dr P.N. Goodfellow of The Imperial Cancer Research Fund, London sent three papers to Africa on his team's researches into the human male-determining gene.

Jim Hodges, formerly Head of English at Dauntsey's School, now secretary of The Old Dauntsean Association, passed correspondence about the book to former students, upon whom the material on the book had been tested.

The Imperial Cancer Research Fund, London permitted me access to their library, where the librarians gave me guidance on useful reference papers in cancer research.

Mrs Nicky Krikorian very greatly assisted in the preparation of the index.

Dr G.H. Jones of The School of Biological Science at Birmingham University was kind enough not only to supply the meiosis photographs used in this book, but was also very helpful in making suggestions about how they should be used together with the captions that go with the photographs.

My son Nicholas found the reference pages for the photographs in the Libraries of London University, while I was in Africa. He was also the link, while I was abroad, between the Book Guild and several individuals and institutions in England.

Dr D.C. Page of the Whitehead Institute for Biomedical Research, in the United States of America, sent papers to Africa on his team's researches into the male determining region of the human Y chromosome.

My wife Penelope typed part of the manuscript and prepared part of the index.

P.G. Pilbeam Esq., Chief Technician at The Department of Biochemistry, University of Cambridge provided the reference to the first experiments on 'interrupted mating' and 'genetic mapping' in bacteria.

The Master and Fellows of St. John's College, Oxford offered very agreeable circumstances to schoolmasters during the summer vacations. The easy access to Oxford libraries for several weeks, during three summers, was invaluable.

Dr A.F. Stimson, Department of Zoology, British Museum (Natural History), London provided information on lizards of the genus *Chemidophorus*.

G.W. Stout, Esq., Headmaster of the International School of Bophuthatswana, Mafikeng read the typescripts of chapters 12 and 14 and made some very useful comments upon them.

Dr M.D. Waterfield of the Imperial Cancer Research Fund, London sent papers on his team's researches into a gene in humans that may be used by blood platelets to induce cell division for repairing wounds. The gene corresponds to an amino-acid sequence in a polypeptide found in a virus that causes cancer in monkeys.

To all those who have responded and corresponded so quickly and helpfully in the preparation of this book, I am extremely grateful.

W.J.C. Roberts,
1994

PREFACE

It has been my privilege to prepare candidates each year for the Oxford and Cambridge entrance examinations in Biology. Teaching up-to-date information suitable for able students has made it necessary to cast a wide net for gathering material at a time when there has been a rapid surge of biochemical discovery. The second half of the twentieth century has seen the fascinating unravelling of several principles of biology. This time therefore seems appropriate to attempt a comprehensive and comprehensible summary of the present state of knowledge about biological inheritance. This book is the result of reading carried out in the libraries of Oxford University and at the Imperial Cancer Research Fund Library in London. It is hoped that all the main principles of genetics are contained in it and that they may be communicated without difficulty to students who are starting university biology courses. In the United Kingdom the material is also aimed at the best A-level candidates, especially those who wish to try for places at the Universities of Oxford and Cambridge.

Some proofs of the discoveries of genetic principles are stated in an historical context in Chapter 1. Each of the other chapters summarises topics that are fundamental for an understanding of the subject of inheritance.

'Genetic engineering' coupled with the prevention and cures of cancer are two areas of biological research that could lead to great improvement in the future for mankind. However there are also some sinister implications of genetic engineering. The ethical and moral problems raised by the sudden ability of geneticists to make genes to their own specifications are briefly discussed at the beginning of Chapter 8, the chapter which discusses the brilliant technological use of 'genetic engineering'. The biology of cancer also has a chapter of its own.

Inevitably some of the language in the text includes terms that are rather specialised. Some familiarity with the language used for elementary genetics would therefore be an advantage for readers of this book. The Dictionary of Genetics by R.C. King (Oxford University Press), The Penguin Dictionary of Biology and The Penguin Medical Encyclopedia are very useful books for biologists.

My aim has been to use descriptive language that is plain, without too much reliance on specialised jargon. I have tried to reduce the essential points of genetics to easily understood principles. The text is presented in the order I use for my own teaching and this order of material is that which I would choose for my own learning, if I was beginning a proper study of genetics. It may be that the genetic crosses given in Chapters 5 and 7 could be learned in accordance with the needs of a student's course, rather than together where they appear in this book.

Simple illustrations are used, some of which include different colours to emphasise important points.

W.J.C.R.

INTRODUCTION

It is useful to know that the cells found in living things throughout the world are broadly divided into two groups.

PROKARYOTIC CELLS

Prokaryotic cells form simple, single cell, living things such as bacteria, blue/greens and one known species of green alga – Prochloron, a symbiont with sea squirts. They have few membranes and few sub-cellular organelles. Bacteria have been particularly useful for genetic investigations, having only one chromosome in the form of a circle. The chromosome has proved to be fairly accessible to scientists and quite easy to purify. Some of the best advances in our understanding of the material of inheritance have come from bacterial DNA.

Another advantage of using bacteria for genetic studies is that they reproduce very quickly, which allows many generations to be studied in a short time.

EUKARYOTIC CELLS

The cells of plants, fungi and animals are much more complex than those of prokaryotes. It has therefore been very difficult to be sure of the results obtained from experiments on the material of inheritance in eukaryotes. There have been severe problems of accessibility and difficulties in purifying cell constituents.

In addition to the two basic cells types there also exist:-

VIRUSES (also called PHAGES when they invade and infect bacteria)

Viruses are not true cells because they depend on their host for making many of their essential metabolic requirements. They do not therefore fall into the category of either prokaryotic or eukaryotic cells. The material of inheritance is not always DNA in viruses. In many plant viruses the material of inheritance is RNA, and RNA is also the material of inheritance in some animal viruses.

It is also useful to know the term used to describe the full complement of DNA in a cell. This is the GENOME and the genome comprises the chromosomes and extra-chromosomal DNA. Chromosomes are large DNA molecules containing most of the genes. Extra-chromosomal DNA usually takes the form of small circles, containing only a few genes. The ways that chromosomal genes and extra-chromosomal genes control inheritance are described in some detail in this book.

1

CLASSIC EXPERIMENTAL WORK IN GENETICS

Rational individuals of our own species have existed for hundreds of thousands of years with a vocal language for expressing curiosity and discovery. Some human societies have collected written records of thoughts and discoveries in their libraries for thousands of years. Yet it has only been in the last two centuries that any significant progress in the understanding of biological inheritance has been made. Even now, at the end of the twentieth century, discoveries are being made in genetics which are forcing scientists to reappraise previous dogma and to search again for the real mechanics by which inheritance is governed in complex living things. Even the broad groups into which living things are divided for classification have yet to be decided for certain. In the matter of classification the detailed analysis of the molecular components of cells, taken from many species, is now leading quite quickly towards a systematic approach to classification. The human genetic system itself is under intense scrutiny and, within a decade or two, many of the puzzling complexities of biological inheritance will have been unravelled.

This book describes some fundamental principles of genetics. The topics have been chosen to give students a broad base from which further, more detailed studies could be made.

It is necessary straightaway to distinguish between visually perceived differences in individual living things, and the study of inheritance at the molecular level in cells. The appearances of living things, or their useful qualities, used to be the only means by which the breeders of animals and plants selected mates or plant crosses that would most likely provide them with new generations of individuals best suited to the breeders needs. Among the earlier famous writers on inheritance, neither Lamarck nor Darwin nor Wallace knew anything of significance about cell biology nor about its regulation of inheritance. However, in the second half of the nineteenth century, Mendel, in Austria, succeeded in becoming the first person to suggest the particulate nature of inheritance in germ cells, and he carried out experiments, now famous, on pea plants to verify his hypothesis. Mendel was therefore the link between simple, visual observations of physical form and modern cellular studies of inheritance. It will be seen from the contents of this book that nearly all of our modern understanding of inheritance comes from cellular studies.

The study of cellular inheritance was started in 1831 with the discovery, by R. Brown, of cell nuclei. The progress of discovery was held back by low magnification microscopy and by a limited knowledge of the chemicals that stain specific biological molecules. It was not until 1881 that E.G. Balbiani saw chromosomes as cross-striped threads in the salivary gland cells of a gnat-like fly (Chironomus), and even then he did not identify them as determining inheritance. It was W. Roux who suggested in 1883, that the chromosomes within the nucleus carry inheritance factors. In the same year E. van Benenden observed the reduction of the round worm (Ascari) diploid number of 4 chromosomes to the haploid number of 2, after meiosis had produced the germ cells. A. Weissman, in 1887, followed up Van Benenden's observations by suggesting that there has to be the regular reduction of the diploid number of chromosomes to the haploid number in the gametes of all sexual species and in 1892 he persuaded many scientists to reject Lamarck's theory of the inheritance of acquired characters and to accept that inherited variations among offspring arise only through particles in the gametes.

Mendel's papers of 1865 were rediscovered, having lain in obscurity for thirty-five years, and his experiments were repeated and corroborated by H. de Vries, C. Correns and E. Tschermak in 1900. In the same year W. Bateson introduced a translation of Mendel's papers to the Royal Society of London.

The ingenuity of biochemists, biophysicists, genetic engineers and ecological geneticists has added a wealth of new knowledge to man's perception of nature during the last forty years. This chapter describes some of the experiments and discoveries that replaced simple perceptions of biology, dependent on man's senses, with classic deductions that eventually led to the deciphering of the genetic code.

The basis for modern discoveries lay in the writings of Darwin, Wallace and Mendel. These three men drew conclusions from their empirical observations of biological inheritance and published their findings in the 1850s and 1860s. Although Darwin and Wallace were not geneticists in the modern sense of the word, it is appropriate to include a summary of their findings among the other classic discoveries that led to the elucidation of the genetic code in the 1960s.

The Austrian monk and abbot Gregor Mendel published his observations of genetic crosses almost exactly a century before the cracking of the genetic code. His experiments were the first to indicate that biological inheritance is governed by 'particles'. It is therefore appropriate to start this brief history of classic genetics experiments with those carried out by Mendel.

a) THE DISCOVERY OF THE PARTICULATE NATURE OF INHERITANCE BY GREGOR MENDEL

Gregor Mendel carried out his plant hybrid experiments in the garden of his monastery near Brünn, at that time in Austria (now Brnö in Czechoslovakia). He published, in 1865, his findings in a local publication – Verh. naturf. Ver. in Brünn – contained in an article entitled 'Observations on Experiments in Plant Hybridisation'. The article contained his belief that the inheritance of characters by individual pea plants is determined by 'particles' in the germ cells. It also described the two Laws of Mendelian Inheritance that are summarised later in this section.

Mendel's Hybrid Experiments

Mendel made the hypothesis that in pure bred plants for the dominant character only the inheritance particle 'A' would be present. In pure bred plants for the recessive character only the inheritance particle 'a' would be present. Like any plant breeder, Mendel took pollen from the anthers of one plant and transferred it to the stigmata of the plants from which he wished to obtain pea seeds. He then used the pea seeds to grow new generations of pea plants.

By crossing two pure bred plants, each being pure bred for different expressions of the same character, Mendel could guarantee that the plants that grew from the seeds he obtained from the pollinated plants would be hybrids.

Whenever he crossed a pure bred dominant plant with a pure

bred recessive plant, all the hybrid offspring of the next generation, the F_1 generation, showed the dominant character.

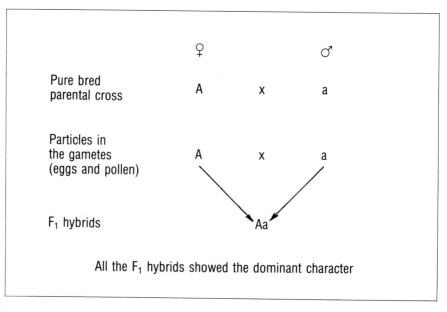

However when hybrid plants were crossed with each other, three-quarters of the subsequent F_2 generation showed the dominant character and one quarter showed the recessive character. Mendel explained these observations as follows.

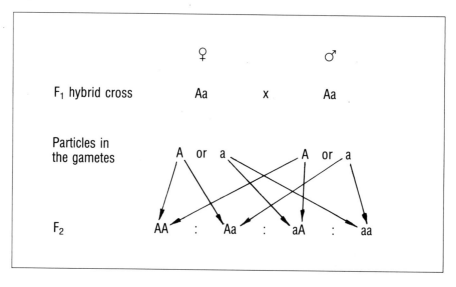

Since Aa is genetically the same as aA, the ratio of particles in the F_2 plants was therefore:-

```
AA  :  Aa  :  aa
 1  :   2  :   1
```

By inspection of the particles present

AA specifies the dominant character
Aa specifies the dominant character
aa specifies the recessive character

Mendel's expected ratio of phenotypes was therefore:-

Dominant character : Recessive character

3 : 1

In order to satisfy himself that his particulate theory was correct, Mendel self-fertilised all of the F_2 plants that showed the dominant character. Two thirds of them produced seeds from which emerged dominant and recessive character plants in the ratio of 3 dominant to 1 recessive. One third of them bred true for the dominant character. These results indicated to him that in the F_2 generation the ratio of pure bred to hybrid plants was:-

```
A   :  Aa
1   :   2
```

When he self-fertilised plants showing the recessive character they all bred true for the recessive character. Mendel had therefore verified his particulate theory of inheritance by demonstrating that the inheritance of particles in the F_2 generation was:-

```
A  :  Aa  :  a
1  :   2  :  1
```

Mendel made a large number of monohybrid crosses. A summary of the results are given below. They give a very close approximation to his predicted phenotype ratios. He used dominant character plants of the F_2 generation to self-pollinate and to demonstrate the 2 : 1 ratio of Aa: A particles in the dominant plants.

Mendel's Monohybrid Crosses

Mendel chose some characters in the pea Pisum sativum:-

1. The form of ripe seeds – 'round' or 'wrinkled'.

2. The colour of the endosperms of seeds – 'pale yellow'; 'bright yellow'; 'orange'; 'green' (all visible through the seed coats).

3. The seed coat colour – 'white' or 'grey/brown'.

4. The shape of the pods – 'smooth' or 'indented'.

5. The colour of unripe pods – 'green' or 'yellow'.

6. The positions of the flowers – 'axial' (on the main stem) or 'terminal' (bunched at the end of the main stem).

7. The differences in the lengths of the stems – 'long' or 'short' (stem length is fairly constant if the plants are subjected to similar environmental conditions).

The hybridisation results in the F_2 generation were:-

Experiment Number

1.		5,474 round	:	1,850 wrinkled
	ratio	2.96	:	1
2.		6,022 yellow	:	2,001 green
	ratio	3.01	:	1
3.		705 grey/brown	:	224 white
	ratio	3.15	:	1
4.		682 smooth	:	229 indented
	ratio	2.98	:	1

5.		428 green	:	152 yellow
	ratio	2.82	:	1
6.		651 axial	:	207 terminal
	ratio	3.14	:	1
7.		787 long	:	277 short
	ratio	2.84	:	1

The average ratio for all Mendel's experiments was 2.98 : 1
or approximately 3 : 1

In addition to substantiating his particulate theory, the experimental results allowed Mendel to make two further deductions from his experiments.

(i) The recessive character can disappear in one generation and then reappear in the next. This allowed him to deduce that inheritance 'particles' are transmitted in the germ cells and that one sort of 'particle' can dominate over the other.

(ii) Neither particle can influence the structure of the other.

Mendel subsequently carried out dihybrid crosses with his pea plants and from these dihybrid crosses emerged results that led to the two famous Rules of Mendelian Inheritance.

Mendel's Dihybrid Crosses

Dihybrid refers to two inherited characters, instead of the one inherited character in monohybrids.

Mendel crossed plants with round, yellow peas with those having green, wrinkled peas. The results he published were taken from 15 plants that produced 556 seeds and were as follows:-

Number of Seeds	Approximate Ratios
315 round and yellow	9
101 wrinkled and yellow	3
108 round and green	3
32 wrinkled and green	1

Mendel made other dihybrid crosses, but the results he obtained from them were not so close to the expected ratios given above. Nonetheless he made use of this experimental data to work out his two Rules of Mendelian Inheritance.

Rule 1 – The Law of Independent Segregation

The characters of colour (yellow or green) and shape (round or wrinkled) in peas can have letters ascribed to the particles that determine the characters. A capital letter implies dominance over the small letter:-

A = round;   a = wrinkled
B = yellow; b = green

Pure bred round, yellow peas would contain the particles:-

AB

Pure bred green, wrinkled peas would contain the particles:-

ab

Mendel predicted which particles would be found in the gametes (pollen and eggs). He also predicted the particles that would be found in the F_1 plants resulting from a cross between a pure bred round, yellow pea plant and a pure bred wrinkled, green plant by using the following scheme.

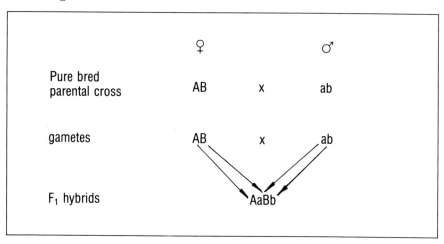

All the F_1 had round, yellow peas. The F_1 plants are heterozygous for both shape and colour, having two different particles for determining each of the two characters. Mendel predicted that the particles for each character would SEGREGATE INDEPENDENTLY when moving into the gametes produced by the F_1 plants. The following diagram shows how this occurs.

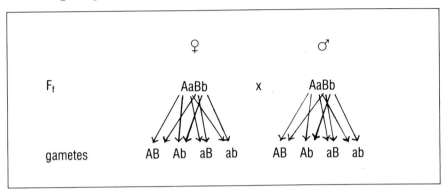

The arrows show that the particles for one gene can move into the gametes with either one of the two particles in the other gene. This is called INDEPENDENT SEGREGATION.

Rule 2 – The Law of Random Assortment

From his experimental results Mendel further worked out that the particles found in any male gamete could randomly associate with particles found in the female gametes. This implies that there is an equal likelihood of a male gamete containing any two inheritance particles fertilising any of the female gametes with either of their two inheritance particles. This is most tidily presented in the form of a table with the particles in the male gametes shown across the top of the table and the female gamete particles shown on the left of the table. The table itself shows the possible associations of the male and female gamete particles by RANDOM ASSORTMENT.

♀ \ ♂	A B	A b	a B	a b
A B	AA BB (1)	AA Bb (1)	Aa BB (1)	Aa Bb (1)
A b	AA bB (1)	AA bb (2)	Aa bB (1)	Aa bb (2)
a B	aA BB (1)	aA Bb (1)	aa BB (3)	aa Bb (3)
a b	aA bB (1)	aA bb (2)	aa bB (3)	aa bb (4)

(Male Gametes across top; Female Gametes down left; F$_2$ Genotypes)

The ratios of the results predicted by this scheme are:-

	① Round Yellow	② Round Green	③ Wrinkled Yellow	④ Wrinkled Green
Ratios predicted by the RANDOM ASSORTMENT table	9	3	3	1

The ratios observed by Mendel were:-

	①	②	③	④
Peas counted	315 :	108 :	101 :	32
Ratios	9.06 :	3.11 :	2.91 :	0.92

This satisfied Mendel that his theory of the particulate nature of inheritance is correct and that his two rules can be applied to some inherited characters in peas that are discontinuous in expression.

b) THE CONTRIBUTION TO THE MODERN INTERPRETATION OF INHERITANCE MADE BY CHARLES DARWIN AND ALFRED WALLACE FOLLOWING ON FROM JEAN-BAPTISTE LAMARCK

The French Zoologist, Professor Lamarck, of the Musée d'Histoire Naturelle, suggested in papers written in 1802, 1809 and 1815 that living things have the ability to alter their form in response to environmental circumstances during their own life-time. He further supposed that these acquired characters can be passed on to offspring, which would inherit a blend of the parental forms at the time of mating.

The way Lamarck might have illustrated his concept was to suggest that giraffes developed long necks because their habit of eating leaves on trees caused them to stretch upwards to feed, so elongating their necks. Each generation would be endowed, by the blending of their parents' characters, with a slightly longer neck than the previous generation, resulting in the evolution of the long-necked giraffes found in Africa now.

Darwin was able to show that Lamarck was incorrect in his theory. Through the good fortune of travelling around the world in the ship HMS Beagle and being a member of the wealthy Wedgwood family Darwin devoted a lifetime to the meticulous study of the inheritance of characters. He knew nothing of Mendel's particulate theory of inheritance and Darwin's work depended entirely on his observations of the phenotypic results of matings.

Darwin's evidence included comparisons between species found on the different islands in the Galapagos group of islands. These island species were also compared with species found on the South American mainland, several hundred miles away. The studies of the finches living on the Galapagos Islands, and of the tortoises there are very well documented. Darwin also drew evidence from the breeders of animals and plants in England to find out how desirable characters could be selected by man.

In very brief and simple terms Darwin's contribution to our understanding of inheritance related to survival or extinction is:-

1. In most species there are many offspring of any pair of mating adults.

2. Although broadly having characters that are a blend of their parents characters, small differences exist between the siblings (brothers and sisters) of any new brood.

3. Most of the offspring die before they reproduce.

4. Those individuals that do survive to find a mate and produce offspring of their own are those best suited to their environment.

5. If changes to the environment occur (Darwin's knowledge of geology and his observations of mountain building in South America indicated that there have been changes in the surface of the earth for a long time, sometimes slowly, sometimes violently) species must have the capacity to produce a proportion of offspring that can cope with the changes, otherwise they become extinct.

6. The isolation of groups of individuals originating from one species by barriers to interbreeding, such as geographical isolation, could lead to different forces of selection acting on each isolated group. Over a period of time natural selection would cause enough divergence of form between the groups so that, even if the isolated groups were later reunited, they would not interbreed and could then be regarded as separate species.

Note: Darwin recognised that some individuals that *are* well adapted to their environment are sometimes killed by unusual, chance events before they have passed on their genes to the next generation. He believed that these chance fatalities have only a small effect on the general rules outlined in 1 to 6 above.

Darwin published his book 'On the Origin of Species' in 1859 having, in 1858, received corroborating evidence from Alfred R. Wallace who had made his observations in the islands of the Malay and Indonesian archipelagoes.

Many of the aspects of inheritance discussed in this book are rather specific and tend to reduce inheritance to the narrow fields of bacterial inheritance, or of biochemistry. It should not be forgotten that in natural populations many characters are controlled by multiple, interacting genes and that the blends and small variations observed by Darwin and Wallace are very much the norm. Mendelian characters which segregate sharply and discontinuously have had to be sought out by scientists, including Mendel himself. The subjects of population genetics and evolutionary genetics are discussed in later chapters.

c) THE PROOF THAT DNA IS THE MATERIAL OF INHERITANCE USING GENETIC TRANSFORMATION EXPERIMENTS IN BACTERIA

For quite a long time during the search for the material of inheritance there were limitations of photographic or magnifying apparatus and of techniques for chemical and biochemical analysis. It is therefore remarkably recently that it was proved that Deoxyribose Nucleic Acid (DNA) is the material of inheritance in most living things (in many plant viruses and in some animal viruses the material of inheritance is Ribose Nucleic Acid – RNA). For many years it was believed that the material of inheritance was protein. The difficulty arose because DNA in eukaryotic cells is very closely associated with protein and it is very difficult to separate the protein from the DNA.

Griffiths' Genetic Transformation Experiments

Using data he had collected between 1922 and 1927 F. Griffiths, at the Ministry of Health's Pathological Laboratory, England, published a paper in the *Journal of Hygiene XXVII, January 1928 pp.113 to 159*. The paper described an experiment in which avirulent bacteria could be TRANSFORMED into virulent bacteria.

The species of bacterium was Diplococcus pneumoniae. When grown on a nutrient agar jelly the virulent strain can be distinguished from the avirulent strain by their different appearances:-

VIRULENT	AVIRULENT
'S' strain (smooth, glistening appearance caused by a surrounding polysaccharide capsule)	'R' strain (rough appearance, lacking the polysaccharide capsule)

Griffiths' experiment was carried out as follows:-

Stage 1

(a) He observed that smooth (virulent) individuals infrequently mutate to rough (avirulent) individuals, but that the reverse mutation does not occur.

(b) He observed no mutations from one virulent strain to another.

Stage 2

(a) He injected heat-killed 'S' bacteria into mice and there was no infection of the mice.

(b) He injected living 'R' bacteria into mice and there was again no infection.

(c) He injected both heat killed 'S' bacteria and living 'R' bacteria into mice together and the recipient mice frequently became infected.

(d) From the mice infected in (c) he isolated living, encapsulated, 'S' bacteria but there was no evidence to suggest that they could be found in (a) or (b).

(e) Griffiths considered the evidence of his experiments against the evidence of Stage 1. He concluded that, because avirulent bacteria never mutate to virulent forms, the living 'R' cells had been TRANSFORMED into living 'S' cells after being in close contact with dead 'S' cells.

(f) When the newly transformed 'S' cells multiplied they transmitted their new character over very many generations, indicating that whatever change had occurred in their inherited character was stable.

The following simplified diagrams show what transformation means in the case of Griffiths' experiment.

Donation of Transforming DNA

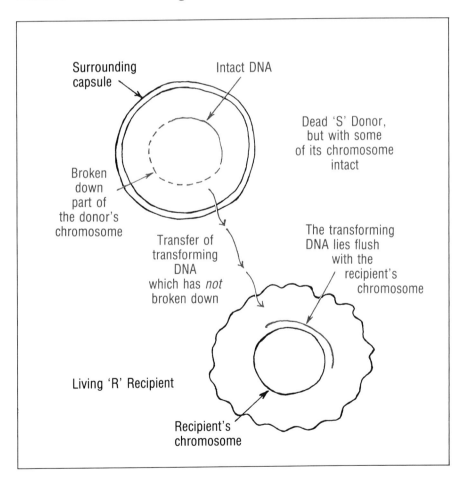

As the dead donor breaks down, part of its chromosome passes out of the capsule and into the recipient. It seems that the recipient bacteria can only take up transforming DNA during one phase of their development when they are said to be COMPETENT. Competence is likely to depend upon 'active' transport (using energy) at reception sites on the recipient's outer surface.

DNA Replication Leading to Transformation

Before the recipient 'R' bacterium can divide to form two daughter bacteria its chromosome has to replicate. The transforming DNA fragment replicates at the same time as the 'R' bacterium's chromosome.

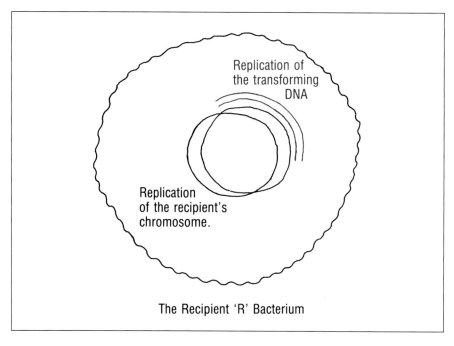

The Recipient 'R' Bacterium

Because the transforming DNA fragment is small, the likelihood of crossing-over occurring between the recipient 'R' bacterium's chromosome and the transforming fragment is also small. However, it does occur in a small proportion of cases. It seems that the expression of the genes contained in the transforming fragment depends upon their incorporation into one or other of the recipient's replicated chromosomes.

The sequence of events is:-

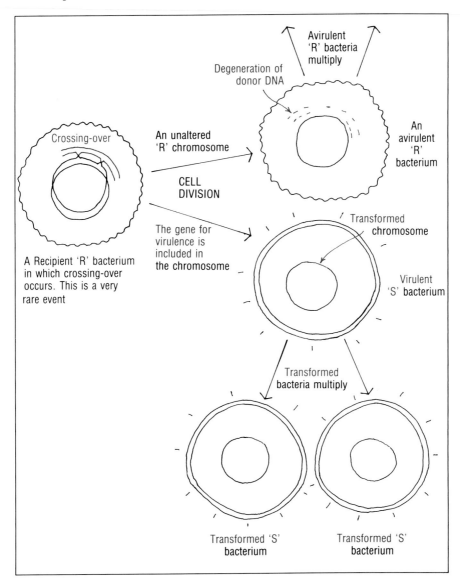

The transformed chromosome is stable and nearly all the cells in subsequent generations will be virulent 'S' bacteria because the TRANSFORMED chromosome now contains the genes for the 'S' characters. Griffiths himself was not aware of the genetic causes of 'transformation' and it took a further sixteen years to identify them.

Experiments Carried out in 1944 by Avery, Macleod and McCarty Proved that DNA is the Material of Inheritance

Although Griffiths' experiment showed that dead cells could pass some of their inherited characters into living cells there was no proof from his experiments about the material which causes transformation.

In 1944 O.T. Avery, C.M. MacLeod and M. McCarty at the Rockefeller Institute, New York caused genetic transformation by introducing highly purified DNA into bacteria.

Bacteria are prokaryotic cells. This allows bacteria to be suitable for transformation experiments because:-

(i) They have DNA which is not closely associated with protein and their DNA can be purified quite easily.

(ii) They have no nuclear membrane and this allows any DNA which penetrates their cell wall to reach the recipient's chromosome without having to cross any more barriers.

In the 1944 experiments, purified DNA which had been extracted from donor strains of bacteria moved into the recipient bacteria to bring about transformation. At that time the yield of transformed cells was only about one in a million.

Subsequent Transformation Experiments

Since 1944 a number of bacteria strains have been used for similar experiments and in some strains the yield of transformed cells is as high as one in a hundred.

The Failure to Transform Bacteria, when DNAase is Present

If the enzyme that breaks down DNA, DNAase, is introduced into the medium in which the purified DNA is found, there is no transformation when the purified DNA in solution is added to the surroundings of the recipient bacteria.

A Modern Understanding of the Mechanism of Genetic Transformation in Bacteria

In recent years it has been discovered that donor bacteria do not, in fact, donate their DNA in the double helix form. It seems that parts of the donor's chromosome produce single-stranded DNA. It is this single-stranded DNA that passes from donor to recipient bacteria. A modern understanding of the events that caused genetic transformation in Griffith's experiments is illustrated below.

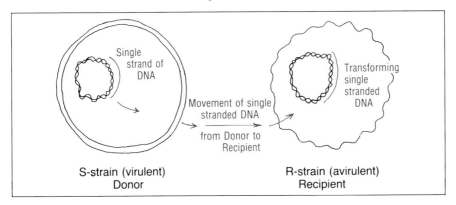

In the Griffiths experiments, the heated donor bacteria disintegrated after releasing their DNA. It is likely that single-stranded DNA was released by the S-strain dead donors and that this single-stranded DNA sometimes entered R-strain recipients. Recipients had the mechanism for base pairing to make double-stranded transforming DNA before cell division. During cell division, crossings-over allowed some of the donors' DNA, now double-stranded, to become incorporated in the recipients' genome, so causing genetic transformation. In experiments described later in this chapter, donors pass their single-stranded DNA to recipient bacteria, but the donors do not generally disintegrate afterwards (pages 59-76).

Radioactive Elements Used in Verifying that DNA is The Material of Inheritance

DNA contains phosphorus but does not contain sulphur. Polypeptides can contain sulphur but never contain phosphorus. A. Hershey and M. Chase showed that radioactive ^{32}P from transforming viral DNA was passed from generation to generation of transformed bacteria. However, radioactive ^{35}S from the polypeptides of transforming viruses was never found in transformed bacteria.

Indirect Evidence Supporting the Proof that DNA is The Material of Inheritance

The Watson-Crick model of DNA structure – see section d) of this chapter – allows:-

1. The semi-conservative replication of DNA (see mitosis and Meiosis – chapter 9).
2. A unique sequence of bases for the synthesis of each polypeptide (see polypeptide synthesis – chapter 2).

Also:-

3. Chromosomes occur singly in gametes but are found in pairs in all other diploid cells.
4. In dividing somatic cells the DNA is seen to replicate before cell division in a way that allows the original number of chromosomes to pass into the two daughter cells.
5. An identical amount of DNA is generally found in the cells of all tissues taken from any 'normal' individual of a species.
6. The ratio of bases:-
$$\frac{\text{Adenine + Thymine}}{\text{Cytosine + Guanine}}$$
 is fairly constant in all individuals of a species.
7. The structure allows viable changes to create new genes, so increasing variation among individuals of a species.
8. In the presence of suitable enzymes, damaged DNA can be repaired quite easily, in a way that replaces the original, damaged gene exactly.

The properties 1 to 8 occur in eukaryotic cells, and several of these properties also occur in bacteria, indicating, but not proving, that DNA is also the material of inheritance in eukaryotic species.

Contrary Evidence

DNA is not the only material of inheritance. Most plant viruses, and some viruses that attack animals and bacteria contain no DNA. Their material of inheritance is RNA.

d) THE STRUCTURE OF DNA

After Avery and his colleagues had proved that DNA is the material of inheritance in 1944 it took nearly a decade to work out the details of its structure. The building components – a pentose sugar (containing five carbon atoms) called deoxyribose, four nitrogenous bases, and a residue of phosphate – were already known.

Dr. Rosalind E. Franklin carried out X-ray diffraction crystallography in London on wet and dry salts of DNA but, although she correctly thought the phosphates point outwards and that there may be a double helix, she was unable to work out the full implications of her X-ray diffraction pictures (*Nature 172, 1953*).

The correct interpretation of the structure of DNA was made by J.D. Watson and F.H.C. Crick, in collaboration with Professor M. Wilkins, at Cambridge University in 1953.

The following series of diagrams shows the simplicity of DNA structure. It is made of a mere six small compounds. These are:-

(i) A sugar called deoxyribose. This has a molecular shape:-

(ii) A phosphoric acid residue:-

(iii) Four bases:- a) Adenine (A)
                b) Thymine (T)
                c) Cytosine (C)
                d) Guanine (G)

The shapes of phosphoric acid, the sugar and of the bases are much simplified – the structure of the bases are shown in more detail later in this section, pages 51-55.

Deoxyribose readily joins phosphoric acid and one base. If we take the base adenine, then the structure formed when deoxyribose reacts with phosphoric acid and adenine is:-

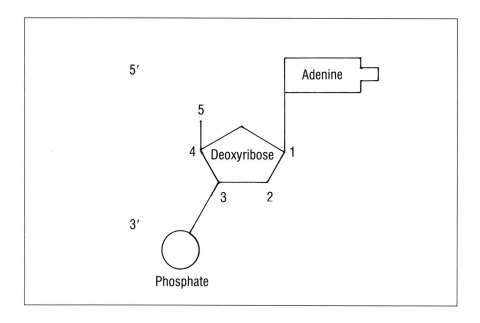

This structure is called a nucleotide unit.

3' and 5' refer to the ends of the nucleotide unit towards which the 3 carbon atom and 5 carbon atom in the deoxyribose sugar point.

Phosphate can then react with a second deoxyribose molecule so that two nucleotide units can be tied together:-

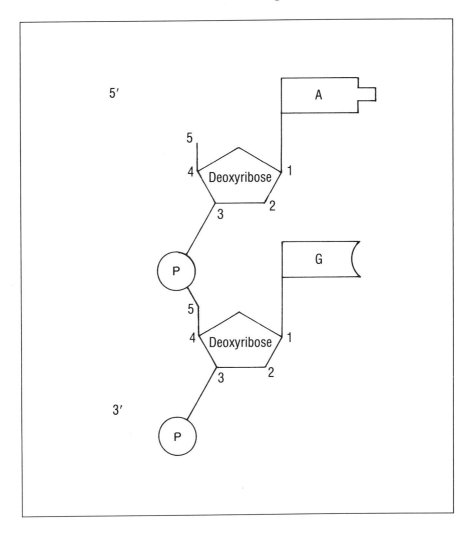

This process can continue so that a long chain of nucleotides is made, as shown overleaf.

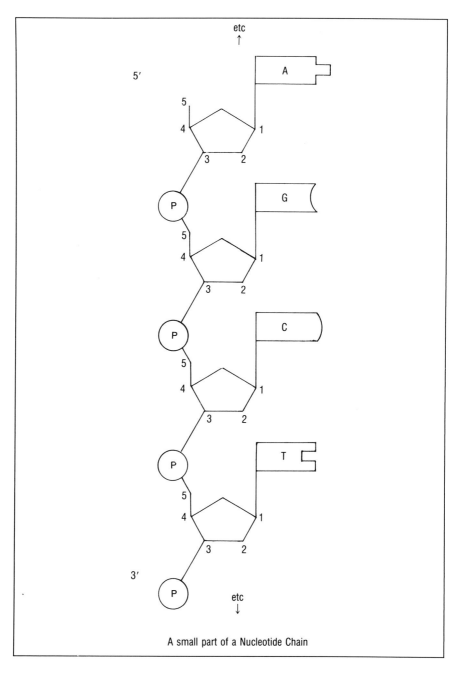

A small part of a Nucleotide Chain

These chains are usually several thousand nucleotide units joined together.

If only one nucleotide unit from each chain is considered, the way the bases pair up is:-

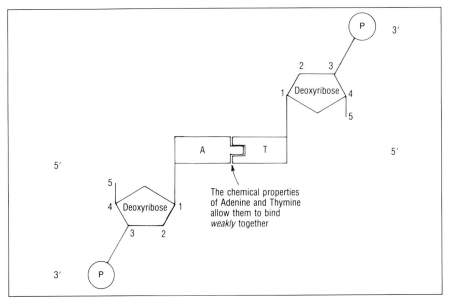

In a similar way the bases Cytosine and Guanine can bind together:-

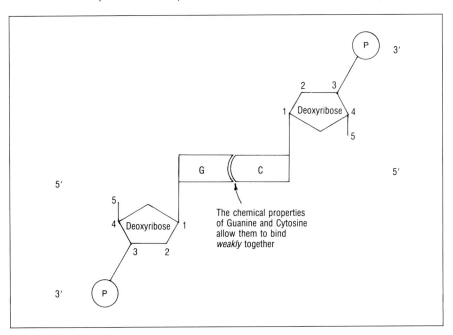

The pairing of bases is very specific. Cytosine can usually only pair with Guanine. Adenine can usually only pair with Thymine.

If more than one nucleotide pair is taken and a small part of a chromosome is considered, the pairing of bases might be:-

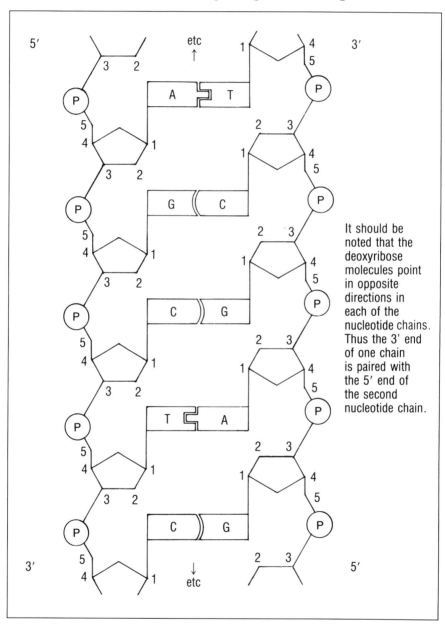

It should be noted that the deoxyribose molecules point in opposite directions in each of the nucleotide chains. Thus the 3' end of one chain is paired with the 5' end of the second nucleotide chain.

Some points to note about the bases are:-

(i) The bases on one nucleotide chain are those that have been selected by natural selection. Thus in every individual of every species the order of bases on every chromosome is very likely to have been inherited from their parent or parents. The exceptions to this general rule are explained in Chapters 10, 11, 12 and 13.

(ii) Whatever the order of bases on one nucleotide chain, the order of bases is pre-selected for the matching nucleotide chain since bases have to pair up with their specific partner (but see tautomerism pages 51-55).

(iii) Bases may appear in any order. The different possible combinations of bases and their order can therefore vary very greatly for any given length of DNA.

(iv) Consider three consecutive paired bases in the last diagram. 64 different linear sequences of bases can be made for these three positions.

'Chromosomes' is the name given to very long, paired chains of nucleotides. In fact the bonding between the atoms of the chains causes the base-bonded chains to curve and the actual form of the chromosomes is that of a double helix.

It was the precise organisation of the components in DNA that caused such excitement in the early 1950s. Teams of research scientists all over the world were striving to establish the exact structure of DNA. It was, of course, two young researchers, the Englishman Francis Crick and the American James Watson, who correctly interpreted the work carried out in London by Rosalind Franklin and Maurice Wilkins. The former complexity of DNA was reduced by these famous scientists to the simple structure we take for granted today.

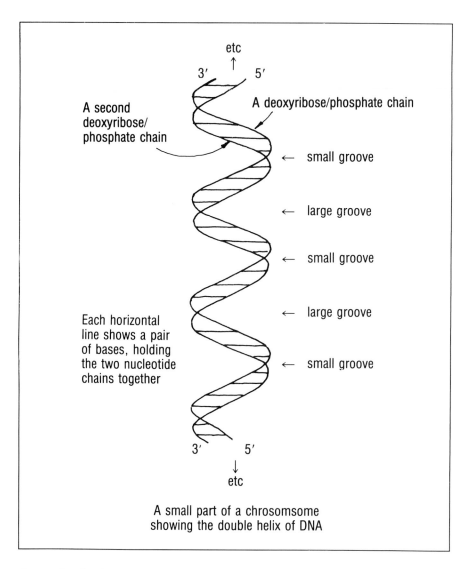

A small part of a chrosomsome
showing the double helix of DNA

Once the basic structure of DNA is understood it is not difficult to explain how it passes the coded messages it holds to the other parts of the cell in which chromosomes are found. This is described in Chapter 2 on polypeptide synthesis.

Base pairing in DNA implies that the number of molecules of Adenine in the chromosomes of a cell should equal the number of molecules of Thymine per complement of cell chromosomes. Similarly the number of molecules of Cytosine in the chromosomes of a cell should equal the number of molecules of Guanine per complement of cell chromosomes.

$$A = T$$
$$C = G$$

But each species has a particular requirement for the bases it needs to maintain its successful genes, and therefore to maintain the successful characters of the species.

Thus the ratio:-

$$\frac{A + T}{C + G}$$

tends to be fairly constant for the individuals of any one species, but varies from species to species.

A More Detailed Look at DNA Structure

It is now necessary to examine the building components of DNA in more detail in order to understand why the structure of DNA has been a suitable vehicle upon which nearly all the species that have inhabited the earth have relied for their success.

The limitations imposed by DNA upon species have also, in a high proportion of cases, led to their extinction. The limitations imposed by DNA structure on the capacity of species to adapt to environmental change are discussed in Chapters 10, 11 and 12.

The Structure and Pairing of the Bases

In order that DNA can best act as the material of inheritance it must be stable enough in its structure to allow a very high proportion of viable genes to be passed from one generation to another without alteration. Chapter 9 on mitosis and meiosis, shows how genes can be passed unaltered from one generation of cells to another, or from one generation of a species to another. It must also have the capacity to introduce changes that confer either a selective advantage to an individual, or a selective disadvantage that is not lethal. The structure of DNA must, in addition, allow sufficient variation in the genotypes of a species so that the species can adapt to changing environments.

The structure and pairing of the bases is a critical feature in both the maintenance of gene stability and in allowing changes to be introduced to create new alleles.

It should be noted that the distances between the carbon atoms of the deoxyribose sugar molecules attached on the outer side of pairs of bases is the same for both pairing bases.

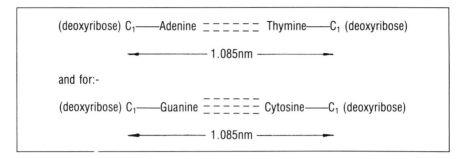

This allows the regular spiral of the double helix of DNA, in which bases fit together easily without having to expend energy to distort the local shape of the DNA molecule.

Another significant feature of the pairing of bases is the type of chemical bonding that holds the bases together. The dotted lines on the base pair diagrams show where 'weak' hydrogen bonds cause $-H$ and $-O$ on opposite bases to attract each other, or $-H$ and $-N$ on opposite bases to attract each other. The fact that these forces of attachment are weak allows the unzipping (under enzymic and dynamic control) of each nucleotide chain from the pairing chain. The bases on each chain then lie 'free', and when they are in this state the unpaired DNA bases can pair up with free bases to make molecules called messenger–RNA, so transcribing the code in a gene into a mobile molecule that can move from the gene on the genome to where polypeptides are made. The details of the processes of translating the genetic code in a gene into a sequence of amino-acids in a polypeptide are shown in Chapter 2.

A third important feature of DNA structure is that the hydrogen atoms on the bases do not require a large input of energy to make them skip from their normal position to another place on their base. This phenomenon is called TAUTOMERIC SHIFT. The following series of diagrams shows how tautomeric shift could cause the 'wrong' base to be recognised for pairing up at cell division, when the replication of DNA takes place (see MITOSIS and MEIOSIS in Chapter 9 and also see Chapter 11). The introduction of a 'wrong' base will henceforward pair up with a 'wrong' base pair. Tautomeric shift therefore introduces a single-base mutation into a gene, and tautomerism is an important property of DNA. Single base, sometimes called 'point', mutations are a fundamental source of variation in DNA and are likely to have been an important cause of variety among living things. This was especially true when there were very simple organisms on earth, at a time when there was no sexual reproduction. The relative importance of single-base mutations compared with large alterations in chromosomes and 'crossings-over' at meiosis and at mitosis is discussed more fully in chapters 9 and 10.

The Theory of Tautomeric Shift

The single plane positions of the atoms in Adenine and Thymine are:-

```
         H
          \
  H        N        N — H ------ O         H   H
   \     /                       \\         \ /
    C = N                         C — C      C
    |        C — C   Hydrogen bonds  \        \
    N       //   \\                   \\       H
   /        C       N ------ H — N     C — H
  H          \                    \   /
              N = C                C — N
                   \              //   \
                    H              O    H

         Adenine                    Thymine
```

The single plane positions of the atoms in Guanine and Cytosine are:-

```
                              H
  H                            \
   \                            /
    C = N       O ------ H — N        H
    |         //               \     /
    N        C — C  Hydrogen bonds  C — C
   /        //   \                  //   \\
  H         C       N — H ------ N      C — H
             \                    \    /
              N = C                C — N
                   \              //   \
                    N — H ------ O     H
                   /
                  H

         Guanine                    Cytosine
```

51

The upper diagrams in the following illustrations show the normal positions of the movable hydrogen atoms (red and starred). The pecked arrows show the positions to which the hydrogen atoms can jump by tautomeric shift. The lower diagrams show the tautomerised forms of each of the four bases, together with the 'wrong' bases with which the tautomers pair up.

The normal form of the bases has been given the name 'Keto'. The tautomers have been given the name 'Enol'. These terms appear again in Chapter 11.

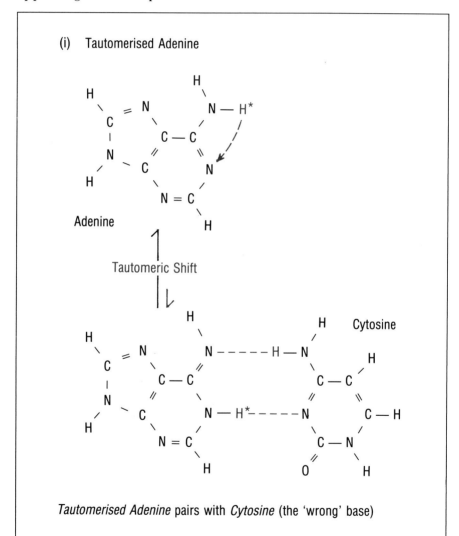

(i) Tautomerised Adenine

Tautomeric Shift

Tautomerised Adenine pairs with *Cytosine* (the 'wrong' base)

(ii) Tautomerised Thymine

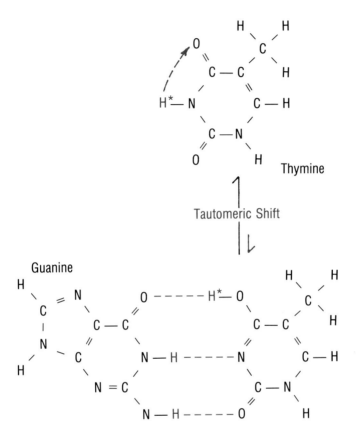

Guanine (the 'wrong' base) pairs with Tautomerised Thymine

(iii) Tautomerised Guanine

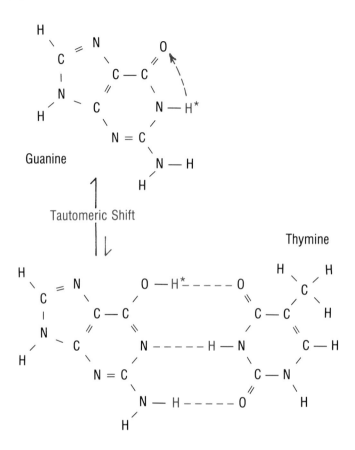

Tautomerised Guanine pairs with *Thymine* (the 'wrong' base)

(iv)   Tautomerised Cytosine

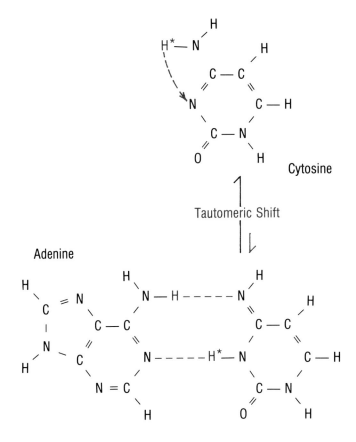

Adenine (the 'wrong' base) pairs with Tautomerised Cytosine

e) THE GENETIC MAPPING OF THE POSITIONS OF BACTERIAL GENES AND CLUSTERS OF GENES BY:-
   (i) INTERRUPTED MATING and by
   (ii) RECOMBINATION FREQUENCIES

The development of techniques to define the precise location and length of genes on the bacterial chromosome led eventually to the proof of the 'One gene – one polypeptide' theory.

G.W. Beadle worked with B. Ephrussi in Paris to show, in 1935, that the vermilion eye colour in Drosophila melanogaster depended on the correct operation of genes to make enzymes. The enzymes drove a biochemical pathway, and if all the enzymes operated correctly the eye pigment was made.

In 1941 the fungus Neurospora was irradiated by G.W. Beadle and E.L. Tatum to produce mutants that suffered from vitamin deficiencies caused by inoperative enzymes (the deficiencies could be defined by adding vitamins to the growth medium).

In 1948 J.Lederberg and E.L. Tatum at Yale University successfully caused RECOMBINATION (the details of recombination are shown later in this section) in strain K12 of the bacterium Escherichia coli.

However, it was not until the development in 1956 by F. Jacob and E.L. Wollman at the Institut Pasteur, Paris and by F.W. Hayes at the Postgraduate Medical School, London, also in 1956, of the 'interrupted mating' technique that the mapping of genes could be carried out in detail. 'Interrupted mating' was also used to prove the 'one gene – one polypeptide' theory (see pages 78-80).

(i) Using 'Interrupted Mating' to Map the Positions of Genes that are Widely Separated on the Bacterial Chromosome

The mapping of the positions and approximate lengths of genes on the chromosome of certain strains of bacteria has been possible by using the techniques described on the following pages.

Mutant Strains of Bacteria

Mutant strains of bacteria are identified by observing the amino-acids that have to be added to minimal medium (MM), a jelly containing only glucose and salts (including an ammonium salt), to allow bacteria to multiply. The amino-acids that have to be added identify the genes that contain inoperative mutations.

In those 'interrupted mating' experiments that were carried out to define the positions of genes (or clusters of genes) used for the syntheses of amino acids the convention is:-

+ = operative gene   (i.e. one working as it would in a thriving colony of wild type bacteria)

− = inoperative gene (i.e. the mutant gene that results in no growth unless the appropriate amino-acid is artificially added to MM).

For example if a bacterium had the following present in its chromosome

$$leu^+ \; trp^+ \; his^+ \; met^- \; thr^-$$

then this mutant strain would not grow on minimal medium. However, if the amino-acids methionine and threonine are added to the minimal medium, the placing of this mutant strain on such an agar will result in the growth of bacterial colonies. Therefore:-
$leu^+ \; trp^+ \; his^+ \; met^- \; thr^-$ + MM = NO growth

but

$leu^+ \; trp^+ \; his^+ \; met^- \; thr^-$ + MM + methionine + threonine = GROWTH (These added amino-acids identify the mutant strain)

Similarly:-
$leu^- \; trp^- \; his^- \; met^+ \; thr^+$ + MM = NO growth

but

$leu^- \; trp^- \; his^- \; met^+ \; thr^+$ + MM + leucine + tryptophan + histidine = GROWTH

Again, the three added amino-acids identify the mutant strain.

The Mating of Different Mutant Strains of a Species of Bacterium called Escherichia coli

If large numbers (10^8 or more) of two different mutant strains of certain bacteria species, such as E. coli (a fairly harmless bacterium that inhabits the human gut) are placed together on minimal medium, the growth of normal bacterial colonies is often observed. It is now known that a mating process exists so that each mutant strain can donate its viable genes in a way that leads to the formation of a chromosome in the offspring that contains a full complement of viable genes. Taking our original example:-

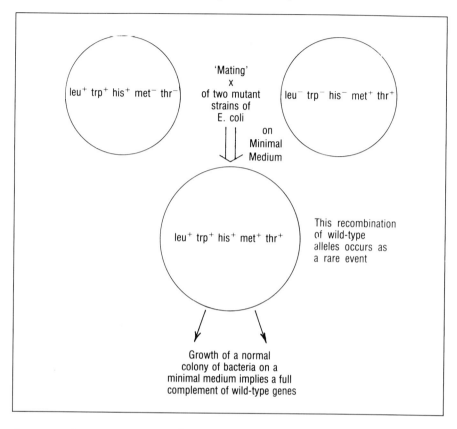

Because there are a very large number of mutant bacteria plated onto the MM there are likely to be a few matings that do give rise to individual bacteria that grow normally into a bacterial colony. The mechanism that gives rise to viable bacteria from mutant strains is RECOMBINATION. Recombination is the term used to describe the stitching together of new combinations of alleles after crossings-over.

The mechanism of recombination was originally envisaged as shown below.

The Mechanism of Recombination

The reason why two mutant strains of E. coli can give rise to normal wild-type offspring is that some of the bacteria act as donors of genetic material and some act as recipients. The process of giving and receiving genetic material (the DNA of the chromosome) can occur if two bacteria lie close enough to each other so that their outer membranes touch each other. The bacterial chromosome is circular and this is shown in the diagram below. The rough positions of the genes for making five amino-acids are shown.

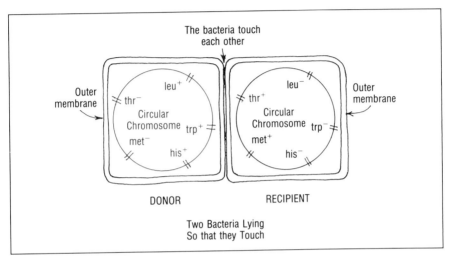

When mating occurs the chromosome of the donor always breaks in the same place between the genes leu$^+$ and thr$^-$ and the broken chromosome can pass through the touching membranes and move into the recipient cell.

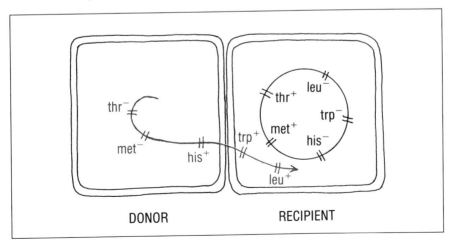

The whole of the chromosome can pass into the recipient so that the donor is emptied of DNA and the recipient contains two chromosomes. Each of the chromosomes can lie flush with each other, with their genes aligned.

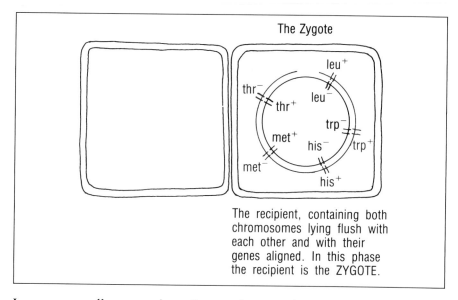

The recipient, containing both chromosomes lying flush with each other and with their genes aligned. In this phase the recipient is the ZYGOTE.

In a very small proportion of cases the two chromosomes can break, at points that lie adjacent, in such a way that 'crossings-over' occur. If crossings-over occur so that the wild-type genes substitute for the mutant genes then the RECOMBINATION of genes in the recipient's chromosomes gives rise to bacteria that have the full complement of wild-type genes and can grow without any need to add amino-acids to minimal medium:

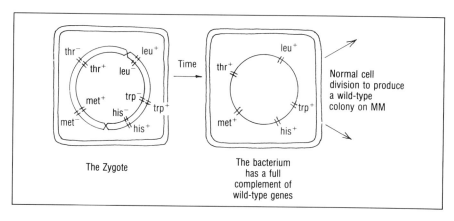

The Zygote   The bacterium has a full complement of wild-type genes   Normal cell division to produce a wild-type colony on MM

In recent years it has been discovered that the DNA transferred from donor to recipient bacteria is single stranded. The single strands of transferred DNA has the second strand of the double helix made to match it in the recipient bacteria.

Crossings-over can then occur in a similar way to those shown on page 39, the modern view of Griffith's transformation experiments. Donors do *not* break down.

Similar processes also apply to the diagrams on pages 63-76.

Genetic Mapping – The Determination of the Precise Positions of Genes by Interrupted Mating

It happens that the process of passing the chromosome from donor to recipient can be interrupted by violent agitation in a 'blender'. Interrupting the donation of the chromosome at different time intervals has led to the precise identification of the relative positions of a large number of genes on the chromosome of E. coli and of other bacterium species.

The 'Interrupted Mating' technique has so far been possible in only a small number of bacterium species. It is only suitable for genes that are quite widely separated on the chromosome.

There is a specific place for breaking the chromosome in the donor bacterium, before mating, in each bacterium strain. Thus the sequence of genes that are passed into the recipient cell is always the same in any one strain. In the case of the E. coli strains already mentioned, the order of genes donated and their relative distances apart can therefore be determined by the 'interrupted mating' process.

The way that interrupted mating allows the defining of the positions of genes relative to each other is as follows.

A mutant strain of bacterium will only grow on agar jelly if deficient amino-acids are added to the jelly, or if the recombinations of genes after 'mating' are such that the full complement of genes are found in the recipient's chromosome.

We can now arrange an agar jelly to have the following properties:-

Amino-acids present: threonine; methionine.
Amino-acids absent: histidine; tryptophan; leucine.

It is also arranged to introduce two mutants of E. coli onto the jelly.

a) The Donor's Genes

thr⁻ met⁻ his⁺ trp⁺ leu⁺

b) The Recipient's Genes

thr⁺ met⁺ his⁻ trp⁻ leu⁻

Interrupted Mating After 5 Minutes

When the amino-acids histidine and tryptophan are added to the jelly, and interruption of the mating takes place within five minutes of plating out the two mutant bacteria strains, there is NO growth. This shows that there is no donation of leu⁺ to allow crossing-over to incorporate it into the recipient's chromosome within the five minute period.

Interrupted Mating After 10 Minutes

However if interruption to mating does not take place until 10 minutes after plating out the two mutant strains of E. coli, then normal growth does sometimes occur. The explanation is that the leu⁺ gene has been passed into the recipient and also incorporated into its chromosome by 'crossings-over'. The recipient therefore has the amino-acids it needs from the following sources:-

From its own genes	Threonine: Methionine		
From the amino-acids added to the minimal medium		Histidine:Tryptophan	
From genes put into the chromosome from crossings-over			Leucine

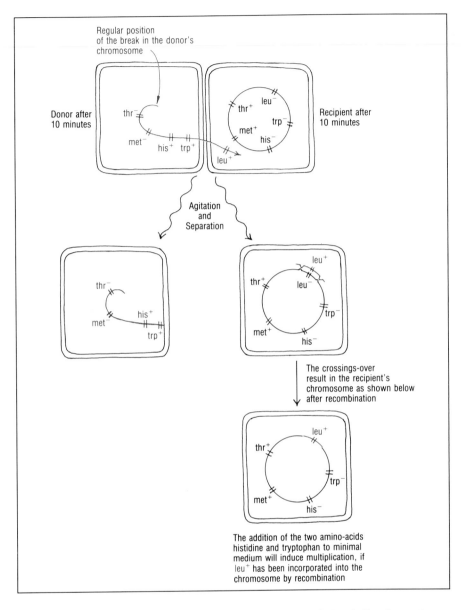

If either histidine or tryptophan are left out of the jelly there is no growth when mating is interrupted after 10 minutes. This shows that leu⁺ is the only gene, among those under study, that is donated within this time.

The modern views of single stranded transforming DNA once again applies. The donor does *not* degenerate.

Interrupted Mating After 35 Minutes

If leucine and histidine are added to the agar jelly then the recipient strain of the bacterium can only grow if the trp$^+$ gene is incorporated into the recipient's chromosome.

From its Own Genes	Threonine	Methionine			
Amino-acids added to MM			Histidine		Leucine
Recombination after crossings-over				Tryptophan	

The time taken for the incorporation of trp$^+$ after recombination has been shown to be 35 minutes.

Interrupted Mating After 50 Minutes

If leucine and tryptophan are added to the agar jelly, growth can only occur when his$^+$ has been incorporated into the chromosome of the recipient by recombination.

From its own genes	Threonine	Methionine			
Amino-acids added to MM				Tryptophan	Leucine
Recombination after crossing-over			Histidine		

The time taken for the incorporation of his$^+$ into the recipients chromosome has been shown to be 50 minutes.

The Significance of the Time when there is Interrupted Mating

It has been established that *the rate at which DNA moves from donor to recipient is roughly constant.*

It has also been established that, for a given strain of E. coli, *the DNA is transferred into the recipient from the same starting point which is always followed by the same order of genes.*

The order of the genes responsible for the three amino-acids under study is therefore:-

If the rate of transfer is constant it can also be deduced that the relative distances between the genes is:-

Between leu and trp = 35−10=25 minutes
Between trp and his = 50−35 = 15 minutes.

The relative distances between the genes are therefore in the ratio:-

The Use of Another Strain of E. coli to Map the Relative Positions of Genes in the Chromosome.

The use of the F factor can be made in Hfr bacteria for regulating specific breaking points and the order of gene transfer in E. coli. Each strain of E. coli is characterised by a different position for the break in the donor chromosome. For example in another strain of E. coli the break occurs between the 'leu' and 'trp' genes. In this case the order of transfer of the genes is altered and is:-

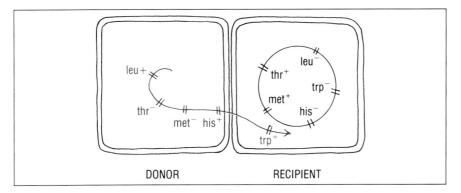

The modern views of single stranded transforming DNA once again apply.

By suitable additions of amino-acids to minimal medium, and by knowing the genotype of the donor for each of the genes shown, it can be shown by interrupted mating that the relative distances between genes can be represented as follows:-

```
    3 ←------------------ 55 mins ----------------→2←15 mins→1←---- 20 mins --→
··· leu ·········································· his ············ trp ···················⟩
```

A succession of maps of this sort can be produced to show the relative positions of a large number of genes on the circular chromosome of E. coli. An example of a genetic map in E. coli is given on page 146.

ii) Mapping the Position of Genes that Lie Close Together by Recombination Frequency

If genes lie close together on the chromosome it may happen that a cross-over will occur between them. It is however very unlikely indeed that more than one cross-over will occur between genes that are close together.

For genes that are far apart, however, there may be several cross-overs between them, and recombination frequency cannot separate how many even numbers of crossings-over or odd numbers of crossings-over there are. Cross-over frequency is only suitable therefore for genes that lie close together. The following diagrams illustrate the problem in determining cross-over frequencies in genes that are far apart.

Any odd numbers of cross-overs gives:-

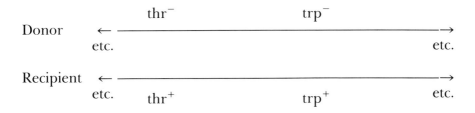

Parts of the donor and recipient chromosomes are shown in both diagrams

Any even number of cross-overs leaves the genes under study linked in the same way as before:-

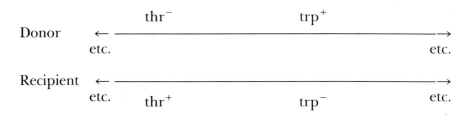

The same parts of the donor and recipient chromosomes are shown

About half the number of crossings-over will be odd numbers and about half will be even-numbers. There is no direct way of discovering the number of crossings-over by this method for genes that are widely separated.

However, if we take genes that *are* close together the probability of a single crossing-over between them is small and the probability of two crossings-over, or more, between such genes is extremely small. For example any crossings-over leading to the recombination of genes in mating E. coli are likely to occur somewhere other than in the DNA between the two genes for threonine and leucine production:-

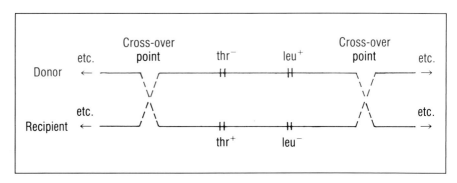

However, in a small proportion of cases crossing-over occurs between these closely situated genes:-

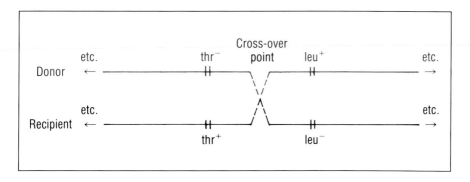

This rare event can be related to the usual situation by calculating the recombination frequency as:-

$$\frac{\text{Total number of thr}^+ \text{ leu}^+ \text{ recombination in the recipient bacteria}}{\text{Total number of thr}^+ \text{ leu}^- \text{ combination in the recipient bacteria}} \times 100\%$$

This recombination frequency ought to give the same % as the calculation:-

$$\frac{\text{Total number of thr}^- \text{ leu}^- \text{ recombination in the recipient bacteria}}{\text{Total number of thr}^+ \text{ leu}^- \text{ combination in the recipient bacteria}} \times 100\%$$

The detection of which recombinant is present can be done using MM and adding suitable amino-acids.

The recombination frequency can be used to determine the proportion of the whole chromosome that lies between the two closely associated genes. It is not a direct measurement however and a recombination frequency of 20% corresponds to about 1% of the total chromosome length. Recombination frequencies as low as 0.1% have been detected, so allowing genes that are very close together to be 'mapped'. This process of mapping very precisely defines both the position and the length of a gene. Each gene can then be associated with the production of a specific cellular component, so proving that one gene (or one gene cluster) is responsible for the production of one polypeptide.

The Mapping of the Positions of Genes in Escherichia coli using Recombination Frequencies and the Capacity of Bacterium Strains to Use a Specific Sugar (Arabinose) As a Source of Energy

It happens that the gene (or cluster of genes) for fermenting (using as a source of energy instead of glucose) the simple sugar called arabinose lies somewhere between the genes that govern the synthesis of the amino-acids methionine and leucine:-

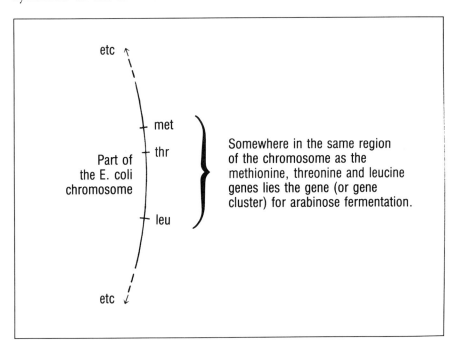

If a:-

$$met^+ \; ara^+ \; leu^- \qquad \text{(recipient)}$$

is mated with a:-

$$met^- \; ara^- \; leu^+ \qquad \text{(donor)}$$

for ten minutes and spread onto a MM agar in which the only addition is glucose, then met^+ leu^+ recombinant bacteria will grow irrespective of whether it is an ara^+ or ara^- that is present, because the bacteria have all the genes needed to use glucose as their source of energy.

If the relative positions and distances from each other of the methionine, arabinose and leucine genes are to be calculated, it is necessary to distinguish between those met^+ leu^+ recombinant bacteria that are ara^+ and those that are ara^-. This can be done by preparing a set of agar plates containing arabinose and no other sugar.

Samples of bacteria are taken from the growing colonies which appear on the original glucose agar plates and bacteria from each colony are spread onto separate plates containing arabinose. If growth occurs on any of the arabinose plates then the bacteria growing on them must be ara^+ leu^+ recombinants. The following diagrams show the two stages:-

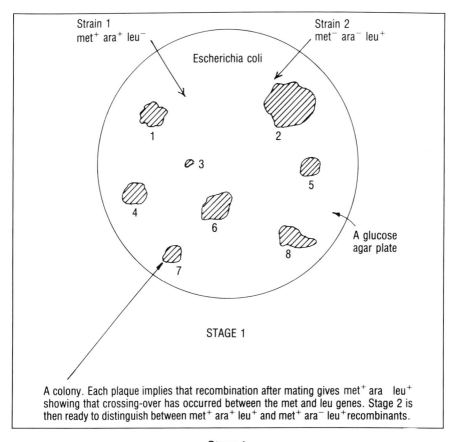

A colony. Each plaque implies that recombination after mating gives met^+ ara leu^+ showing that crossing-over has occurred between the met and leu genes. Stage 2 is then ready to distinguish between met^+ ara^+ leu^+ and met^+ ara^- leu^+ recombinants.

Stage 1
The Matings of the Two Strains are Placed on Several Glucose Plates to Give Many Growth Colonies Allowing Bacteria from Every Colony to be Transferred to Arabinose Plates

The numbers refer to colonies on the glucose plate in stage 1 and to colonies taken from other glucose plates (that were not shown). Growth is shown on only one out of 20 arabinose plates in stage 2, showing that the ara$^+$ leu$^+$ recombination occurred in only one in 20 glucose plate colonies.

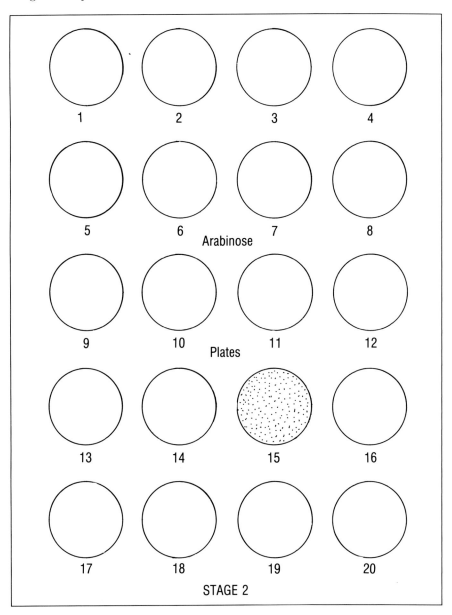

The Recombination Frequency is calculated as:-

$$\frac{\text{Number of ara}^+ \text{ leu}^+ \text{ Recombinants}}{\text{Number of ara}^- \text{ leu}^+ \text{ Recombinants}}$$

ara$^+$ leu$^+$ recombinants are formed when crossing-over between the two bacteria strains occurs as follows:-

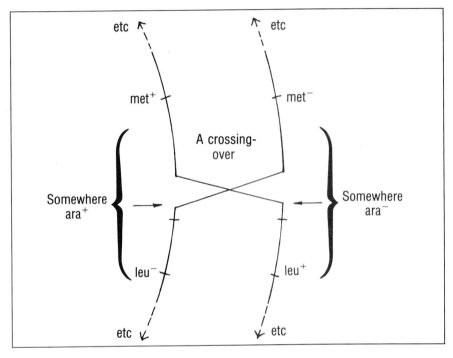

ara$^-$ leu$^+$ recombinants remain if there is no crossing-over between them.

If the recombination frequency is small (5% or so) then the leucine gene is probably very close to the arabinose gene. A recombination frequency of 20% in E. coli has been shown to represent about 1% of the entire chromosome so a recombination frequency of 5% represents a case when two genes are close together.

It is now necessary to carry out a second experiment to determine the distance of the arabinose gene (or gene cluster) from the threonine gene. The following genetic cross between E. coli strains is arranged:-

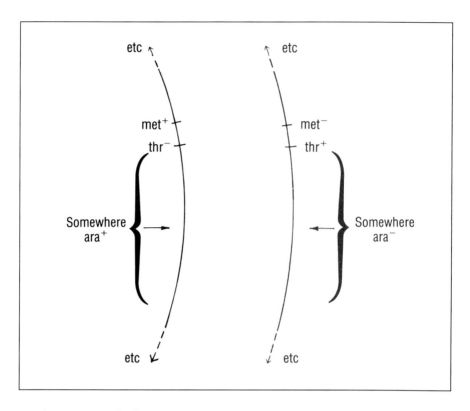

Mating proceeds for ten minutes, then the bacteria are spread on MM containing glucose. Growth will occur provided the bacteria are met$^+$ thr$^+$, irrespective of whether ara$^+$ or ara$^-$ is present, because the energy source is glucose.

In a way similar to that used in the met ara leu experiment (page 71), met$^+$ thr$^+$ bacteria can be plated out from their glucose colonies onto the arabinose plates. In this case the recombination frequency can be calculated as:-

$$\frac{\text{Number of ara}^+ \text{ thr}^+ \text{ recombinants}}{\text{Number of ara}^- \text{ thr}^+ \text{ recombinants}}$$

This frequency is about 15%. The distance from ara to thr is therefore about three times that of ara to leu.

A further recombination frequency experiment can be carried out to calculate the distance between the genes for leucine and threonine:-

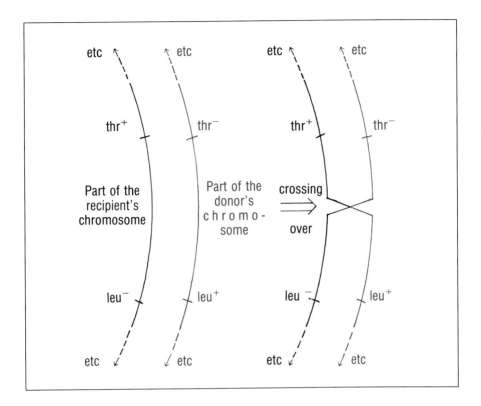

The recombination frequency is:-

$$\frac{\text{Number of thr}^+ \text{ leu}^+ \text{ recombinants}}{\text{Number of thr}^+ \text{ leu}^- \text{ recombinants}}$$

This is about 20%. Thus the threonine gene and leucine gene are further apart than the distance of the arabinose gene from either thr or leu. The arabinose fermentation gene therefore lies between the genes for the synthesis of threonine and leucine. The distances apart of the genes for threonine, arabinose fermentation and leucine are shown in the diagram overleaf:-

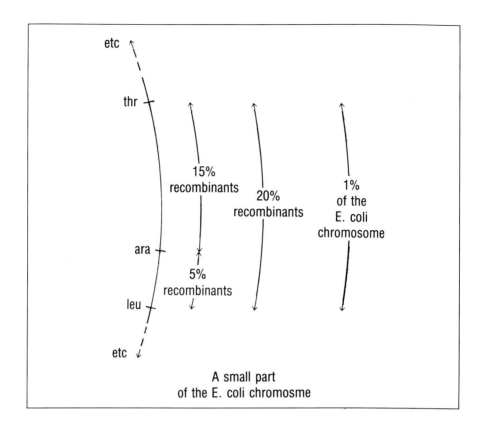

A small part of the E. coli chromosme

Genetic Mapping Using Computer Technology

In recent years the use of computers has allowed the rapid analysis of the order of amino-acids in polypeptide molecules. The rapid analysis of the order of bases in mRNA molecules extracted from living things then allows the order of amino-acids in a polypeptide to be compared with mRNA molecules until a corresponding sequence matches up mRNA with a specific polypeptide.

Large numbers of the mRNA molecule or DNA single strands can be synthesised to include radioactive atoms. These are injected into a medium containing an organism's DNA. Wherever the radioactive mRNA attaches to the DNA, by pairing bases, is found the gene that codes for the mRNA. This gene will, of course, be the gene for the polypeptide under study. This type of genetic mapping is very useful for eukaryotic genomes and overcomes problems presented by complexities such as introns and exons (page 150) and crossings-over in diploids (chapter 9) during meiosis.

Scanning electron micrograph of Escherichia coli, a rod shaped bacterium. Here seen as a dense swirling mass, these bacteria inhabit the human intestine where they are normally harmless.

Magnification x 530

Photograph by Dr. Tony Brain. With permission from The Science Photo Library, London.

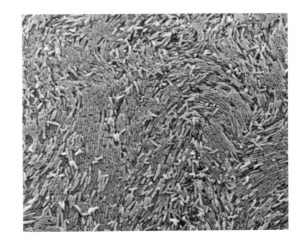

Transmission electron micrograph of an Escherichia coli bacterium that has spewed forth its chromosome following osmotic shock.
The chromosome is a circular double-stranded DNA molecule about 1500 mµ in length (1 mµ is one millionth of a metre). This molecule is packaged into a bacterium that is only 1 mµ in length by extensive folding.

Magnification x 17,000

Photograph by Dr. Gopal Murti. With permission from The Science Photo Library, London.

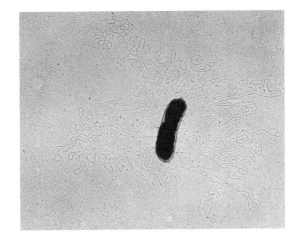

The mapping of the human genome and the identification of the positions of potentially cancerous genes is likely to depend on this type of genetic analysis.

Cotransformation

Cotransformation in bacteria occurs when two genes on a DNA fragment both cause genetic transformation. Of course, the further apart two genes are on the mobile DNA, the lower the likelihood that both of them will lie on the same DNA fragment that enters a competent recipient bacterium. It follows, therefore, that the higher the proportion of cotransformed bacteria, the closer together two genes are likely to be. Cotransformation frequencies are useful for both bacterial genetic mapping and for the genetic mapping of genes in eukaryotes, including human genetic mapping.

f) THE PROOF OF THE 'ONE GENE – ONE ENZYME' THEORY

It is commonly found that the genes responsible for all the enzymes needed for catalysing a biochemical pathway are clustered together in a continuous length of DNA. In some cases biochemical pathways have more than one cluster of genes. An example is the genes responsible for making the nine enzymes that catalyse the synthesis of the amino-acid arginine:-

> One cluster of four genes
> One cluster of two genes
> Three isolated genes, quite widely separated.

The genetic mapping of the positions of the genes governing the synthesis of leucine, tryptophan, histidine, methionine and threonine, discussed in e), in fact mapped the positions of clusters of genes, each cluster controlling the synthesis of the enzymes needed for the biosynthetic pathway leading to production of each of these five amino-acids.

The three genes that govern the production of enzymes for the synthesis of the amino-acid SERINE have been used to prove the 'one gene – one enzyme' theory. The three genes are quite widely separated on the chromosome, and it is this fact that has made it possible to prove the theory.

The biochemical pathway and the three enzymes involved in catalysing the reactions are shown in the following diagram.

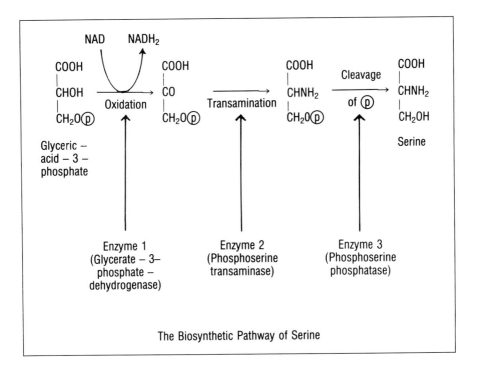

The Biosynthetic Pathway of Serine

The mutant strains of the bacterium E. coli that cannot make serine are subdivided into:-

 (i) Those mutants that cannot make enzyme 1
 (ii) Those mutants that cannot make enzyme 2
 (iii) Those mutants that cannot make enzyme 3

Studies of these three mutants have allowed the positions of the three genes to be 'mapped' on the chromosome. Mapping by interrupted mating has shown that the three regions of DNA responsible for the production of Enzymes 1, 2 and 3 lie quite far apart on the chromosome.

By contrast, the mapping of mutations that cause the loss of any one enzyme shows that all the mutations lie very close to each other. The map positions for each of the lengths of DNA responsible for the production of one enzyme show that the lengths of DNA for each enzyme occupy only about 0.0002 of the total chromosome length. The three small regions responsible for the production of each of the three enzymes are described as three genes.

The enzymes that catalyse the biosynthesis of serine are polypeptides. The mapping of the small lengths of DNA on the E. coli chromosome has shown that one gene does indeed determine the production of one enzyme for each of the biosynthetic conversions in the manufacture of serine.

The Proof Slightly Modified

The 'one gene – one enzyme' proof for the enzymes in serine synthesis is a common feature of those bacterial genes that have been studied so far. However some enzymes are complex molecules that are made of more than one polypeptide. For example one of the enzymes needed for the synthesis of the amino-acid tryptophan has two components. Mapping the genes responsible for each component shows that the genes are distinctly separated from each other on the chromosome. One gene leads to the synthesis of the α sub-unit. The second gene leads to the synthesis of the β sub-unit. In this case two genes are needed for one enzyme.

Cases have recently come to light where there are overlapping genes. Chapter 4 discusses overlapping genes and other variations in the ways in which genes are expressed.

g) THE GENETIC TRIPLICATE CODE

The most significant work demonstrating that the genetic code uses three bases to code for one amino-acid (the triplicate code) was done in three centres of research:-

(i)   At the National Institute of Health, Bethesda, Maryland, USA by M.W. Nirenberg and H. Matthaei.

(ii) At the New York University School of Medicine by S. Ochoa.

(iii) At the Medical Research Council Laboratory of Molecular Biology at Cambridge, England by F.H.C. Crick, Mrs L. Barnett, S. Brenner, R.J. Watts-Tobin and L. Shulman.

A summary of the techniques used to establish the triplicate code is:

Species used:- T4 virus which has 200,000 base pairs in its DNA

Host for the virus:- Strains of E. coli bacterium

Genes examined:-	rIIA and rIIB. Both genes are needed to operate correctly if the virus is to multiply in the host bacterium. There are less than 1,000 base pairs per gene.
Other points:-	If only a very small proportion of the amino-acids in a polypeptide are altered by genetic changes the function of the polypeptide (which may be an enzyme) need not be much impaired. However, if many of the amino-acids in a polypeptide are changed, the function of the polypeptide will be completely damaged.
Method of causing genetic change:-	Chemicals called ACRIDINES can both add or delete bases in DNA. They cause adjacent bases to separate from each other by the exact width of one base molecule.

In mRNA, the base URACIL (U) replaces the THYMINE (T) of DNA. A full explanation of the production of messenger-RNA (mRNA) from DNA is given in Chapter 2.

An examination of the processes of gene transcription (pages 105-107 of polypeptide synthesis) shows that when mRNA is made at a gene the paired bases on DNA unzip. It seems that the sequence of bases on one strand of the unpaired bases of the gene is 'blocked', while the second strand has its sequence of bases 'transcribed' into the sequence of pairing bases that are put into the mRNA molecule. It is this sequence of bases in the DNA used for transcription that is now considered.

A small part of the base sequence in DNA used for transcription from the unblocked DNA strand was:-

etc. ←A T T G C A T C G A C C T→ etc.

The genetic code is read from left to right.
The sequence of pairing bases that go into the mRNA at this part of the gene is shown below in red:-

etc. ←A T T G C A T C G A C C T→ etc. DNA

etc. ←U A A C G U A G C U G G A→ etc. mRNA

Single Base Changes Induced by the Chemical Acridine

Whenever there are free bases present, the introduction of ACRIDINES can cause the replacement of a base in the DNA, or the addition of a base into the sequence of bases. ACRIDINES can also cause the removal of a single base from a point in the DNA chain. The following scheme shows the consequences of each of these possibilities, if each was to occur in different experiments at the same point in the DNA chain.

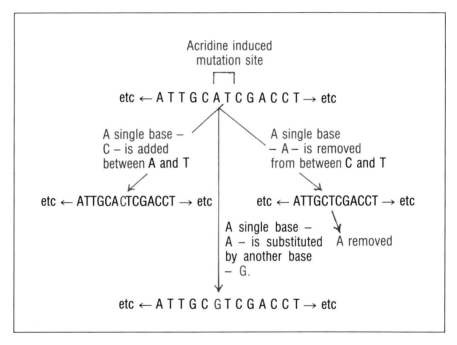

The consequences for the viability of the function of the polypeptide coded by the genes are:-

(i) **One Base Added:**
The virus was unable to multiply

(ii) **One Base Removed:**
The virus was unable to multiply.

(iii) **One Base Changed:**
The virus could multiply

Changing Two Adjacent Bases in the Gene

When acridine was used to introduce, change or delete two adjacent bases at a place on the gene the results were similar to the single base changes:-

(i) **Two Bases Added**
The virus was unable to multiply.

(ii) **Two Bases Removed**
The virus was unable to multiply

(iii) **Two Bases Changed**
There was the multiplication of the viruses.

Changing Three Adjacent Bases in the Gene

Whether the three adjacent bases were added, deleted or changed the gene lost little of its function.

Crick's Interpretation of these Results

The results suggested to Crick that the genetic code is made up of a multiple of 3 bases (probably 3 itself) coding for each amino-acid put in sequence into a polypeptide (the details of how this is done is shown in Chapter 2).

His reasoning was that the total loss of function by the genes under study in the T4 virus was caused because the introduction or loss of 1 or 2 bases caused a shift in the way the bases would be interpreted for the building of an enzyme. However the gain or loss of 3 bases would only cause a local addition or deletion of one amino-acid.

(i) 1, 2 or 3 Bases Added

The following scheme shows the consequences for polypeptide synthesis of adding 1, 2 or 3 bases at a point on the DNA.

In the example given overleaf, the 'genetic code' is read from *left to right*. Only the small part of the strand of DNA close to the mutation point is shown. When DNA is transcribed into a sequence of bases in mRNA, one strand of the DNA is 'blocked'. The second strand, which is the one that is illustrated, is used for transcription. The details of this process are shown in Chapter 2.

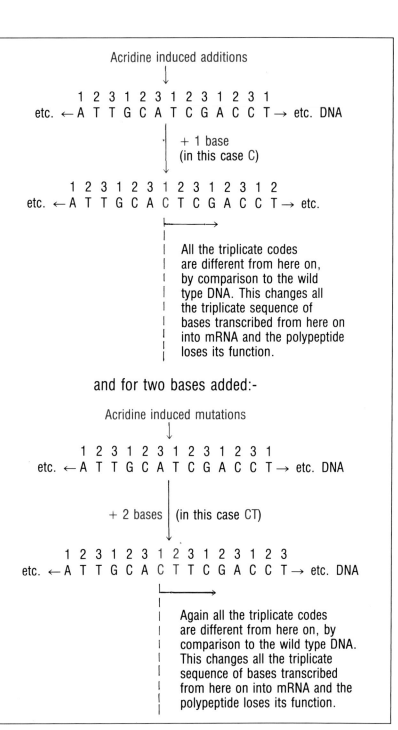

However when three bases are added there is an additional triplicate code introduced into the DNA, but all the other codes continue to be read in their original form.

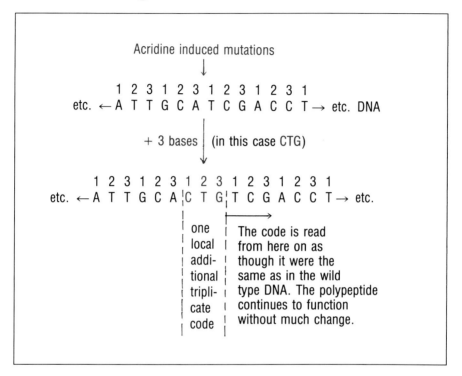

(ii) 1, 2 or 3 Bases Deleted

The deletion of one or two bases causes all the triplicate codes to the right of the deletions to be changed, so destroying the function of the polypeptide coded for by the DNA.

The deletion of three bases causes a local omission of one amino-acid, but elsewhere the polypeptide sequence of amino-acids is unaltered.

The results obtained from these experiments, adding or deleting bases, convinced Crick that the genetic code is a triplicate code.

(iii) Substituting 1, 2, or 3 Bases

The modifications made to the functional properties of polypeptides are not much altered if bases different from those found in the wild type are substituted at a small number of adjacent positions.

Chapter 11 deals with base substitutions in some detail.

Having established by 1962 that the genetic code is a triplicate code, it was then necessary to specify the sequences of triplicate bases that code for each of the naturally occurring amino-acids.

h) HOW THE GENETIC CODE WAS WORKED OUT USING FRAMSHIFT MUTATIONS

F.H.C. Crick suggested, in his 1961 article on the triplicate genetic code, that the lysozyme gene of the T4 virus might be suitable for cracking the genetic code. The genetic code ascribes a particular amino-acid to each of the 64 possible triplicate combinations of four different bases. By convention, the triplicate code relates the amino-acids to the base sequence on messenger-RNA (mRNA). Messenger RNA carries exactly the same sequence of bases as the 'blocked' sequence of bases in the double helix of the DNA gene from which mRNA is transcribed. The way this is done is explained at length in Chapter 2. The ascribing of amino-acids to messenger RNA base sequences therefore indirectly describes the base sequence in DNA. The only difference is that the base uracil in mRNA substitutes for the base thymine in DNA.

Two teams of collaborators, one at the University of Oregon, Eugene, USA (George Streisinger, Yoshimi Okada, Joyce Emrich and Judith Newton), and the second at the University of Osaka, Japan (Akira Tsingita, Eric Terzaghi and M. Inouye) worked out most of the genetic code by studying the changes in amino-acid sequences caused by adding or removing bases at known points in the lysozyme gene of T4 phage.

The alterations to the RNA base sequences were consequent upon the changes in the viral gene. An important fact has to be understood at this stage of the discussion. DNA has two strands. The order of bases that appears in messenger-RNA is the order of bases that appear in the 'blocked' strand of DNA. The 'blocked' strand of DNA is the strand that is *not* used to determine the order of bases in messenger-RNA. However, with U substituting for T in mRNA, the following diagram shows how the order of bases in DNA corresponds to the order of bases in mRNA.

Mutant genes in the 'blocked' strand of the lysozyme genes were analysed for their base sequences. These DNA base sequences were compared with the mRNA base sequences transcribed from the mutant genes. Both the sequences in the blocked strands of mutant DNA and in the mutant derived mRNA could then be compared with the changes in the order of amino-acids in the lysozyme polypeptide.

An analysis was then carried out to compare the base sequences of amino-acids that corresponded to the region of the gene under study. A full list of amino-acids and their mRNA codons is given on page 89.

One of the regions of the lysozyme gene's transcribed mRNA had the following sequence, from left to right, in the wild type.

```
etc. ← T A A C G T A G C T G G A → etc. DNA that is blocked
etc. ← A T T G C A T C G A C C T → etc. DNA used for transcription
etc. ← U A A C G U A G C U G G A → etc. mRNA transcribed from The DNA
```

Scientists who analyse DNA can search for a base sequence that matches the base sequence in a messenger RNA molecule. Although they have found a 'blocked' sequence of bases in the DNA, they nonetheless know they have found the gene that corresponds to a particular mRNA. This fact allowed the two teams mentioned above to crack the genetic code, which was finally worked out in its entirety by 1966.

For example, in an analysed short sequence of the wild-type mRNA bases, the corresponding order of amino-acids was:-

		later removed ↑						
mRNA base sequence	AC.	AAA	AGU	CCA	UCA	CUU	AAU	GC.
amino-acids	Thr	Lys	Ser	Pro	Ser	Leu	Asn	Ala

In one analysis of a mutant virus, there was the removal of the third base from the former codon for lysine. By reading the code from left to right, serine's first base, adenine, now becomes the third base of the codon to its left. All the codons to the right of lysine are changed. However, if a new base is inserted into the sequence, let us say guanine to the right of uracil in the former Asn codon, there is the return to the normal wild type reading frame from here onwards to the right:-

						inserted ↓	
AC.	AAA	GUC	CAU	CAC	UUA	AUG	GC.
Thr	Lys	Val	His	His	Leu	Met	Ala

A large number of analyses of the amino-acid sequences produced from the lysosyme mutant genes of T4 showed that the genetic code is DEGENERATE. For example, in the two genetic code sequences shown above, it seems surprising that the same amino-acid, leucine, corresponds to two different triplicate codes in the mRNA. In fact, many of the twenty naturally occurring amino-acids have more than one triplicate code. This is because there are 64 possible triplicate codes that can arise from four different bases. All of these 64 triplicate codes have been found in genes.

B.F.C. Clark and K.A. Marker at Cambridge University worked out that AUG is the genetic code for starting the synthesis of a polypeptide (GUG may also be used as an initiation codon). The termination codons were discovered by S. Brenner at Cambridge University and A. Garen at Yale University to be UAG, UAA and UGA.

A glance at the genetic code table quickly reveals the degenerate nature of the genetic code.

The full genetic code consisting of 64 triplicate codons had been fully worked out by 1966. It should be noted that the mRNA codons correspond to the equivalent sequence of bases on the BLOCKED strand of DNA (U substituting for T).

The details of DNA transcription into messenger-RNA and the translation of codons in messenger-RNA into amino-acid sequences in polypeptides are given in Chapter 2.

The full genetic code is:

UUU	Phenylalanine	UCU	Serine	UAU	Tyrosine	UGU	Cysteine
UUC		UCC		UAC		UGC	
UUA	Leucine	UCA		UAA	Termination	UGA	Termination
UUG		UCG		UAG	Termination	UGG	Tryptophan
CUU	Leucine	CCU	Proline	CAU	Histidine	CGU	Arginine
CUC		CCC		CAC		CGC	
CUA		CCA		CAA	Glutamine	CGA	
CUG		CCG		CAG		CGG	
AUU	Isoleucine	ACU	Threonine	AAU	Asparagine	AGU	Serine
AUC		ACC		AAC		AGC	
AUA		ACA		AAA	Lysine	AGA	Arginine
AUG*	Methionine	ACG		AAG		AGG	
GUU	Valine	GCU	Alanine	GAU	Aspartic Acid	GGU	Glycine
GUC		GCC		GAC		GGC	
GUA		GCA		GAA	Glutamic Acid	GGA	
GUG		GCG		CAG		GGG	

* The way the initiation codon for methionine is used to start polypeptide chains is discussed in Chapter 2.

By 1966 the genetic code had therefore been discovered. It should be noted that the code is not quite universal. The code in the DNA of some eukaryotic cell organelles is different from the code in viruses – see mitochondrial inheritance, page 211.

Chapter 1 Questions

1. What were the most significant steps taken to advance man's knowledge of genetic inheritance?
2. What experiments allowed Gregor Mendel to formulate:-
   i. the Law of Independent Segregation?
   ii. the Law of Random Assortment?
3. Were Lamarck, Darwin and Wallace really geneticists?
4. How was it proved that the material of inheritance in bacteria is DNA?
5. What is the indirect evidence supporting the notion that DNA is the material of inheritance in higher organisms?
6. What makes DNA an ideal material for determining inheritance?
7. What are the normal rules for base pairing in DNA?
8. How does tautomeric shift change the normal base pairing rules?
9. How is genetic mapping carried out in a haploid species by:-
   i. interrupted mating?
   ii. recombination frequencies?
10. Of what value are contransformation frequencies?
11. How was the 'one gene – one polypeptide' theory verified?
12. How is genetic mapping carried out in diploids?
13. What were the indications that the genetic code is a triplicate code?
14. How was the genetic code cracked?

2

THE TRANSCRIPTION AND TRANSLATION OF THE TRIPLICATE DNA CODES INTO A SEQUENCE OF AMINO-ACIDS IN POLYPEPTIDES

The chemical reactions of the biochemical pathways that endow 'life' to the collections of molecules contained in cells are usually inter-related. A large number of precisely controlled reactions have to take place in each living cell so that it can take in the nutrients it needs, in order to provide itself with energy-rich molecules, and so that it can reduce the poisonous or osmotic consequences of some of the chemical reactions. There is also an enormous variety of specialised structural or functional properties that confer specific character on each cell. The specialised properties of cells largely depend on the collection of polypeptides found in them.

The structures of widespread living materials such as muscles and membranes are fashioned by their polypeptides. The rates of biochemical reactions often depend on polypeptide enzymes. Structures made of sugars, salts, polypeptides or fats depend for their synthesis on enzymes. Chemical messenger molecules such as hormones are often polypeptides. Essential cell metabolism, such as the release of energy from chemicals, depends on polypeptides acting as oxidation and reduction molecules as part of the biochemical pathway for energy release. The carriage of oxygen in blood is increased by the polypeptide haemoglobin which holds oxygen onto the surfaces of red blood cells. A very wide range of characters found in the living cells of each species, in individuals of each species and in specialised parts of individuals are dependent upon their complement of polypeptides. Both specific character, and life itself, are therefore conferred by cellular sets of polypeptides.

During the long period of life on this planet some polypeptides have been handed on from primitive species to become widespread among a large number of species that descended from them. Other polypeptides became confined to only a small number of species or to only one species. The specialisation of cells within individuals restricted some polypeptides to local organs or structures in only a small proportion of individuals or only one species. The elucidation of the triplicate genetic code and its relationship with particular amino-acids has, in recent years, allowed the study of amino-acid sequences in polypeptide molecules to corroborate the fossil evidence relating the different species. The topic of amino-acid sequences in polypeptides is discussed in Chapter 11.

The evidence from microfossils found in the strata of very ancient rocks shows that living cells have existed on earth for at least three-and-a-half thousand million years. If only four bases have been used in the material of inheritance since life began, it is generally believed that a single or a duplicate genetic code would have been too restricting in the determination of amino-acid sequences put into polypeptides. There would not have been enough variation in phenotypes resulting from mutations in these unicellular, haploid asexual cells using a single or duplicate code. By contrast a quadruplicate code would present problems of control by introducing too many possible quadruplicate combinations, leading to a degree of complexity that evolution was unable to accommodate.

Although it is not at all certain, the consensus view is that the change from a single to a duplicate, or from a duplicate to a triplicate code, would be very disruptive from an evolutionary point of view. The evidence from the organelles of eukaryotic cells, such as mitochondria, which are believed to be very ancient in their origin and which contain their own DNA, shows that the genetic code in mitochondria is triplicate, but slightly different from the triplicate code discovered in viruses and in bacteria.

Before the relationship between the triplicate code in DNA and the sequence of bases in polypeptides can be studied it is necessary to examine some of the properties of polypeptides.

Polypeptides

The heading 'polypeptides' is chosen in preference to 'proteins' or 'enzymes' because all molecules having amino-acids in them joined together by 'PEPTIDE BONDS' are POLYPEPTIDES. However, the properties of proteins and of enzymes sometimes include factors other than sequences of amino-acids, such as the agglutination of more than one sort of polypeptide by bonds other than 'peptide bonds' (see Quaternary Structure, pages 100-101).

During the long period of evolution, natural selection has chosen only twenty amino-acids as the building units for all the polypeptides studied so far. However the twenty naturally occuring amino-acids can be joined together in an enormous number of different possible sequences to make polypeptides. Polypeptides vary in molecular weight from about 5,000 to many millions.

The Naturally Occurring Amino-Acids

The general formula of the naturally occurring amino-acids is:-

$$\text{(amino group) } H_2N - \underset{\underset{H}{|}}{\overset{\overset{R(\text{varies})}{|}}{C^*}} - COOH \text{ (acid group)}$$

C*. This atom has four different chemical groups attached to it. It is therefore an 'asymmetric' carbon atom. The orientation in space of the four groups is specific for each of the naturally occurring amino-acids. This has been established by putting each of them into aqueous solution and passing polarised light through the solution. All the naturally occurring amino-acids rotate polarised light to the left and they are therefore called the L-amino-acids.

R is one of the twenty different chemical groups that indentify each naturally occurring amino-acid.

The functional properties of polypeptides often depend on their solubility in the 'watery medium' found in cells. Each amino-acid has a different solubility in water and the number of each amino-acid, together with their positions within a polypeptide, determines the solubility of each polypeptide. It is therefore useful to categorise the amino-acids according to their solubility.

The Amino-Acids

(i) Hydrophobic (non-polar and not very soluble in water)

Eight of the naturally occurring amino-acids are non-polar and therefore hydrophobic:-

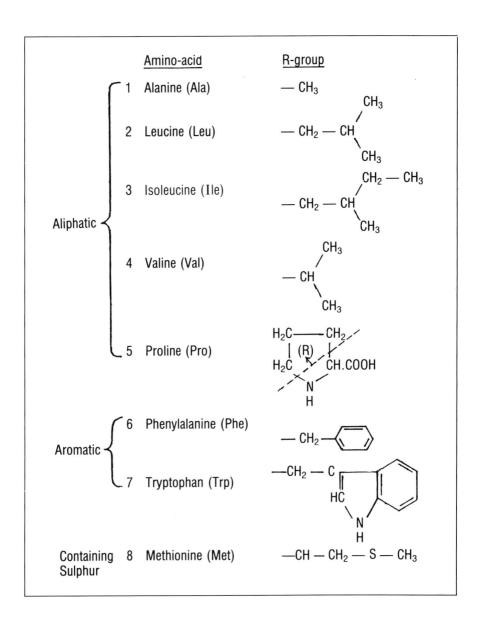

(ii) Uncharged Polar (fairly soluble in water)

Although there is no overall charge on the seven amino-acids in this group their polar localities make these amino-acids quite soluble when they are included in polypeptides:-

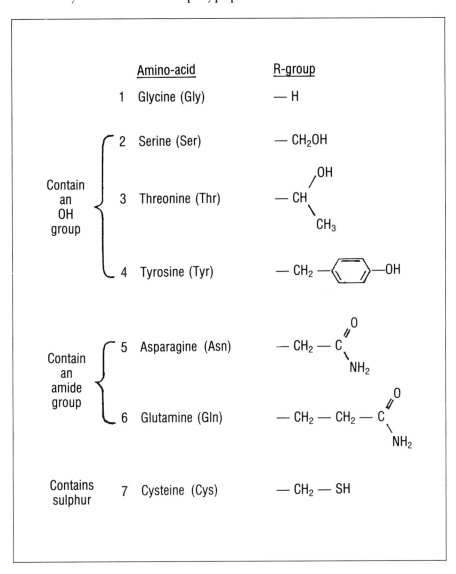

(iii) Charged Polar (very soluble in water)

The five amino-acids of this group carry an overall electrostatic charge and this allows them to be soluble in water.

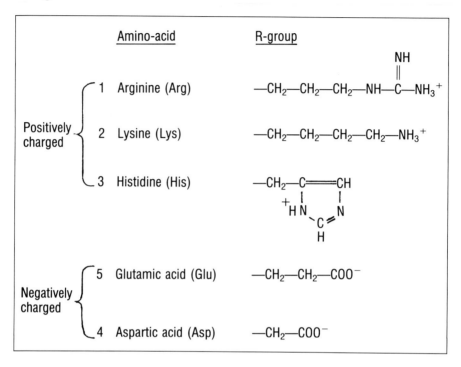

The Structure of Polypeptides

Under suitable conditions (some details are described on page 119) amino-acids can join together when the amino- group of one amino-acid reacts with the carboxyl- group of a second amino-acid. In simplified terms this can be shown as:-

Amino-acid 1 + **Amino-acid 2**

(OH and H *Excluded by reacting together*)

↓ The formation of a peptide bond

A DIPEPTIDE (+ H_2O)

The result of the reaction shows that the two amino-acids become joined by a peptide bond. Water is a by-product of the reaction.

It will be seen in the section on polypeptide synthesis (pages 113-123) that the genetic code determines a precise order of amino-acids that are joined together to make the polypeptides in cells. It has also been demonstrated that the genetic code in genes and the order of amino-acids in polypeptides are COLINEAR.

The Shape of Polypeptide Molecules

The shape of polypeptides has to be precise in that part of them that contains the 'functional' region of the molecule. For example if the function of a polypeptide is to be an enzyme it will only be able to lower the activation energy of the chemicals it catalyses if the shape of the enzyme in its functional region is precisely that which alters the bonding of one or other (or both) of the reacting chemicals to make the reaction easier. Some lack of precision in the non-functional regions of a polypeptide can occur without necessarily damaging the function of a polypeptide. Chapter 11 discusses some aspects of functional and non-functional regions of polypeptides.

Most polypeptides in living cells are surrounded by water. Water, by being a polar molecule, alters the shape of polypeptides so that the sequence of amino-acids found in living cells are those that have been put there by natural selection acting on their capacity to function correctly in a watery medium.

Polypeptides fall into two broad categories.

Fibrous Polypeptides

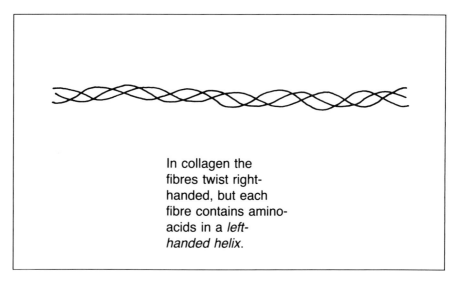

In collagen the fibres twist right-handed, but each fibre contains amino-acids in a *left-handed helix*.

Fibrous polypeptides are made of individual, filamentous chains of amino-acids joined laterally to form a stable, insoluble structure. Examples are myosin in muscle and the spindle fibres in dividing cells.

Globular Polypeptides

This category simply defines by shape all those polypeptides that are not long and thin. Among the multitude of examples are haemoglobin and many enzymes.

In the case of haemoglobin there are three amino-acids contained in side chains.

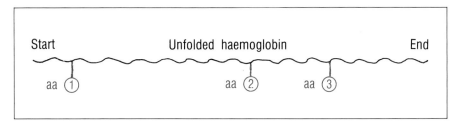

These three amino-acids become essential to the active site of the molecule, where oxygen is carried, when it takes up its folded, tertiary structure:-

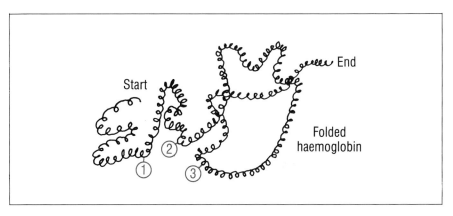

Both filamentous and globular polypeptides depend for their shape upon their amino-acids, and the order of them. The following systematic structural description shows how some of the chemical groups of the amino-acids in water help to determine the functional shapes of polypeptides:-

(a) Primary Structure

This is the linear sequence of amino-acids joined together by peptide bonds.

(b) Secondary Structure

When amino-acids form long chains they naturally twist left-handed, even if the strands so formed then twist right-handed to form the α-helix that characterises many polypeptides.

(c) Tertiary Structure

Superimposed upon the α-helix extensive folding occurs to give the complex structures found in many polypeptides.

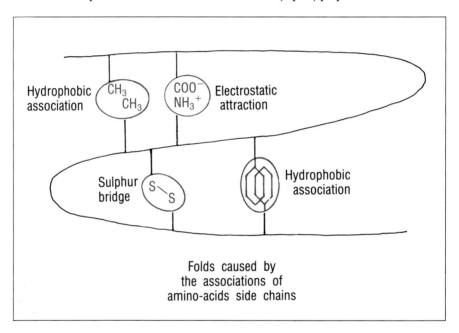

(d) Quaternary Structure

Some polypeptides have to associate with other polypeptides before they can function correctly. Two examples are given.

(i) Haemoglobin

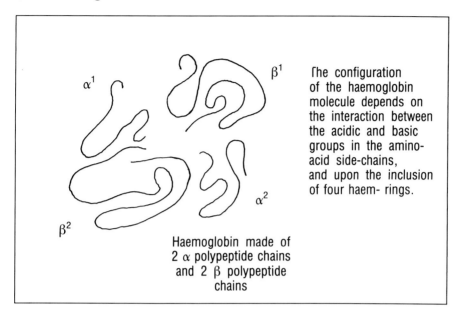

The configuration of the haemoglobin molecule depends on the interaction between the acidic and basic groups in the amino-acid side-chains, and upon the inclusion of four haem- rings.

Haemoglobin made of 2 α polypeptide chains and 2 β polypeptide chains

(ii) Tryptophan Synthesis Enzyme

One of the enzymes needed for the biosynthesis of the amino-acid tryptophan must have both the α and the β polypeptide constituents present in order to operate as a catalyst.

The vast range of polypeptides found in living things ultimately depend upon the triplicate codes in genes for their correct structure. The colinearity of the triplicate codes in genes and the order of amino-acids in polypeptides was proved using the collected work of a large number of scientists working in widely separated research centres.

Introduction to Polypeptide Synthesis

The following simplified scheme shows the essential processes of polypeptide synthesis:-

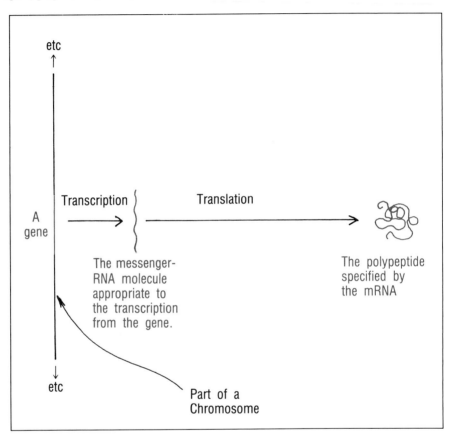

It is now necessary to make a comparison between ribonucleic acid and deoxyribonucleic acid, and to examine the processes of polypeptide synthesis in more detail.

Ribonucleic Acid (RNA)

Ribonucleic acid differs from deoxyribonucleic acid in several ways:

(i) Atomic Structure

a) There is one atom of oxygen more in the pentose sugar ribose by comparison to deoxyribose.

The numbers 1 to 5 refer to the carbon atoms.

b) The base URACIL is found in ribonucleic acid and replaces the thymine of DNA. Uracil pairs with adenine:-

Uracil differs from thymine as shown below:-

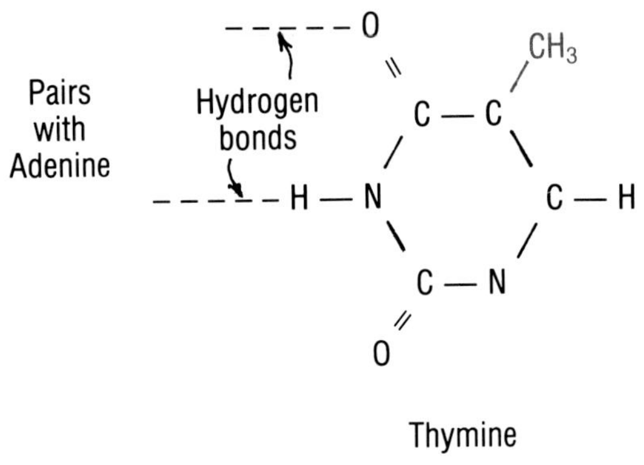

```
                         O            CH₃
                      ╲╲            ╱
  Pairs    Hydrogen    C — C
  with      bonds     ╱       ╲
  Adenine            H — N      C — H
                        ╲      ╱
                         C — N
                        ╱╱
                       O

                    Thymine
```

(ii) Molecular Structure

The differences in the atomic structures of the sugar and of uracil allow the molecular structure of RNA to be distinctly different from DNA. RNA has *no* double helix and the bases are often largely unpaired. Base pairing does occur in some types of RNA and confers a degree of rigidity to their structures (see transfer-RNA, page 110). However much of RNA is single-stranded.

(iii) Size and Mobility

RNA molecules are usually very much smaller than DNA molecules so allowing RNA molecules to be much more mobile than those of DNA.

(iv) Chemical Stability

Different types of RNA differ in their molecular stability. In general, however, RNA molecules exist for only a short time by comparison to the DNA of the cells in which RNA is made.

(v) Synthesis

When DNA is synthesised it usually involves the replication of the whole genome (Chapter 9). By contrast the synthesis of any one RNA molecule involves only the bases of DNA found within the gene specific for the synthesis of that particular RNA molecule. In general the synthesis of DNA is an 'all or nothing' process whereas the synthesis of the many RNA molecules needed in a cell is under dynamic and enzymic control, depending on the immediate needs of the cell. An exception occurs when locally damaged DNA is being repaired. Other exceptions are also known.

Polypeptide Synthesis

Much of the evidence for the processes of polypeptide synthesis comes from studies of the bacterium Escherichia coli. Although it is thought that polypeptide synthesis in all cells follows the same pattern as in E. coli, the description of the processes on the following pages refers largely to studies in this one bacterium species.

It is now known that the codes contained in bacterial genes are transcribed into intermediary, mobile molecules called 'messenger-RNA'. Messenger-RNA (mRNA) molecules can move easily to the sites of polypeptide synthesis, if they need to do so. In bacteria the site of polypeptide synthesis is often very close to its determining gene in the DNA, so mRNA in bacteria does not have to move far before its codons are translated into a polypeptide.

The TRANSCRIPTION of a Sequence of Bases in DNA into a Sequence of Bases in mRNA

mRNA is synthesised when double-stranded DNA separates to form two single strands. One of the single strands is thought to be 'blocked'. The other single strand has a sequence of unpaired bases available for pairing with bases of ribose nucleotides which freely diffuse in the vicinity of the DNA.

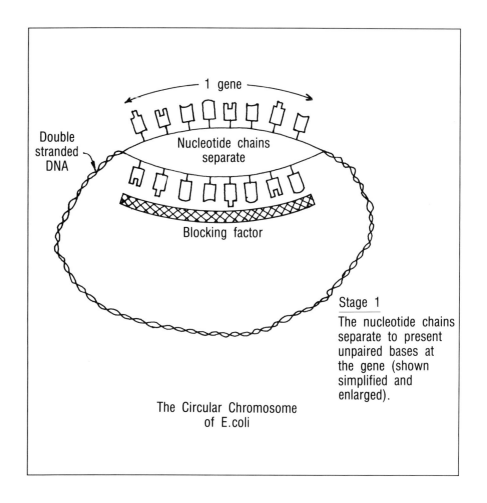

The freely diffusing nucleotides are then formed into an RNA chain called messenger-RNA. Enzymes are needed to separate the DNA into single strands and to join adjacent nucleotides along the RNA chain. The simplified diagram on the facing page shows how mRNA is synthesised. The number of bases shown in the diagram is much smaller than the number of bases in a gene.

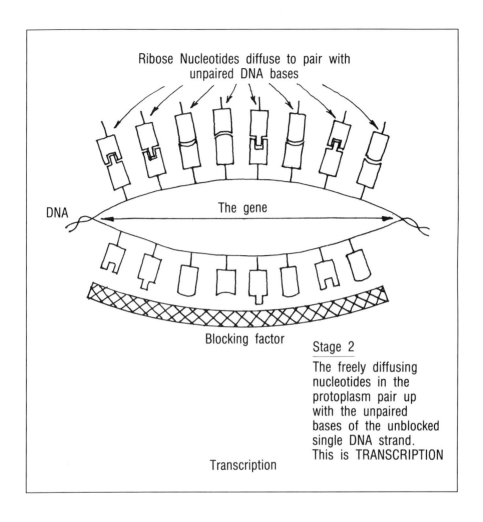

Nucleotides are then joined together with the help of the enzyme RNA polymerase:

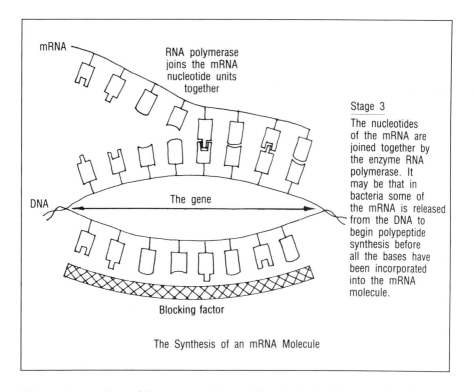

The Synthesis of an mRNA Molecule

Thus the order of bases on the unblocked single strand of DNA determines the sequence of bases on the mRNA. When the mRNA is completely broken away from the DNA the sequence of bases in mRNA can be used later for determining the order of amino-acids in polypeptides.

The theory that messenger-RNA carries the transcribed code was put forward more or less simultaneously by F. Jacob and J. Monod at the Institut Pasteur, Paris, by F. Gros at Harvard University and by S. Brenner and M. Meselson at the California Institute of Technology in 1960/61.

Other RNA Molecules

While the messenger-RNA contains the codes for determining the sequence of amino-acids in polypeptides, there are two other types of RNA needed for polypeptide synthesis.

(a) Ribosomal-RNA (rRNA)

Unlike mRNA, which is synthesised and degraded very quickly, ribosomal-RNA is much more stable. It is thought that there are only three different types of rRNA molecules in E. coli. This seems to imply that ribosomes themselves are not very specific in their associations with mRNA for polypeptide synthesis. The ribosomes/mRNA relationship is illustrated on page 121. In E. coli the ribosomes are made of two parts. Two of the rRNAs are found in the bigger part and one rRNA is found in the smaller part. The exact function of rRNA molecules is not yet clear but when they are removed the ribosomes become non-functional. Ribosome synthesis is the subject of Chapter 3.

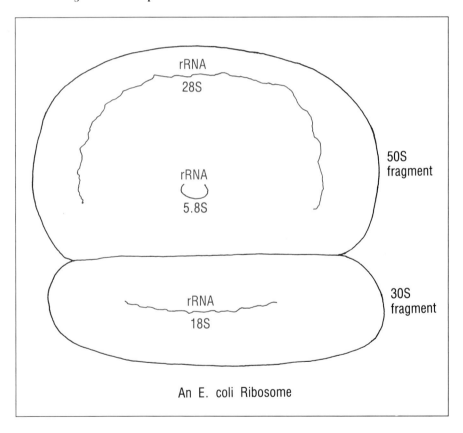

An E. coli Ribosome

(b) Transfer-RNA (tRNA)

The function of transfer-RNAs is to bind their specific amino-acids to their 3' end and to take the amino-acid to a specific site on the messenger-RNA. There is a small part of each tRNA, the ANTI-CODON, which can recognise the CODON on the mRNA, which in turn specifies the incorporation of one particular amino-acid. Each tRNA is fairly specific for one particular amino-acid (but see the Wobble Hypothesis page 125).

A simplified example of a transfer-RNA is shown in the next diagram.

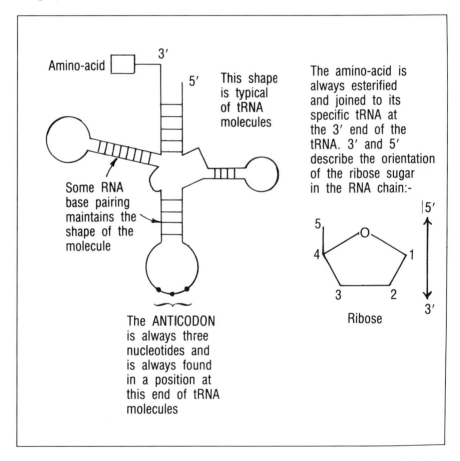

tRNAs contain about 70 to 80 nucleotides, some of which include bases other than A, G, C or U. The ANTICODON in tRNA is recognised in the presence of a ribosome by a CODON (see page 115) in mRNA. Thus the codon in mRNA, also three nucleotides, determines which amino-acid shall be included in a polypeptide molecule at the position of the codon. The way this is done is shown on page 119.

Ribosomes

In E. coli nearly all the ribosomes have a sedimentation constant (the layer in which they are found after centifugation, and measured in Svedberg units) of 70S. Their molecular weight is about 2.6 million (Hydrogen=1).

Each ribosome is made of two sub-units. One is of sedimentation constant 50S and the other of 30S. Sedimentation constants are *not* directly additive; they simply refer to the relative positions of molecules after centifugation.

The 50S sub-unit contains two molecules of ribosomal-RNA and about 35 different polypeptides with a molecular weight about 1.8 million altogether.

The 30S sub-unit contains one molecule of ribosomal-RNA and about 20 molecules of different polypeptides, molecular weight about 800,000 altogether.

None of the polypeptides found in each sub-unit are found in the other.

It seems that, when the purified molecules of all ribosomal polypeptides and all ribosomal-RNAs are mixed together, ribosomes are self-assembling in water.

The Function of Ribosomes

Ribosomes can capture and carry transfer-RNA anticodons to codons on the messenger-RNA. Because each anticodon automatically implies a specific amino-acid attached to the transfer-RNA, the genetic code can be transcribed from DNA into mRNA and thereby translated into a particular sequence of amino-acids which, in turn, specifies the structure and function of the polypeptide specified by a gene.

Genes

By comparison to the total length of a chromosome the length of DNA which codes for any one polypeptide is very small. Because many polypeptides are used as enzymes, the small region of DNA within which mutations lead to the loss of a single enzyme is defined as a gene. This led to the 'one gene-one enzyme' hypothesis.

Since some enzymes consist of the association of more than one polypeptide chain any changes in the structure of either sub-unit leads to the loss of enzyme activity. Thus a more accurate description of the gene is therefore:-

'One gene – one polypeptide'.

The proof of this has already been described in Chapter 1. However, it has recently become evident that there are several complications in defining a gene. This subject is discussed in Chapter 4. Nonetheless, for the bacteria in which fundamental genetics were studied, there is a very large proportion of cases in which the 'one gene – one polypeptide' hypothesis is true.

It is now appropriate to show how the base codons in genes transcribed into mRNA are made to be COLINEAR with the amino-acids in polypeptides.

Coding for Amino-acids on the mRNA

There are four possible bases at each base position along the mRNA molecule – A, G, C, or U (adenine, guanine, cytosine or uracil).

It has been proved that the mRNA codons for determining which amino-acids are joined to the polypeptide chain are contained in sequences of three adjacent, consecutive nucleotides. (The genetic code table for mRNA is given at the end of Chapter 1).

There are sixty-four possible arrangements of three (out of four) nucleotides on the mRNA. It is now known that sixty-one of these code for amino-acids. There is more than one combination of three nucleotides for nearly all the amino-acids, so making the genetic code DEGENERATE. The name given to each combination of three nucleotides on messenger-RNA is a 'codon'. The remaining three codons specify the termination of the polypeptide chain, making 64 codons altogether.

There is no 'punctuation' in the *chemical* form of the codons so they form a continuous series of bases.

It has now been shown that polypeptide chains are synthesised in a sequence of amino-acids starting at their amino- ends. It has also been shown that the 5′ end of the mRNA is near the start codon for the amino- end of the polypeptide. It happens that the 5′ end of the mRNA is made first at the transcription site of DNA; so in prokaryotic cells polypeptide synthesis may start before the mRNA molecule has broken completely free of the DNA.

It should be noted that the triplicate genetic code associates each of 61 possible triplicate codons in mRNA with one of the amino-acids or with a termination codon. *The order of bases in mRNA corresponds to the order of bases on the BLOCKED strand of DNA, U substituting for T in mRNA.*

Colinearity therefore passes indirectly from the blocked strand of DNA to mRNA and finally to the order of amino-acids in the polypeptide.

Starting the Synthesis of a Polypeptide – TRANSLATION

The codon on mRNA which starts the synthesis of a polypeptide chain is near, but not at, the physical 5' end of the mRNA.

It is thought that methionine is always the first amino-acid to be put into every polypeptide chain. It happens that there is only one codon for methionine – AUG. However methionine is an amino-acid that also appears in polypeptides in places other than at the amino- end of the polypeptide. There must therefore be specific mechanisms for allowing the initiating AUG codon to recognise the initiating methionine tRNA anticodon. The tRNA for subsequent AUG codons is modified so that mid-chain methionine is not attached to the initiating AUG codon. Methionine can react to form peptide bonds with both its amino- group and its carboxyl- group. The amino- side of *initiating* methionine is 'blocked' so that no amino-acids can be added at that side of it.

One further fact needs to be explained. Methionine does not appear at the amino- end of all completed polypeptides. It is thought, therefore, that the initiating methionine, together with other amino-acids attached to it by a series of peptide bonds, is often lost by cleavage from the polypeptide. This cleavage may occur before completion of the polypeptide or after the polypeptide has completely dissociated from the ribosome/mRNA complex.

The initiation of a polypeptide chain takes place as shown in the following diagrams.

(i) A 30S ribosome sub-unit attaches to the mRNA at the initiating codon for methionine

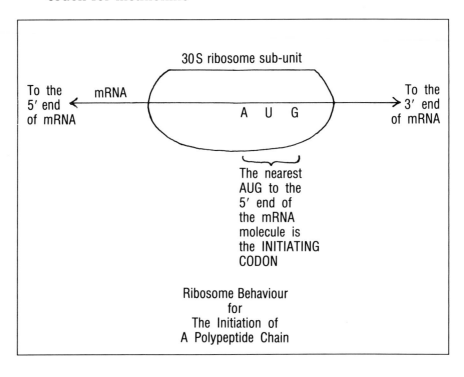

(ii) The form of the transfer-RNA which brings methionine to the initiating codon is specific for the initiating codon. The initiating transfer-RNA complex is recognised by the initiating codon, and the anticodon of the tRNA is paired with the codon on the mRNA.

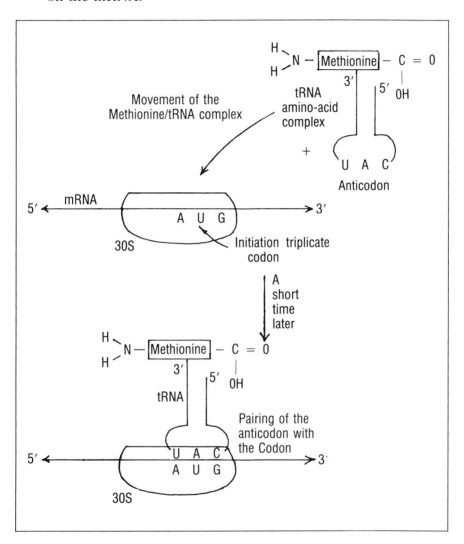

(iii) A 50S ribosome sub-unit now binds to the complex and the initiation process is complete.

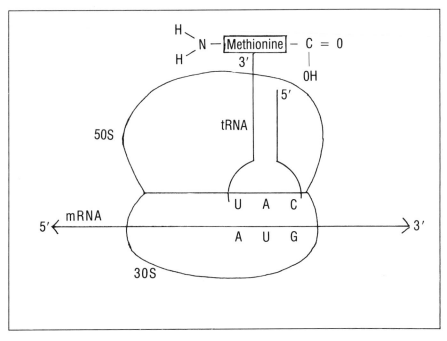

Elongation of the Polypeptide Chain

Messenger-RNA has no *molecular* 'punctuation' and consists of an uninterrupted sequence of nucleotides. For the purposes of illustration it is convenient to introduce 'punctuation' to define the beginning and end of codons.

The real:-

To 5' end ← A A U U A C U G U A A A U A G → To 3' end

is therefore rewritten:-

To 5' end ←. A A U . U A C . U G U . A A A . U A G . → To 3' end
            Asn   Tyr   Cys   Asn   Termination

(The amino-acids corresponding to the codons are written below them).

At any one time at least six nucleotides are covered by the 30S sub-unit of the ribosome. Elongation of the polypeptide chain proceeds as follows:-

(i) At the instant that a ribosome moves to cover a new codon, one of the two codons covered by it will already have a peptide or polypeptide extending from it. The other codon at this instant will have nothing extending from it.
The site where, at any instant, there is a peptide extending from it, is called the P (Peptide) site. The empty site is available for the attachment of an amino-acid appropriate to the codon and is called the A (Amino-acid) site.
In the same way that the initiating transfer-RNA required cofactors for introducing the molecule into the 30S ribosome sub-unit, so too are elongation factors necessary for introducing the tRNA of amino-acids into the 30S ribosome sub-unit, and for allowing the amino- group of the new amino-acid to react with the carboxyl- end of the peptide which has already been formed.

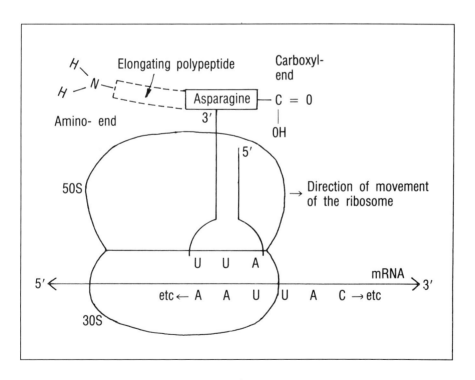

At the instant the ribosome moves to cover the next codon (UAC), the P site has a peptide extending from it and the A site is empty.

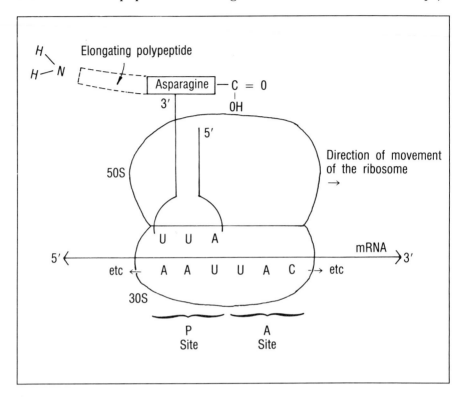

(ii) The transfer-RNA specific for the codon UAC (tyrosine transfer-RNA) makes use of elongation factors and the energy in GTP (a molecule similar to ATP) for introducing the tyrosine transfer-RNA into the ribosome. This allows the pairing of codon and anticodon and for the reaction between the amino- end of tyrosine and the carboxyl- end of the peptide, in this case the carboxyl- end of asparagine.

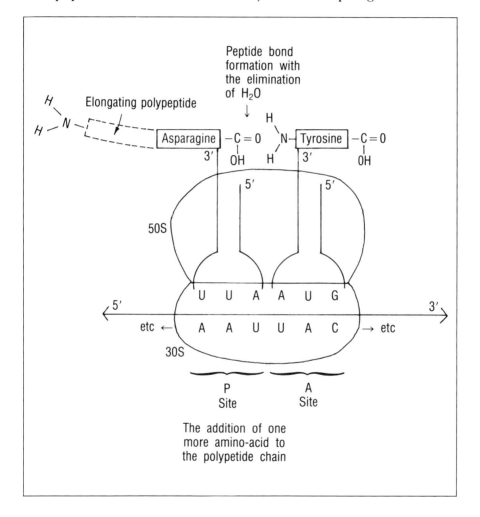

Ending a Polypeptide Chain

UAA, UAG and UGA are the codons for ending a polypeptide chain.

Wherever these termination codons appear along the mRNA they switch on a polypeptide release factor when they are covered by a ribosome in the A site. The release factor releases the polypeptide from the transfer-RNA held at the P site.

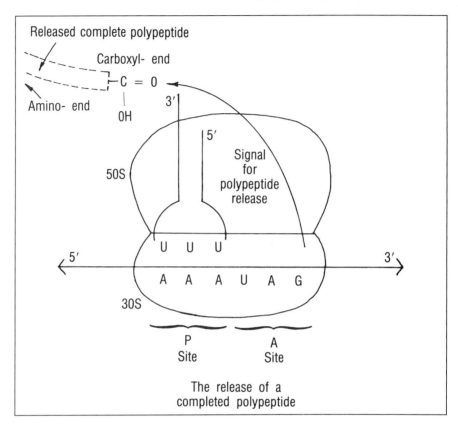

The release of a completed polypeptide

Ribosome Movement Along mRNA

Electron microscope photographs show that ribosomes are packed close together along the strands of mRNA in both prokaryotic and eukaryotic cells.

A general characteristic of ribosomes is that they can move past the transfer-RNA at the 'A site' as soon as the transfer-RNA at the 'P site' is released. It is thought that the ribosome changes shape near the 'A site' tRNA to allow movement past it. Once the ribosome has moved to the termination codon it breaks free from the messenger-RNA and may be used for the synthesis of another polypeptide. The ribosome movement diagram on the facing page shows how ribosome recycle.

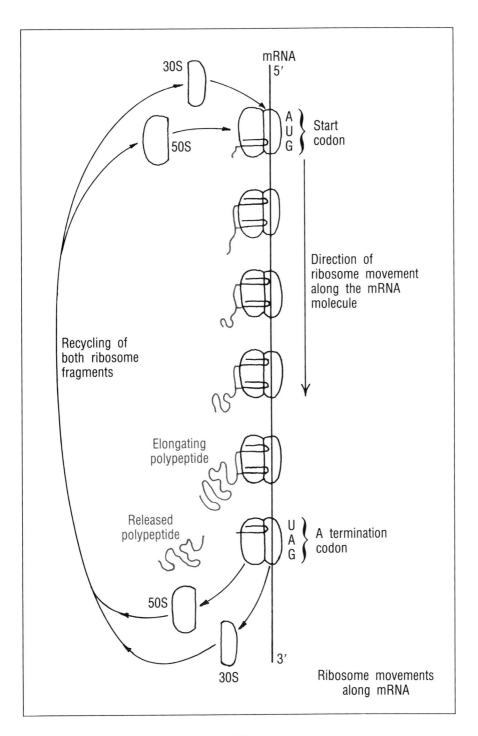

Summary of Polypeptide Synthesis

The following diagram (which is not drawn to scale) shows a schematic summary

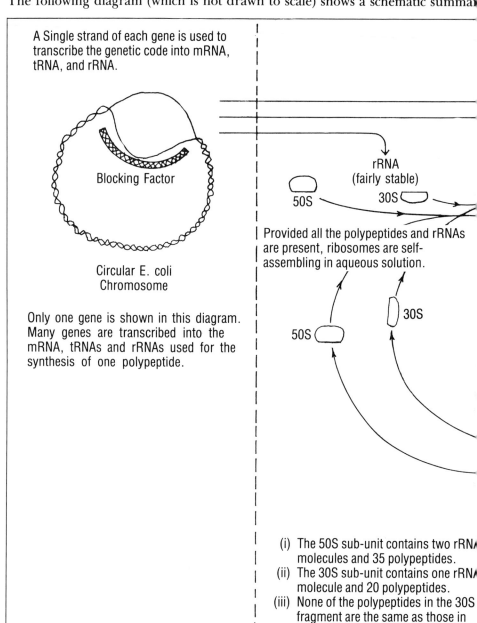

A Single strand of each gene is used to transcribe the genetic code into mRNA, tRNA, and rRNA.

Blocking Factor

Circular E. coli Chromosome

Only one gene is shown in this diagram. Many genes are transcribed into the mRNA, tRNAs and rRNAs used for the synthesis of one polypeptide.

rRNA (fairly stable)

50S   30S

Provided all the polypeptides and rRNAs are present, ribosomes are self-assembling in aqueous solution.

30S

50S

(i) The 50S sub-unit contains two rRNA molecules and 35 polypeptides.
(ii) The 30S sub-unit contains one rRNA molecule and 20 polypeptides.
(iii) None of the polypeptides in the 30S fragment are the same as those in the 50S fragment of a ribosome.

of polypeptide synthesis in E. coli:–

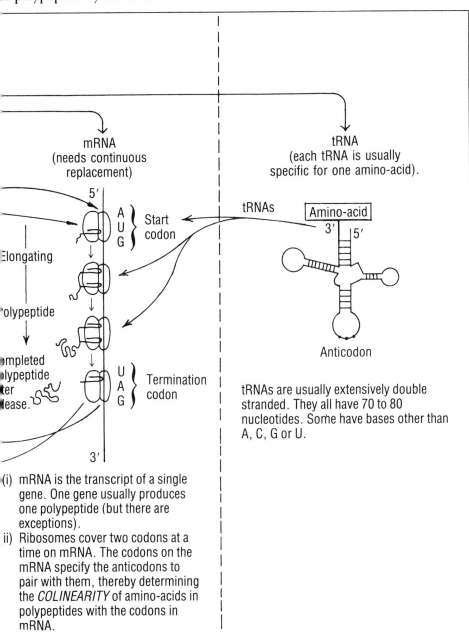

(i) mRNA is the transcript of a single gene. One gene usually produces one polypeptide (but there are exceptions).
(ii) Ribosomes cover two codons at a time on mRNA. The codons on the mRNA specify the anticodons to pair with them, thereby determining the *COLINEARITY* of amino-acids in polypeptides with the codons in mRNA.

tRNAs are usually extensively double stranded. They all have 70 to 80 nucleotides. Some have bases other than A, C, G or U.

Polypeptide Synthesis in Eukaryotic Cells

Whereas there is considerable polypeptide synthesis in prokaryotes close to the bacterial DNA, in eukaryotic cells most of the polypeptide synthesis is carried out at sites quite far removed from the genes of the chromosomes in the nucleus. Electron microscope photographs of eukaryotic cells show that there is a highly branched membranous system, the endoplasmic reticulum, that has a large number of ribosomes close to it on the cytoplasm side of the membranes. It is thought that the endoplasmic reticulum provides a channel, through which the RNA molecules made in the nucleus can move to the ribosomes where polypeptide synthesis occurs close to the endoplasmic reticulum. Pores in the nuclear membrane lead into the channels of the endoplasmic reticulum. Even if there is little resistance to the movement of RNA molecules in the channels of the endoplasmic reticulum, they are quite large molecules and have to cross the endoplasmic reticulum membrane to associate with the ribosomes in the cytoplasm. How they do this has yet to be explained.

The synthesis of eukaryotic polypeptides at the ribosomes is similar to polypeptide synthesis in prokaryotes.

The following diagram shows the relationship between the synthesis of RNA molecules in the nucleus and in the nucleolus, their transport and the site of polypeptide synthesis:

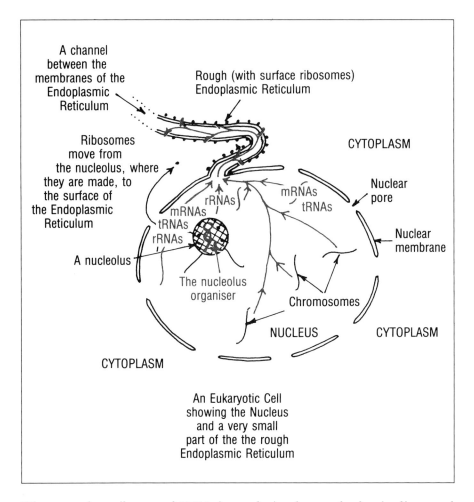

An Eukaryotic Cell showing the Nucleus and a very small part of the the rough Endoplasmic Reticulum

The way that ribosomal-RNA is made in the nucleolus is discussed in Chapter 3.

Ambiguity in Codon/Anticodon Recognition- The Wobble Hypothesis

F.H.C. Crick put forward his 'Wobble Hypothesis' in 1966. This hypothesis states that when the codons of mRNA come into close contact with the anticodons in tRNAs there are small changes in the shape of both mRNAs and tRNAs in the locality of their associations. This sometimes allows the 'wrong' base to be recognised on the codon, especially in the position of the third codon base:-

tRNA 1st anticodon base	mRNA 3rd codon base
U	can pair up with A or G
G	can pair up with U or C
I	can pair up with A, C or U.

(i) I is inosine, a rarely found base that occurs in tRNAs. It is synthesised from adenine but also has some properties of guanine. It can pair up with any of the three bases shown in the table.

(ii) It is thought that A on tRNAs always pairs up with U and that C on tRNAs always pairs up with G, irrespective of local distortion to mRNA and tRNA.

Ambiguity of translation of the genetic code allows some variation among the polypeptide molecules produced by the translation of the sequence of tripicate codes in a single gene. Such changes in translation have consequences for an individual but are much less important for evolution than the single base mutations that occur in DNA caused by tautomeric shift (pages 51-55 and Chapter 11) during DNA synthesis.

The Central Dogma

F.H.C. Crick summarised the processes that have been described in this chapter in his 'central dogma', which was made in 1970. By this time the experimental work on polypeptide synthesis in Escherichia coli and in viruses gave fairly clear proofs of the transcription and translation of the coded information in the genes of DNA. In his 'central dogma' Crick included the facts that DNA can, in certain circumstances, provide codes that control both DNA and RNA synthesis and, in special circumstances, can also control polypeptide synthesis without using RNA as an intermediary.

In the presence of specific enzymes, RNA can be used as the template for DNA synthesis. Chapter 8 on genetic engineering discusses the usefulness of this reverse transcription process. Chapter 13 dicusses some of the complexities of DNA/RNA synthesis in viruses and in other very small particles.

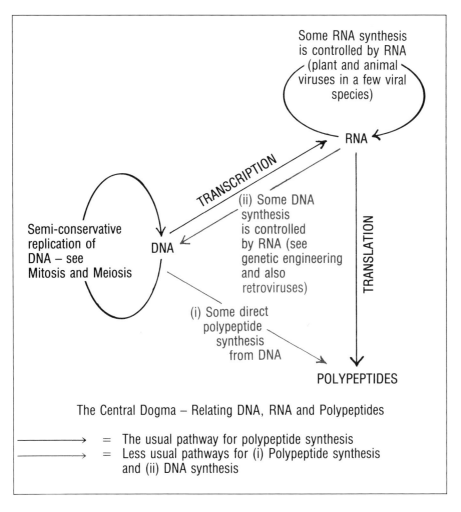

The Central Dogma – Relating DNA, RNA and Polypeptides

⟶ = The usual pathway for polypeptide synthesis
⟶ = Less usual pathways for (i) Polypeptide synthesis and (ii) DNA synthesis

Recent Discoveries about Eukaryotic Species

Chapter 4 discusses recent discoveries about gene transcription and translation in eukaryotic species. In particular, the section describing 'introns' and 'exons' needs to be understood.

Further complications about gene expression, caused by altering the sequence of bases when DNA is broken and rejoined in mobile genes are discussed in some detail in Chapter 14. Chapter 14 also describes a process called 'splicing' in which mRNA molecules are altered after they have been synthesised.

In several ways, therefore, the simplicity of prokaryotic polypeptide synthesis, as just described in this chapter, needs to be revised when eukaryotic polypeptide synthesis is described. It is now known, for example, that eukaryotic initiation codons and termination codons are contained within 'islands'. Within these islands are repeating CpG sequences. It is also known that some eukaryotic polypeptides are made more complex than those in prokaryotes by the presence of sugar residues.

Non-glycosylated Polypeptides in Eukaryotes

The cytosol is the watery, non organelle part of the cytoplasm in eukaryotic cells. Most polypeptide constituents of enzymes are synthesised in the cytosol by way of ribosomes, mRNAs and tRNAs. These soluble polypeptides do not have sugar side-chains added to them. Such polypeptides are described as 'non-glycosylated'. They are similar, therefore, to the polypeptides synthesised in prokaryotes.

Glycopolypeptides in Eukaryotes

The endoplasmic reticulum of eukaryotes is a widespread series of passages that pass through the cytosol. The membranes around the E.R. separate the contents of the so-called lumen (the watery medium inside the E.R.) from the cytosol. The E.R. membranes also connect the nucleus to other important organelles in such a way that nuclear products and E.R. lumen products can move from organelle to organelle without entering the cytosol.

Most polypeptides produced within the lumen of the endoplasmic reticulum of eukaryotic cells are glycopolypeptides. The polypeptide part of glycopolypeptides are synthesised in the conventional way using ribosomes, mRNAs and tRNAs. Various sugary chemicals are then added.

The majority of these glycopolypeptides have an oligosaccharide (a chain of sugar molecules) joined on to the $-NH_2$ group on the side chain of an asparagine residue in the polypeptide. These small sugars include glucose, mannose and N-acetylglucosamine. They are N- linked oligosaccharides.

In fewer cases, oligosaccharides are linked to the -OH group on a side chain of residues of serine, threonine or hydroxylysine. These are O- linked oligosaccharides.

The oligosaccharides are synthesised by the metabolism of the cytosol. They are then transported across the membrane of the endoplasmic reticulum into the lumen, where they bind to the appropriate amino-acid residues.

Glycopolypeptides are transported in the lumen of the endoplasmic reticulum to an organelle called the Golgi Apparatus. There is considerable modification to the glycopolypeptides during their passage to the Golgi and there is further modification to them once they are contained within this organelle.

It also seems that many non-glycosylated polypeptides travel to the Golgi Apparatus and are modified within it. The Golgi therefore acts as a modifying agent for an enormous number of genetically controlled products. Essentially it is a storehouse for gene controlled products. These products can be secreted either intra-cellularly or extra-cellularly when the Golgi is stimulated to do so.

So far as glycopolypeptides are concerned, mature glycopolypeptides can have complex oligosaccharides or oligosaccharides rich in mannose attached to them. Both types of oligosaccharide can attach to the same polypeptide simultaneously.

The functions of the products secreted by the Golgi Apparatus are more appropriately described in a biochemistry book than in one that concentrates on genetics.

Chapter 2 Questions

1. What confers the property of 'life' to cells?
2. What are polypeptides?
3. What categories of polypeptides are there?
4. How many types of RNA are there?
5. What are the functions of RNA molecules?
6. How long do RNA molecules exist?
7. How do the life spans of RNA molecules compare with the life span of DNA?
8. What are ribosomes?
9. What are the functions of ribosomes?
10. How does the initiation of a polypeptide molecule occur?
11. How does the elongation of a polypeptide occur?
12. How does the termination of a polypeptide molecule occur?
13. Summarise the processes of polypeptide synthesis.
14. How is polypeptide synthesis organised in eukaryotes?
15. Summarise the 'central dogma', describing the relationship between polypeptides, RNAs and DNA.
16. What are glycopolypeptides? Where are they synthesised and what happens to them after synthesis?

3

RIBOSOME SYNTHESIS IN THE NUCLEOLI OF EUKARYOTIC CELLS

Reference was made, in the chapter on polypeptide synthesis, to RIBOSOMES, which are needed for translating the codons in mRNA into a sequence of amino-acids in polypeptides. In eukaryotic cells there are specialised parts of the nucleus in each cell, the NUCLEOLI, in which some of the important components of ribosomes are made. Some of the polypeptides used for making ribosomes are, however, encoded in the genes of the nuclear chromosomes *outside* the nucleolus.

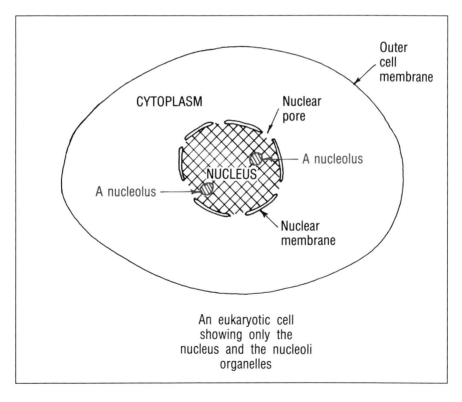

An eukaryotic cell showing only the nucleus and the nucleoli organelles

The nucleoli are very rich in RNA and they respond to stains in a different way from the rest of the nucleus, where DNA predominates.

There are usually a specific number of nucleoli per nucleus (2, 4 and 6 are common), but these can fuse together after cell division to form less numerous large nucleoli.

An analysis of the contents of nucleoli generally gives:-

    80% protein

    10% RNA

    10% DNA

In Xenopus laevis (the African horned toad) about 0.3% of all the genes in the genome are found in the nucleoli. This probably represents about 600 genes per haploid genome and all or nearly all of these 600 genes encode the production of ribosomal RNA molecules.

The number of nucleoli per cell in a species can vary, and has been shown, in some species, to be inherited. Xenopus laevis populations contains individuals with:-

    2 : 1 : 0 (nucleoli/cell)
    25 50 25 (% of population)

In this species it has been shown that individuals with 0 nucleoli per cell can make mRNA and tRNA, but they cannot make rRNA. Polypeptide synthesis is therefore inhibited in these individuals and they die. Individuals with 1 or 2 nucleoli develop normally.

Ultrastructure of Nucleoli

A nucleoprotein called chromatin surrounds the nucleolus, but this is *not* the same as the normal protein/lipid membranes found in cells. The surrounding chromatin engulfs a specific part of each of two homologous chromosomes, leaving parts of the same chromosome pair outside the nucleolus.

Fibrils have been observed associated with the engulfed DNA. At other places in the nucleolus granules have been observed.

The following simplified diagram shows the ultrastructure:

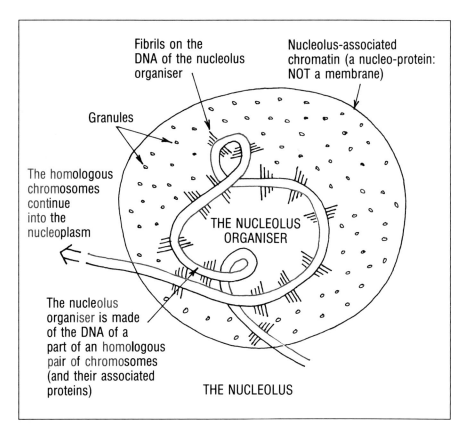

The engulfed part of the DNA is called the NUCLEOLUS ORGANISER. If one of the homologous chromosomes in the nucleolus is examined in more detail, the fibrils associated with the DNA make a Christmas Tree shape, repeated several times:-

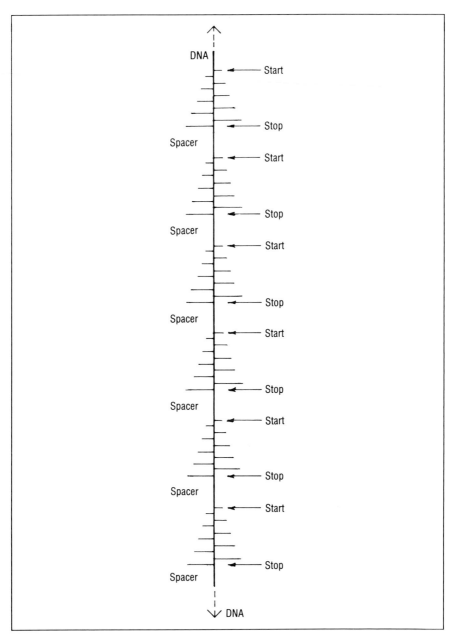

The fibrils are increasing lengths of rRNA. The transcription of the repeating genes in the nucleolus organiser produces identical rRNA molecules, each of which contains three different rRNA fragments, later used for making ribosomes:-

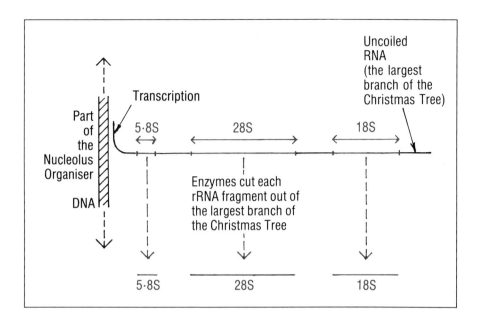

The Synthesis of Ribosomes from rRNA and Polypeptides

Eukaryotic cells have slightly larger ribosomes in their cytoplasm than those found in bacteria. The two components of eukaryotic, cytoplasmic ribosomes have sedimentation constants of 60S and 40S (by comparison to the 50S and 30S of bacteria)

The 60S and 40S components of ribosomes are made from rRNA molecules associating with about 70 different polypeptides. Some of the polypeptides are made in the nucleoplasm from where they move into the nucleoli.

Although the 5.8S, the 28S and the 18S rRNA fragments are made in the nucleolus, there is also a 5S rRNA fragment found in eukaryotic ribosomes and this is made by the transcription of a nuclear gene outside the nucleolus.

In summary, the ribosomes of eukaryotic cells are made of two fragments, as shown in the following diagram:-

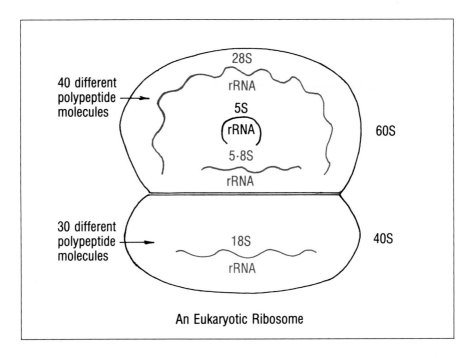

An Eukaryotic Ribosome

When the 40S sub-unit combines with the 60S sub-unit, their combined sedimentation constant is 80S (sedimentation is not necessarily additive), and 80S is the sedimentation constant of ribosomes in the cytoplasm of eukaryotic cells.

A Scheme Showing the Synthesis of Ribosomes in Eukaryotic Cells

The diagram on page 137 shows that the synthesis of ribosomes in eukaryotes depends on products made in the nucleolus, in the nucleus and in the cytoplasm.

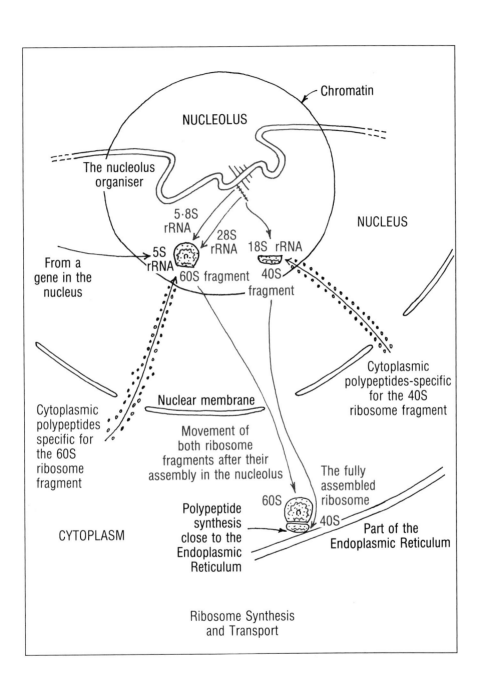

Ribosome Synthesis and Transport

The Rate of Ribosome Production

Differences occur in the activity level of cells, the most common reasons being cell division during growth and development, and when living things do work. Because ribosomes break down after about 10 days, there may not be enough of them to cope with an increased demand for the synthesis of polypeptides. Steroid hormones are produced at times of cellular stress to increase the production of rRNA polymerase (an enzyme used for rRNA synthesis in the nucleolus).

rRNA polymerase increases rRNA production by:-

   (i) Causing the nucleolus organiser to uncoil more.
   (ii) Making easier the initiation processes of rRNA synthesis.
   (iii) Assisting the joining together of nucleotides.

Some cells (animal oöcytes) can make extra genes, in the form of circular DNA, in their nucleoli. The circular, extra DNA is separate from the nucleolus organiser and it is made at times when rapid polypeptide synthesis is needed, requiring extra ribosomes.

The Relationship Between the Activity of the Nucleolus and Other Factors That Control Gene Expression

The activity of the nucleolus is not only associated with the increased requirement for ribosomes at times of increased cell activity, but it is also related to cell differentiation. The relationship between rRNA synthesis, mRNA synthesis, hormones and inducers or repressors is discussed in the chapter on cell differentiation, Chapter 14.

The Replacement of Missing Nucleolus Genes in Mutants During Cell Division

The synthesis of DNA during mitosis (both sexes) and meiosis (males only) has been shown to insert missing genes into the nucleolus organiser in Drosophila species. A mutant fly zygote (the fertilized egg) can thereby be changed into a wild type adult during mitosis, and the synthesis of extra DNA in mutant males during the production of male gametes from testis mother cells by meiosis allows the inheritance of the offspring to be the wild type for nucleolus organiser genes.

The Number of rRNA Genes Related to The Inherited Number of Nucleoli In Each Cell, and Related to The Activity Levels of Cells

In Xenopus laevis tadpoles with two nucleoli per nucleus, there are about 1,600 nucleolar genes per cell, each gene coding for rRNA. Tadpoles with one nucleolus per cell have about 800 genes per cell coding for rRNA, and those with no nucleolus have none.

Nucleoli become larger and make extra rRNA genes when the activity level of a cell is high.

The Non-specific Recognition of mRNA by Ribosomes

Ribosomes made by the nucleoli in eukaryotic cells seem to be identical to each other and it is thought that they can be used for the translation of the codons in *all* mRNA molecules into polypeptides.

The Nucleoli at Cell Division

The nucleoli break down at the same time as the nuclear membrane breaks down during cell division, but the chemical contents of the nucleoli may stay closely associated during the processes of cell division.

At anaphase of mitosis (photograph in Chapter 9), it has been observed that there are a number of secondary constrictions (thin regions of the shortened chromosomes, other than the primary constrictions which attach to the spindle). The number and size of the secondary constrictions coincide with the number and size of the nucleoli in the cells in which they have been observed. The significance of this is not yet clear.

The Transport of Ribosomes to Where They are Used in the Cytoplasm

The mechanism for transporting ribosomes through the chromatin around the nucleolus and through the nuclear pores is not yet known.

When they reach the cytoplasm, the ribosomes can float free either as two separate sub-units or combined together. Ribosomes are usually found close to the endoplasmic reticulum and it is close to the endoplasmic reticulum that polypeptide synthesis occurs.

Transmission electron micrograph of an animal cell (a B-lymphocyte white blood cell) taken from guinea pig bone marrow. The nucleolus can be seen near the centre of the nucleus. The dark regions close to the inner membrane of the nucleus is heterochromatin in these plasma cells, a typical position in such cells. Heterochromatin is a complex of DNA and specialised polypeptides (it is described in chapter 4).

The endoplasmic reticulum is the widespread membranous system dispersed widely throughout the cytoplasm. On the cytosol side of the E.R. can be seen dark dots. Each dot is a ribosome, the several dots giving the name 'rough' endoplasmic reticulum to the E.R. of this animal eukaryotic cell. (Other cell structures – organelles – also appear in the micrograph, but they do not concern us here).

In the micrograph, the separation of the membranes on either side of the lumen of the E.R. is considerable. This swollen E.R. shows that the lumen is full of transcribed and translated molecules, in particular the antibodies produced by this white blood cell.

*Photograph by Don Fawcett.
With permission from the Science Photo Library, London.*

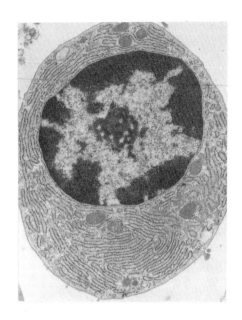

Pancreatic acinar cell fixed with osmium and stained with lead hydroxide. The E.R. is *not* swollen in this photo, suggesting less transcription and translation than in the photo above.

From The Cell, *by D.W. Fawcett W.B. Saunders Co., with permission.*

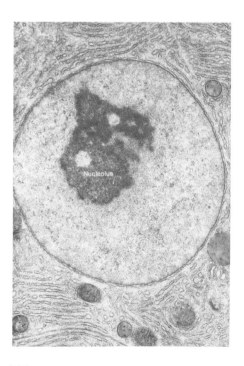

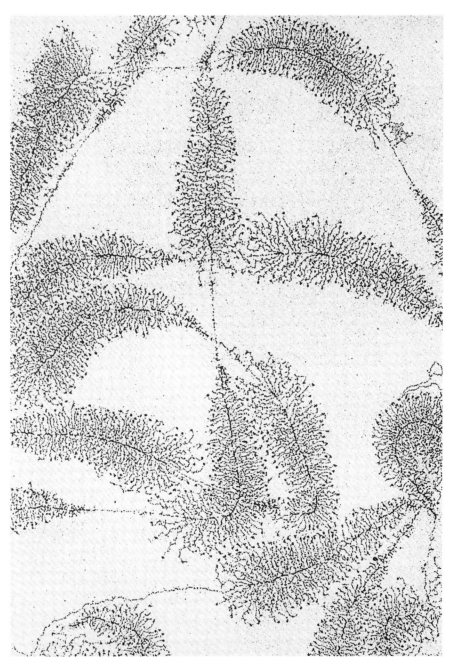

Electron micrograph of a portion of a dissociated nucleolar core isolated from an oöcyte of *Notophthalmos viridescens*. Christmas Trees are clearly visible. (Micrograph courtesy of Oscar Miller, from Miller and Beatty, Science 164: 955–957, 1969.)

Human rRNA Genes

Many human cells contain about ten million ribosomes. Multiple copies of each of the four different rRNA genes are needed if this very large number of ribosomes is to be produced quickly for use in new cells after cell division. Each human, haploid genome contains about 200 rRNA genes, distributed on the five chromosome pairs 13, 14, 15, 21 and 22. The arrangements for including parts of all five of these chromosomes within the nucleolus to form the nucleolus organiser are not known in detail. Nor is the relationship between the nucleolus organiser regions on homologous chromosomes in diploids, such as humans, understood in their coordinated functions.

Chapter 3 Questions

1. What is the ultrastructure of a nucleolus?
2. What is the function of a nucleolus organiser?
3. What are the components of:-
    i. A prokaryotic ribosome?
    ii. An eukaryotic ribosome?
    iii. Ribosomes found in organelles such as mitochondria and chloroplasts?
4. Summarise the production processes of ribosome assembly.
5. Describe the translocation of ribosome components in eukaryotic cells.
6. What governs the rate of ribosome synthesis?
7. What happens to nucleoli when cells divide?
8. Does the number of nucleoli per nucleus vary among individuals of a species?
9. What is known about human rRNA genes?

4

GENES, SUPER-GENES, ALLELES AND SUPER ALLELES, WITH SOME COMMENTS ON HOW GENES ARE CONTROLLED

Genes

Genes are mostly found in chromosomes, plastid DNA and plasmids. Genes are also found in a variety of mobile genetic elements. Plasmids are discussed at some length in Chapter 8. Plastids are discussed in Chapter 6. Mobile genetic elements are described in Chapter 13. This Chapter considers genes in chromosomes.

The cells of living things contain either single chromosomes (haploid cells such as bacteria or the gametes of diploids), or the chromosomes occur in homologous pairs (diploids) or the chromosomes are found in higher homologous multiples than two of the haploid number (polyploids). The significance of haploidy, diploidy and polyploidy are discussed in the chapters on mutations and evolutionary genetics.

Genes are partly defined by their *location* on single chromosomes (haploids), on pairs of chromosomes (diploids) or on the multiple chromosomes of polyploids.

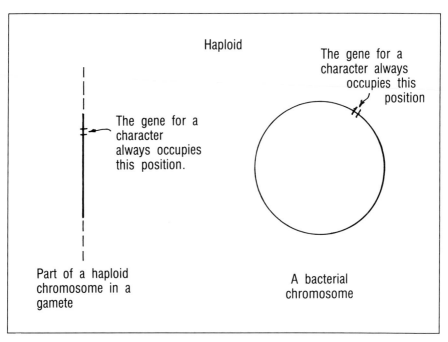

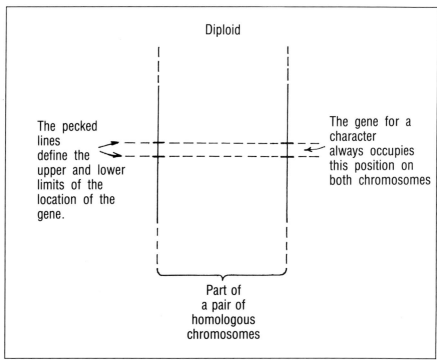

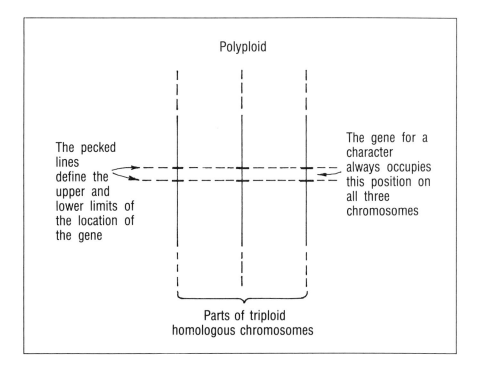

Examples to Show The Positions of Genes in a Haploid Species and in a Diploid Species

Some gene positions on the circular chromosome of the haploid species Escherichia coli and on the pair 1 chromosomes in the diploid species Drosophila melanogaster are shown in the two diagrams overleaf.

(i) Haploid – Escherichia coli

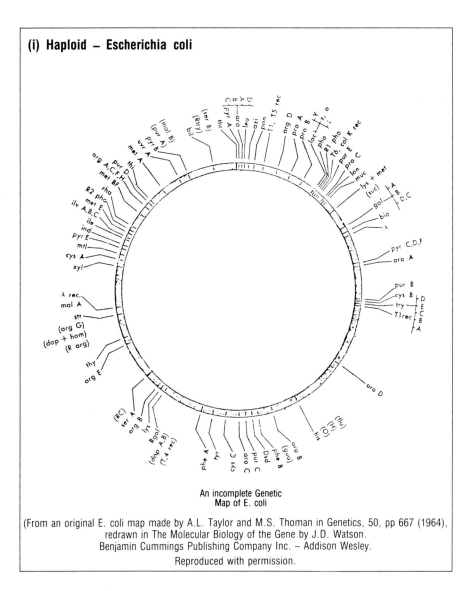

An incomplete Genetic Map of E. coli

(From an original E. coli map made by A.L. Taylor and M.S. Thoman in Genetics, 50, pp 667 (1964), redrawn in The Molecular Biology of the Gene by J.D. Watson.
Benjamin Cummings Publishing Company Inc. – Addison Wesley.
Reproduced with permission.

The gene positions in the E. coli diagram are for genes that control biochemical processes. The D. melanogaster genes overleaf are those that control visually perceived characters.

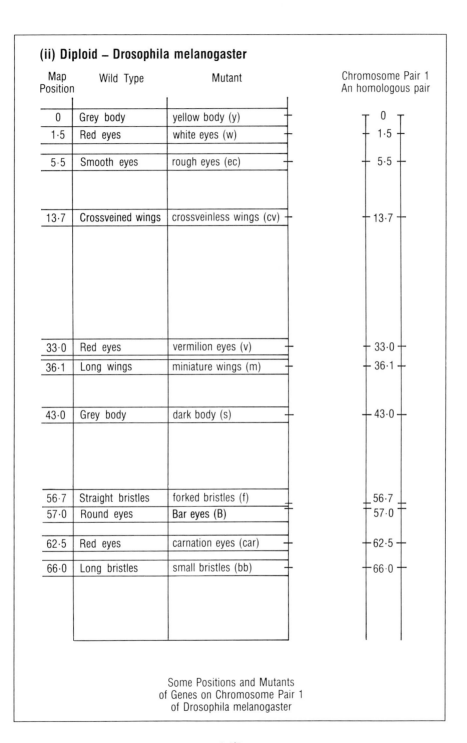

Some Positions and Mutants
of Genes on Chromosome Pair 1
of Drosophila melanogaster

An Attempt to Define a Gene

Each gene:-

> (i) Can be defined by its *position* on a chromosome (haploids) on a pair of homologous chromosomes (diploids), or on homologous multiples of chromosomes (polyploids). A gene can therefore consist of more than one piece of DNA in diploids (two pieces) and in polyploids (more than 2 pieces). It should be noted that the actual composition of bases in a gene is called an ALLELE (page 155).
>
> (ii) Is responsible for the production of *one* polypeptide. Although several alleles may be present in polyploids, or two alleles present in diploids, it is common that one allele dominates over the other so that only one polypeptide is produced for each gene:-

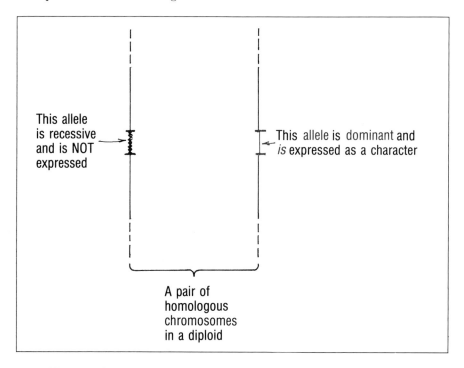

> (iii) Consists of a linear set of mutable sites. The mutable sites are the bases of DNA.
>
> (iv) Each mutable site can exist in four different forms, each form being one of the four bases found in DNA.

Problems Related to the Definition of a Gene

(a) Crossings-over

Crossings-over occur commonly at meiosis in diploids, during recombination in haploids, and instances are known of crossings-over at mitosis. It is tempting to believe that crossings-over always occur between genes, rather than within them. Experiments using the virus T4, growing in the bacterium E. coli, have clearly shown that crossings-over can occur within genes. Crossings-over points cannot therefore be used for defining the positions of genes.

(b) Overlapping Genes

Evidence has come from the virus φX 174 to show that the same sequence of bases in DNA can be used for two (at least) overlapping genes.

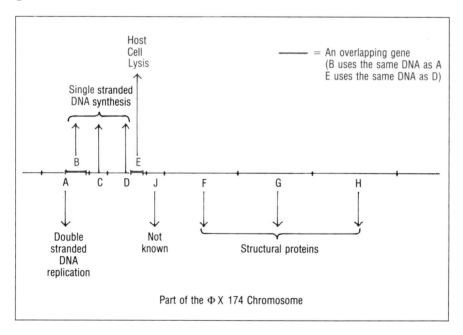

Overlapping genes have also been found in eukaryotes such as mice. Overlapping therefore makes the definition of a gene by 'position' not entirely satisfactory.

(c) 'Redundant' DNA Within Genes

In some eukaryotic and viral DNA it has been shown that 'redundant' loops occur along the length of a gene in such a way that the coding for mRNA ignores the loops and only a limited part, perhaps as little as 5% of the DNA is used to code for the transcription of mRNA.

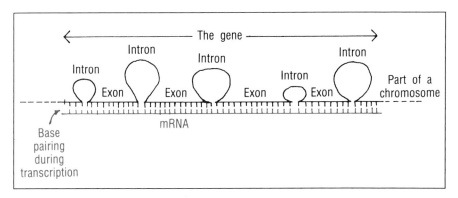

At the boundaries between introns and exons there are similar, perhaps identical nucleotide sequences on the DNA. Small particles – associations of RNA and proteins – are found where introns meet exons and the RNA has a base sequence complementary to the DNA base sequence in this region:-

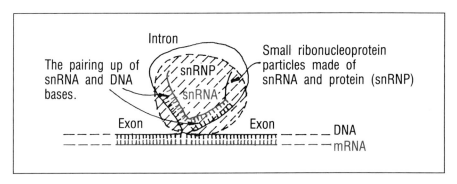

The sequence of bases in many genes are therefore interrupted before transcription occurs.

A specific example of an exon has been found in the short arm of the human Y chromosome. Some details of the gene containing the exon, which plays a significant part in human sex determination, are given in Chapter 16.

(d) Incomplete Dominance

In some diploid genes there is incomplete dominance of one allele over the other. In such instances one gene may be responsible for the production of more than one polypeptide because each allele produces a different polypeptide and each polypeptide is expressed in the phenotype. An example is flower colour in sweet peas.

	♀		♂
P	RR (red)	x	WW (white)
gametes	R	x	W
F_1		RW (pink)	

In this case the presence of both red and white alleles in F_1 heterozygotes causes the production of both red and white pigments, the blend of which is pink.

e) Codominance

An example is the genetics of human blood groups. If I^A is the allele for blood group A and if I^B is the allele for blood group B then:-

$I^A I^A$ gives blood group A
$I^B I^B$ gives blood group B
$I^A I^B$ gives blood group AB

Blood group AB is a distinct blood group and is *not* a blend of blood groups A and B.

f) Mobile Genes

Several examples of mobile genes are known.

i. DNA can move from the genome of an invading virus such as Bacteriophage λ to become incorporated in the chromosome of the bacterium Escherichia coli.

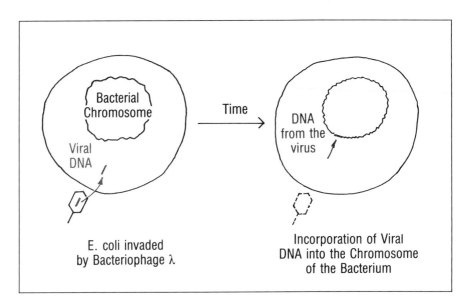

E. coli invaded by Bacteriophage λ

Incorporation of Viral DNA into the Chromosome of the Bacterium

ii. An example of a bacterial plasmid gene in E. coli that moves to become incorporated into the chromosome of the same individual bacterium is illustrated on page 220. The example shows that the expression of the mobile DNA is different when it is in the plasmid by comparison to when it is in the chromosome.

iii. An example is given on page 489 of a mobile gene in humans that changes from being an undamaging gene, when it is in one chromosome, to become a cancerous gene when it transfers to a different chromosome.

iv. The translocation of genes from place to place on the genome is fairly common in some species. The mobility of genes in this way can alter the base sequence in DNA where the break occurs, because mobile DNA from another part of the genome can join on to the broken end of the DNA. Mobile genes also alter the base sequence where they become attached in their new position. The alterations to base sequences in this way are one of the ways for creating genetic variation. Chapters 8 to 14 discuss further aspects of genetic change.

v. Mobile genes in complex, genetic systems, such as the immune system in mammals, allow the shuffling of a small number of genes to control the synthesis of a very large number of different polypeptides. An example to show how mobile genes, in the immune system sets of genes in mice, greatly increase the variety of antibodies available for fighting disease is given in Chapter 14.

vi. Transposons are fragments of DNA that can move into and out of chromosomes, sometimes becoming incorporated in one place, sometimes in another.

The relationship of mobile genetic elements with their host cells' genomes, of course including host cell chromosomes, is the subject of Chapter 13.

g) The Modification of mRNA Between Transcription and Translation – SPLICING

Some genes are known which give rise to several different polypeptides, all of which depend for their synthesis on the same base sequence in DNA. It seems that there can be the modification of some mRNA molecules between transcription and translation. The modification process is known as 'SPLICING'. Splicing is known to occur in genes of the large virus, Adenovirus, that infects animal cells. Another example is that of the splicing of the mRNA transcript of the ovalbumin gene in chickens. Splicing has also been observed in the production of mammalian antibodies. An example of antibody splicing is given on page 468.

h) Changes in Gene Expression Caused by Gene Amplification

The manufacture of large numbers of repeated, identical genes along a length of DNA is known to occur from time to time. It is known that there are consequences for the differentiation of cells. In some cases repetition, also called gene amplification, is known to cause cancer.

i) Deletion of A Gene

The deletion of regulatory genes can sometimes remove repressor genes. In some cases the absence of such repressor genes can allow the derepression of cancer genes.

It is important to note that, although mutations are the ultimate source of new alleles, it is NOT the arising of new alleles in diploid species, which already have a proportion of genes with alternative or several alleles, that has the most significant influence on differences between individuals of the species. Mutations have to be studied together with crossings-over at meiosis, genetic drift and natural selection. The interactions of these four factors are discussed in Chapters 9, 10, 11 and 12.

Super-Genes

It is thought that a sizeable proportion of genes lie adjacent to other genes involved in the expression of a phenotypic character. For example the genes for making all the enzymes needed for a biosynthetic pathway often lie beside one another on the DNA. In E. coli the amino-acid histidine has ten adjacent genes that are used for its synthesis. The positions of the genes along the DNA do not necessarily occur in the same order in which the enzymes are used in catalysing the reactions of the biosynthetic pathway.

In eukaryotic species some inherited characters have discontinuous expression, with no intermediate forms. Discontinuous phenotypic expression is called 'genetic polymorphism'. Single genes with clearly dominant and recessive alleles commonly produce polymorphism. There are, in addition, several known examples of super-genes that have clearly dominant and recessive super-alleles giving rise to polymorphism:-

> (i) The light and dark forms of the peppered moth, Biston betularia.
> (ii) The male and female projections in primroses, Primula vulgaris.
> (iii) The shell colour in the snail Cepaea nemoralis.
> (iv) O and A blood groups in humans.

The Multiple Effects of Super-Genes

Genes usually have multiple effects because their polypeptide products operate in unison with, or opposed to other genes. Sometimes the multiple effects are dependent, in part at least, on closely linked genes in super-genes:-

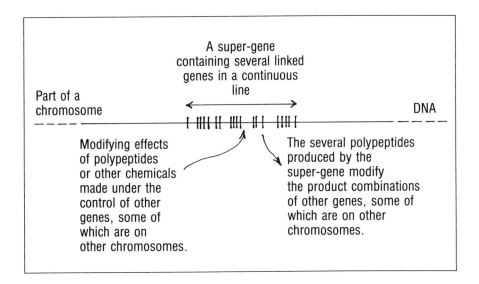

Alleles

The term 'alleles' refers to a distinct order of bases found in the DNA of a gene. Diploids will therefore have two alleles for each gene. The alleles may be the same or they may be different. The ways their alleles are inherited by an individual are shown in the genetic crosses of Chapters 5 and 7.

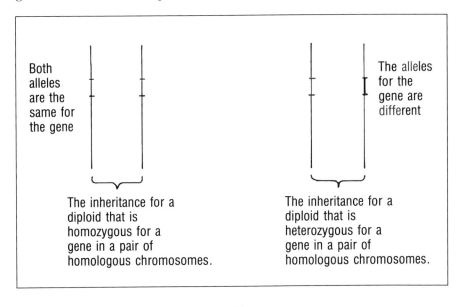

If 'gene' defines 'position', the term 'allele' defines 'what' is in the gene.

Mutations, either as single base substitutions or as larger changes to chromosomes (see Chapter 10 which describes some of the evolutionary implications of mutations), are the ultimate causes of creating new alleles. Genetic drift and natural selection has allowed a number of stable alleles to exist for a proportion of the genes in each species. The multiple effects of the alleles inherited by an individual give it its characters.

It is important to note that, although mutations are the ultimate source of new alleles, *it is NOT the arising of new alleles in diploid species that is the most significant influence on differences between individuals*. A careful study of the effects of crossings-over at meiosis, chapter 9, needs to be made to gauge the relative importance of mutation and crossings-over for creating variation.

Super-alleles

If a viable mutation occurs within a super-gene there may be multiple effects:-

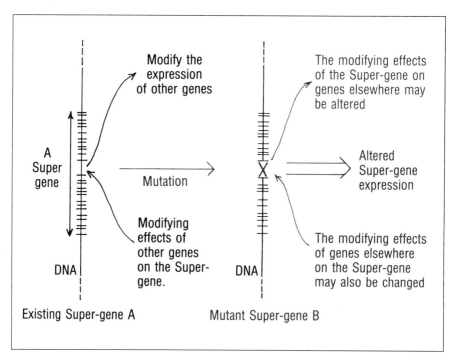

For the mutation in B to cause polymorphism there has to be the dominance of either A or B over the other super-allele. An examination of the dark and light forms of the peppered moth shows the difference between the wing markings of this species. The dark super-allele dominates over the light super-allele in this case. When the British Biston betularia is crossed with a closely related North American moth, the multiple effects are altered to break down the dominant effect of the dark super-allele so that forms intermediate between dark and light are found among the offspring.

The Control of Genes

The rate at which cells need to produce each of the polypeptides encoded in their genes depends upon:-

(i) The existing polypeptide concentration for each gene.
(ii) The concentration of the end product of a biosynthetic pathway for which several genes produce enzymes.
(iii) The stage of the life-cycle of the cell (resting or dividing).
(iv) Damage to the cell.
(v) Environmental factors.

Each of these factors operate together with genetic controls for switching genes on and off.

Genetic Control of Genes

Two types of molecule, synthesised under genetic control, assist in the 'switching off' and 'switching on' of genes.

(a) Repressors

Repressors are polypeptides which regulate the production of mRNA by binding at specific mRNA transcription sites on DNA close to the mRNA gene. Repressors stop mRNA synthesis, so stopping the production of a gene's polypeptide.

Repressors are often inactive on their own and need to be associated with a co-repressor to make them active:-

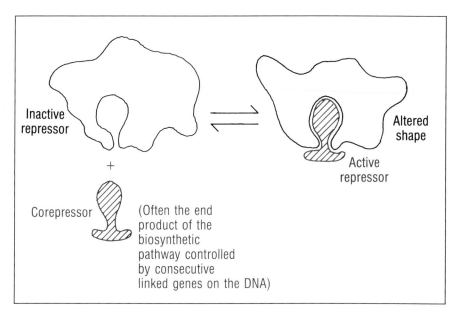

The chemical configuration of the active repressor causes it to bind to part of a gene to stop the synthesis of the gene's mRNA:-

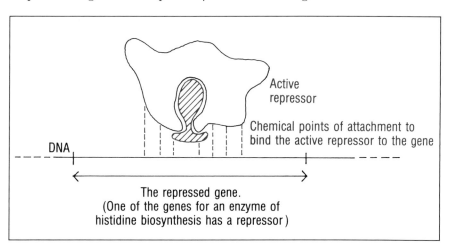

(b) Inducers

Inducers increase the rate of polypeptide production. They work by associating with repressors to cause the repressors to dissociate from a gene, thereby 'switching on' the gene. The site of attachment of a repressor is called '*The Operator*'. The following diagrams show the relationship between an Inducer, a Repressor, an Operator and a gene:-

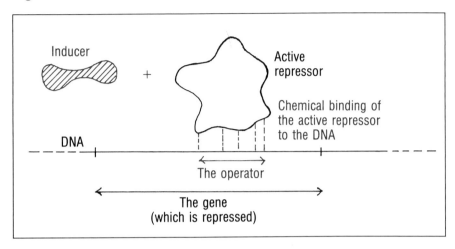

The inducer then associates with the repressor to deactivate it:-

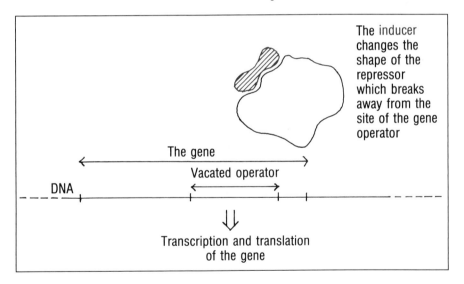

An Example to Show How Inducers and Repressors Control the Operation of a Gene

In the 'lac' region of the chromosome of Escherichia coli, there are genes that code for the enzymes of a biosynthetic pathway and some of these genes are controlled by an inducer and a repressor:-

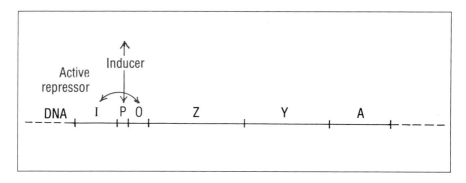

I controls the production of the active repressor.

The inducer is produced from a gene beyond the region of DNA shown in the diagram. P is the promoter.

O is the operator.

The genes of the biosynthetic pathway are A, Y and Z.

A is a gene that functions without being influenced by the promoter P or the operator O.

P has to be associated with an inducer molecule before the promoter can switch on mRNA transcription at genes Y and Z.

O has to be associated with the active repressor from I before the repressor causes the operator to switch off the promoter, so stopping mRNA transcription by Y and Z.

Chapter 14 also discusses gene control.

Zinc Fingers in Gene-controlling Polypeptides

It was discovered in 1987 that certain polypeptides have protruding 'fingers', often associated with the heavy metal zinc, which readily associate with DNA. There seem to be specific DNA loci where the zinc-finger polypeptides bind in order to carry out their functions as inducers or repressors. Zinc-finger polypeptides probably have a considerable influence on the differentiated expression of genes.

Differentiation is discussed in Chapter 14.

An example of a zinc-finger polypeptide and its influence on sex determination in humans is given in Chapter 16.

Genes in the Chromosomes of Plants and Animals

The single, circular chromosome of bacteria operates in a straightforward way to produce the mRNAs needed for making the polypeptides of the cell. However the form of DNA in eukaryotic cells is much more complex. The DNA is closely associated with large numbers of different polypeptide molecules.

The polypeptides can be divided into two groups according to size – histones and non-histones. It is thought that the polypeptides of histone size form complexes with DNA called NUCLEOSOMES, between which there are non-complex regions of DNA where there is only one histone.

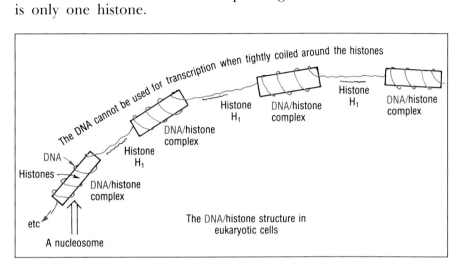

The DNA/histone structure in eukaryotic cells

It seems that RNA transcription from eukaryotic chromosomes only occurs in the NUCLEOSOME regions, and then only when the DNA dissociates from the nucleosome:-

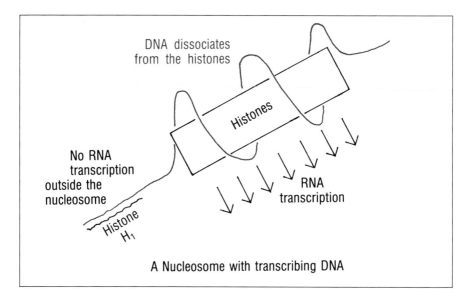

A Nucleosome with transcribing DNA

It is likely that nearly all eukaryotic chromosomal genes are found in multiple copies arranged linearly quite close to each other on the DNA, separated from each other by spacer regions:-

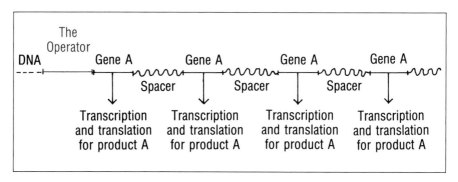

Multiple genes, dissociated from the histones, may be controlled by one operator region. It is estimated that only between 5% and 10% of animal DNA is used for transcription. This contradiction of the economy of nature, in not taking the least energy consuming pathways, has yet to be fully explained for evolution. One advantage is that, whenever genes are duplicated, or occur in multiple copies, there can be the modification of a proportion of the genes by mutations without destroying the ability of unaltered genes to produce the polypeptides needed by individuals. 'Trial and error' mutations, modifying one or other copies of multiple genes, can therefore occur without damaging all copies of each vital gene. The example of myoglobin and haemoglobin is given in Chapter 11.

Redundant DNA and spacer regions generally allow more 'trial and error' in modifying one or other copy of each existing multiple gene, redundant region or spacer region, without damaging all copies of each vital gene.

In brief, eukaryotic DNA can be described as containing single copy genes, slightly repetitive DNA and highly repetitive DNA. The proportions of each type of gene in each species is not yet known.

The Advantage of Arranging DNA in Nucleosomes

The arrangements of DNA, chromosomal polypeptides and RNA in nucleosome chains are collectively called CHROMATIN. One considerable advantage of cells in eukaryotic species is that chromatin, and histones in particular, cause the super-coiling of DNA. Super-coiling very greatly reduces the physical length occupied by DNA. DNA can therefore be contained in much smaller spaces than would otherwise be the case. If DNA had its super-coiling removed, it would be 100,000 times as long as it is in chromatin.

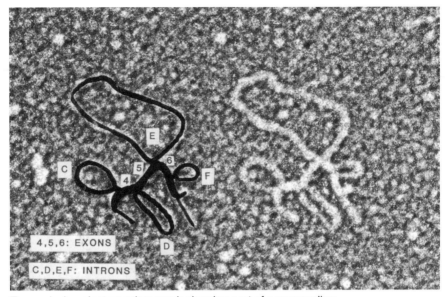

Transmission electron micrograph showing part of a mammalian gene. The introns C, D, E and F are seen as loops. Exons are marked 4, 5 and 6 and can be seen to form a continuous sequence.

Magnification x 131,000

Photograph by P. A. McTurk.
With permission from,
The Science Photo Library, London.

Part of protein framework and surrounding DNA from a histone-depleted metaphase chromosome.

From *The Cell*, by D.W. Fawcett, with permission of W.B. Saunders & Co.

Chapter 4 Questions

1. What is the difference between a gene and an allele?
2. What is a super-gene?
3. What is a super-allele?
4. Attempt to define a gene.
5. Why are simple definitions of a gene unsatisfactory?
6. Write notes on:-
    i. Crossings-over.
    ii. Overlapping genes.
    iii. Introns and exons.
    iv. Incomplete Dominance.
    v. Mobile genes.
    vi. Splicing.
    vii. Gene amplification.
7. What are 'zinc fingers'?
8. How do 'zinc fingers' assist in controlling the repression or induction of gene expression?
9. What are histones?
10. How is eukaryotic DNA arranged to associate with histones?
11. What is chromatin?
12. Into what three catagories of repetitiousness does DNA usually fall?
13. Of what advantage to cells is the super-coiling of DNA in chromatin?

5

INHERITANCE DETERMINED BY CHROMOSOMAL DNA

Genetic crosses cause an unnecessary degree of difficulty to some students. This difficulty can easily be overcome provided the genes and alleles are visualised on the chromosomes and on the extra-chromosomal DNA. For the genetic crosses of this Chapter, and in Chapters 7, 14 and 16, much of the presentation includes diagrams of the DNA in which the inheritance particles are found. This visual approach to genetic crosses, together with standard methods for drawing genetic cross tables, makes the understanding of genetic crosses very easy. Students should approach the subject in the knowledge that it can be understood easily and that solutions to genetic problems can usually be found by following very simple procedures.

In order to study the ways that the alleles or super-alleles in genes and super-genes determine the phenotype of an individual it is necessary to divide genetic inheritance into:-

> i. The expression of alleles in chromosomes.
> ii. The expression of alleles in extra-chromosomal DNA.
> iii. The modification of allele expression by chemicals in cells.

Chapter 6 describes extra-chromosomal inheritance and some examples of the expression of extra-chromosomal alleles are then given in Chapter 7. An example of the modification of gene expression by cytoplasmic chemicals is given in Chapter 14.

Genetic Crosses Depending on Chromosomal Genes

In species that depend on DNA for their inheritance, most of their genes and super-genes are on the chromosomes. Whenever it is the chromosomal genes that control characters it is possible to make predictions about the results of genetic crosses, depending on which alleles or super-alleles are found in the chromosomes.

Haploid organisms such as bacteria have only one allele for each character, so each gene is expressed according to the single allele present. Diploid species contain pairs of chromosomes so there are usually two alleles for each of the chromosomal genes. The expression of each gene in diploids depends on the dominant/recessive relationship of the two alleles present. Polyploid inheritance is more complex and depends on the expression of three or more alleles at each gene.

The alleles that are found in a gene define the GENOTYPE. The way that a gene is expressed as a character is called the PHENOTYPE. For example in the diploid fruit fly species Drosophila melanogaster there may be a dominant allele and a recessive allele present in the same gene on a pair of homologous chromosomes:-

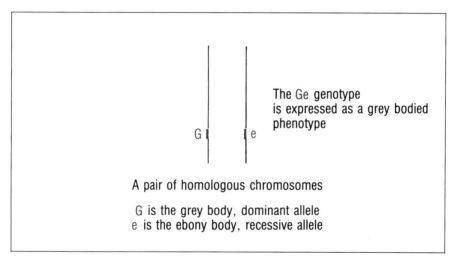

The Ge genotype is expressed as a grey bodied phenotype

A pair of homologous chromosomes

G is the grey body, dominant allele
e is the ebony body, recessive allele

Homologous chromosomes in diploid species are pairs of chromosomes upon which lie equivalent genes. One chromosome of an homologous pair comes from the male gamete and one comes from the female gamete. For example the diploid number (2n) of chromosomes in Drosophila melanogaster is eight. The diploid number is derived from the contribution of four chromosomes from each gamete at fertilisation.

In the diagram overleaf the pair of homologous chromosomes containing the body colour gene show the position of the gene in red.

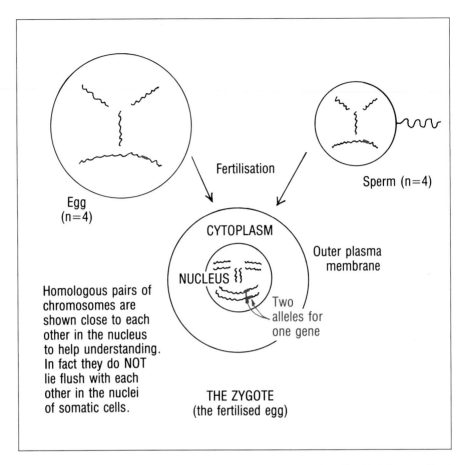

Some details of the processes of the type of cell division called MEIOSIS, which causes the production of gametes by the reproductive organs, are described in Chapter 9. In order to understand the genetic crosses in this chapter it is only necessary to understand that, when gametes are produced in diploids by the sex organs, the diploid number of chromosomes is reduced to the haploid number. If only one pair of chromosomes are illustrated, the following diagram shows that only one allele per gene is passed into each gamete. This point is crucial for understanding the genetic crosses of this chapter.

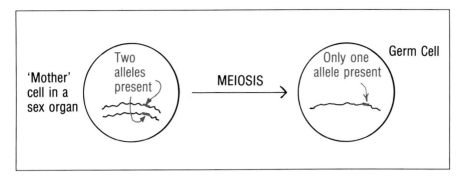

Which of the two alleles goes into a gamete from the mother cell is a matter of chance. A large number of mother cells produce gametes during the life-time of an individual. An equal number of the two alleles present in the mother cells goes into the gametes, one allele per gamete. It is for this reason that the allele which goes into any gamete produced by an heterozygous parent, such as Ge, can be *either* G *or* e.

Returning to the Drosophila melanogaster body colour cross mentioned earlier, the expression of the body colour gene depends on which alleles are inherited by an individual from the gametes of its parents. For example a pure bred, grey body fly is homozygous for the grey body allele:-

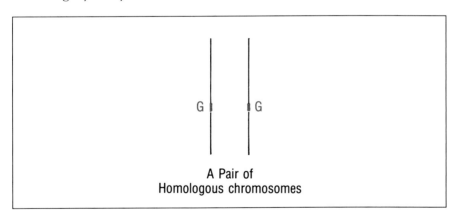

In a similar way, a pure bred ebony body fly is homozygous for the ebony body allele.

If a cross were arranged between a pure bred, grey body female and a pure bred, ebony body male the following scheme predicts the genotypes and phenotypes of the first filial (F_1) and second filial (F_2) generations.

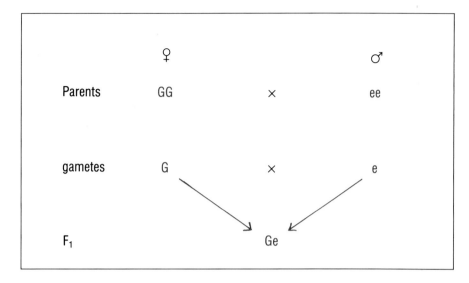

Any fly which inherits the dominant allele G in its body colour gene will be grey bodied. Flies with the genotype Ge in the F_1 generation will therefore all be grey bodied. There will be roughly equal numbers of male and female, grey bodied flies in the F_1 generation. If F_1 flies mate (matings between brothers and sisters are very common in natural populations of many species), the genetics of their cross to produce the F_2 generation is:-

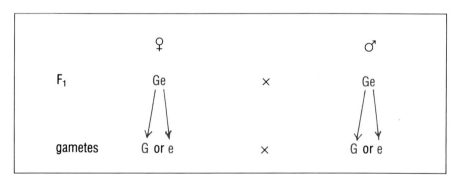

In order to predict the genotypes and phenotypes of the F_2 generation it is best to draw a table showing the alleles from the female gametes on the left and the alleles from the male gametes across the top of the table. The genotypes of the F_2 generation are shown to the South East of the double lines. An examination of the alleles present can easily be used to predict the phenotype ratio of the F_2 generation.

♀ \ ♂	G	e
G	GG	Ge
e	eG	ee

F_2 genotypes

The F_2 genotype ratio is:-

GG : Ge : ee
1 : 2 : 1

All flies containing the dominant G allele will be grey. Double recessive ee flies will be ebony. The phenotype ratio in the F_2 will therefore be:-

Grey body : ebony body
3 : 1

For many genetic crosses it is conventional to describe phenotype using a description in plain English. Genotypes, by contrast, are described in symbols that represent the alleles. In this chapter the convention is largely applied. One further convention needs to be mentioned. Whenever the allele that is predominantly expressed in wild populations is present in an individual, the symbol for the allele is +. Thus the grey body allele is written as + because this allele is expressed as grey body flies more frequently than the expression of any other body colour in wild populations of Drosophila melanogaster. The expression of the + allele is called the 'wild type'. The term 'wild type' applies to the predominant expression of each and every gene. For example the 'wild type' wing shape for Drosophila melanogaster is long wing. When discussing the wing shape gene the long wing allele is written +. Similarly the allele that determines the wild type eye colour, which is red eye, may also be written +. The use of the + allele appears frequently later in this chapter.

It is also common to use the convention of using a single capital letter for a dominant allele and a small letter for the recessive allele. This convention also appears frequently later in the chapter.

Before starting the genetic crosses it is necessary to point out some general observations about genetic inheritance.

It should be remembered that the expression of a gene can be as straightforward as the synthesis of a chemical compound in cells.

The manufacture of each chemical compound in a biochemical pathway must have an enzyme to catalyse its production, and each enzyme has a gene that specifies its synthesis. By contrast some genes and super-genes interact with many other genes to control characters such as height in humans. For a character such as height, many gene-controlled cellular products interact. Some of the chemicals are produced by specialised cells in one part of the body and transported in the bloodstream to bones to cause the continuous variation in height in the human population. Continuous variation and discontinuous variation are discussed in a mathematical context in Appendix 2.

It should also be remembered that the expression of a gene often depends on the environment. For example green plants have the genes needed for the synthesis of the green chlorophyll molecules they use for photosynthesis. However, plants kept in the dark become pale in colour, indicating that the environmental factor 'light' is needed to allow the chlorophyll genes to function correctly.

The study of chromosomal inheritance has been carried out using many species of plants and animals. One species that has been outstanding in providing information about the ways inheritance works is the fruit fly Drosophila melanogaster.

The Study of Genetics Using Drosophila melanogaster

Drosophila = lover of dew
melanogaster = black belly

Seventeen species of fruit fly have been identified in Britain, and several species in addition to these are found in other parts of the world. Among the advantages of using the species D. melanogaster for genetic crosses are:

1. The flies breed rapidly and their genetics can be studied over several generations within a short lapse of time.

2. The diploid number of chromosomes is only 8, so the small number of chromosomes help in the assigning of genes to their linkage groups on only 4 pairs of chromosomes (i.e. which chromosomes contain which genes can be easily identified).

3. Many of their genes segregate sharply to give phenotypes that conform to simple, theoretical ratios.

4. Because it is a sexual species, their sex linkage is quite easy to study.

5. There is no crossing-over in male meiosis, so autosomal linkage and genetic mapping by recombination frequency is relatively easy.

The Life Cycle of D. Melanogaster

(a) Females

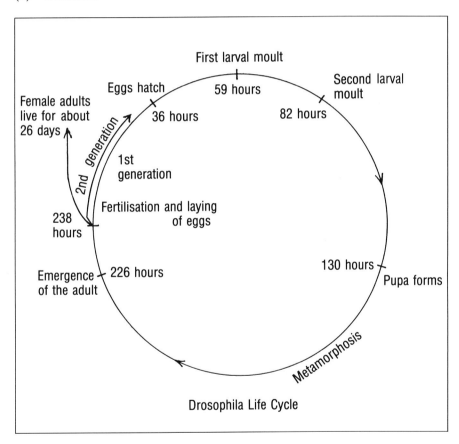

Drosophila Life Cycle

All the times shown are those after the fertilisation of the eggs. The generation time is close to ten days, but varies according to the temperature. An increase in temperature reduces the generation time.

It is shown by the life-cycle diagram that females can mate twelve hours after emergence from the pupa. Therefore, in order to guarantee VIRGIN FEMALES for genetic crosses, females must be separated from males of their generation WITHIN TWELVE HOURS of emergence from the pupae.

(b) Males

The life cycle of males differs from females only in that:-

They produce fertile sperms twelve hours after emergence from the pupa.
They usually live longer than the females – about thirty-three days.

The Diet of D. melanogaster

The species feeds mainly on yeast. There are several suitable foods that provide the requirements of D. melanogaster, which are:-

- (i) Provision of yeast as food for larvae and adults.
- (ii) A medium for larvae to burrow into.
- (iii) A fairly solid material to prevent drowning or glueing of larvae, pupae or adults.

Biological suppliers provide suitable materials and instructions for setting up Drosophila crosses.

Flies for Genetics Experiments

The initial stock of flies should be obtained from approved biological suppliers. This should guarantee the correct genotype and that flies are free from mites or disease. It is important that all factors that might alter the results of genetic crosses, other than the genes themselves, are eliminated.

Adult Morphology versus Other Life Cycle Stages

There are complications that affect the genetics of the larvae and pupae. It is generally desirable, therefore, to use adult characters for genetics experiments with this species.

Drosophila Crosses

There are two conventions used in this book for genetic crosses that may help students in their studies of genetics:-

(i) Even for the simplest of crosses it is advantageous to use a table. The male gametes are shown across the top of the table and female gametes at the left hand side of the table.

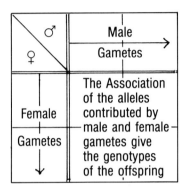

(ii) The chromosomes are shown as a straight line with the position of a gene shown at the appropriate place along its length. The symbol used at the position of the gene shows which allele is present in the gene on the chromosome:-

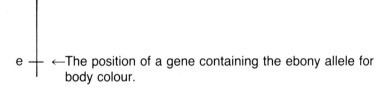

Whenever the symbol + appears beside the position of a gene, the + symbolises the 'wild type' allele. This allele is the one that is most common among individuals in a wild population. The wild type allele is usually dominant and its expression therefore produces the most common phenotype for the gene in wild populations.

The following schemes show how to predict the genotypes and phenotypes of the offspring resulting from Drosophila crosses.

(i) Monohybrid Cross

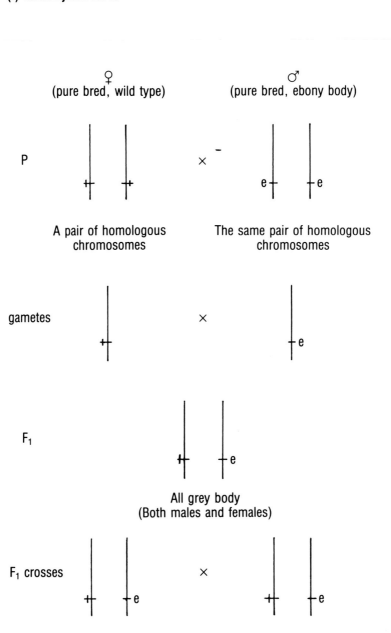

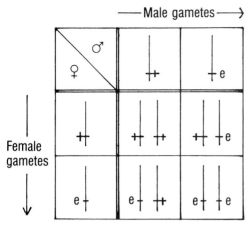

F₂ genotypes
(the female contribution to the genotype of the offspring are shown to the left of the contribution from the male. This applies in all the tables).

The table shows that the ratio of genotypes in the F₂ generation is:-

```
  + +     :    + e    :    e e
   1      :     2     :     1
```

The F₂ phenotype ratio is:-

```
       Grey         :         Ebony
        3           :           1
```

Using a similar procedure, it can be seen that a reciprocal cross leads to the same genotypes in the F_1 and F_2 generations:-

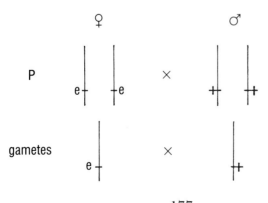

The (so-called) Back-cross of The Monohybrid

If ever an individual has the phenotype determined by the dominant allele, it is impossible by inspection to tell if the individual is homozygous dominant, or heterozygous:-

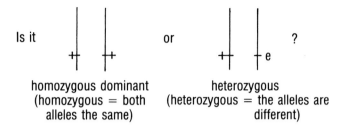

Individuals of the F_2 generation of the monohybrid cross contain some individuals of both sexes that are homozygous dominant and some that are heterozygous. In order to distinguish between them they are 'back-crossed' with the double recessive homozygote from the parental generation. The two possibilities from such a cross depend on the genotypes (what alleles are present) of the F_2 individuals that are back-crossed:-

(i) If the F_2 individual is homozygous dominant (the female in this case)

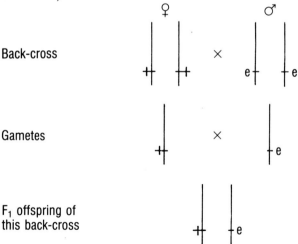

Back-cross

Gametes

F_1 offspring of this back-cross

All the F_1 will show the dominant character (grey body in this case)

(ii) If the F₂ individual is heterozygous (the female in this case)

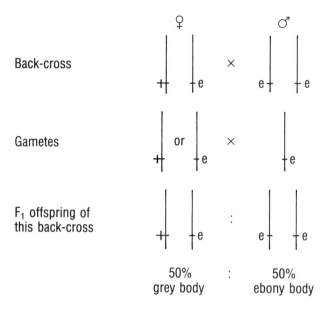

By using a back-cross with a parental double recessive in this way, it is possible to distinguish between a pure bred dominant and a heterozygote.

Dihybrid Cross – NO Autosomal Linkage

The gene for wing shape (long or dumpy) is on chromosome 2 and the gene for body colour (grey or ebony) is on chromosome 3.
 The following cross is made, as shown overleaf.

P

Gametes

F₁

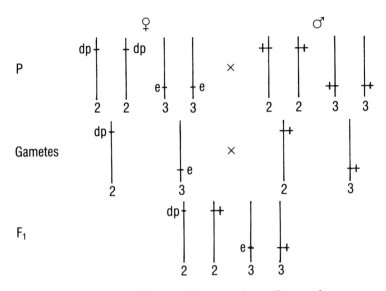

All these flies have long wings and grey bodies

It is now appropriate to introduce a convention that allows the symbols that define the genotype to be used without having to draw the chromosomes. If the genes for two characters are found on different chromosomes (pair 2 and pair 3 in this case) an oblique stroke can be used to signify that the alleles for each gene are on different chromosomes. Continuing with the dihybrid cross examples:-

F₁ gametes

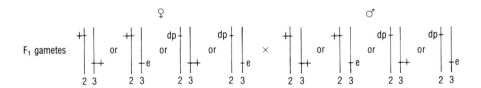

This cross can be simplified as follows:-

F₁ gametes +/+ or +/e or dp/+ or dp/e x +/+ or +/e or dp/+ or dp/e

A table can now be used to predict the F₂ genotypes:-

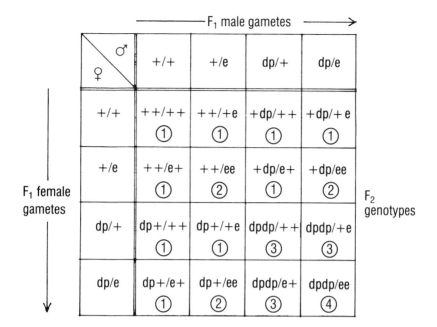

Numbers in circles represent phenotypes.

The phenotypes are:-

	①	②	③	④
	long/grey :	long/ebony :	dumpy/grey :	dumpy/ebony
phenotype ratios	9 :	3 :	3 :	1

Autosomal Linkage

Autosomes are all the chromosomes other than the sex chromosomes. Two genes can be shown to be on the same chromosome by using the fact that crossings-over occur in female Drosophila, but NOT in males. When genes are on the same chromosome they are said to be 'linked'.

Mapping The Positions of Autosomal Linked Genes (genes on the same non-sex chromosomes) in D. melanogaster

Provided genes lie fairly close together on a chromosome there is the possibility, in female Drosophila, of either no crossing-over between the genes or of one crossing-over between the genes, but it is very unlikely that there will be two or more crossings-over between them.

In this species the genes for ebony body and curled wing are both on the same chromosome – pair 3. If a pure bred (for both genes) wild-type fly is mated with a pure bred (for both genes) double recessive, the offspring are heterozygous for both genes:-

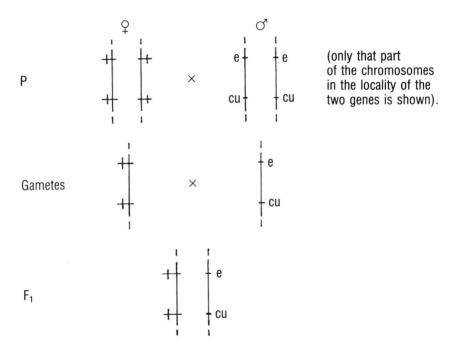

(only that part of the chromosomes in the locality of the two genes is shown).

A female F_1 heterozygote for both genes is crossed with a pure bred double recessive male for both genes (this is a back-cross with the male parent):-

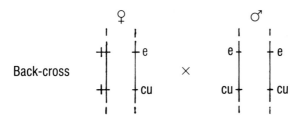

Back-cross

Two different results of this cross can occur, depending on whether or not crossing-over occurs between the two genes during egg production.

i. There is no crossing over between the genes

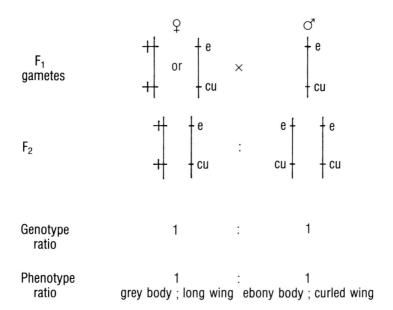

F_1 gametes			
F_2			
Genotype ratio	1	:	1
Phenotype ratio	1	:	1
	grey body ; long wing		ebony body ; curled wing

183

ii. A single crossing-over occurs between the two genes

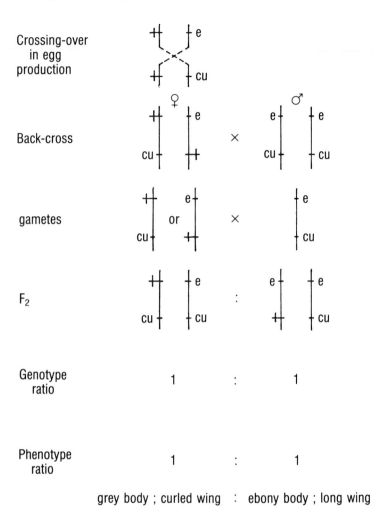

In practice, the progeny of a cross involving linked genes will produce some flies of each category:-

Grey body Curled wing	Grey body Long wing	Ebony body Curled wing	Ebony body Long wing
37	: 211	: 203	: 33

The further apart the genes, the higher the proportion of phenotypes produced from a single crossing-over between the genes. Genetic mapping for Drosophila, using similar recombination frequencies techniques to those used for mapping the positions of genes in bacteria, pages 67-76, has been possible using cross-over values to produce the chromosome map shown on page 147. Maps for the other Drosophila chromosomes are produced in a similar way.

It should be noted that this method of genetic mapping cannot be used for those species in which crossing-over occurs in both sexes. Genetic mapping is made even more complex when there are several pairs of chromosomes. The 23 pairs of human chromosomes, with crossings-over in both sexes, has made the mapping of human genes quite difficult. In humans it is completely unacceptable to carry out experimental crosses and it has therefore been difficult to identify the chromosome pair upon which the gene for any particular character is found, let alone the position of the gene on the chromosome. Chapter 16, which discusses human genetics, shows that there is still a very high proportion of questions about human genetics that are unanswered, including many that influence the health of an individual.

Sex Linkage

For Drosophila melanogaster, chromosome pair 1 are the sex chromosomes. The gene determining whether the eye colour of the flies is to be red or white is found only on the X chromosome. Thus males have only one, unpaired allele for determining red or white eye colour.

i. Cross a White Eye Female with a Wild-type (Red Eye) Male

The red eye allele is dominant over the white eye allele.

Parents

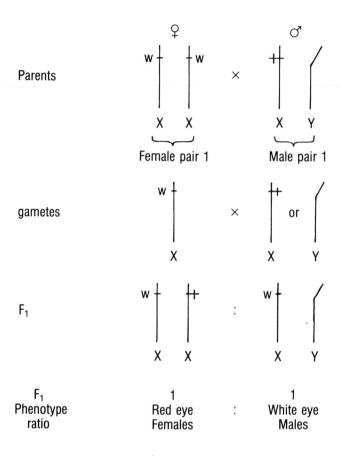

An F_1 male is then crossed with an F_1 female:-

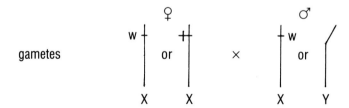

186

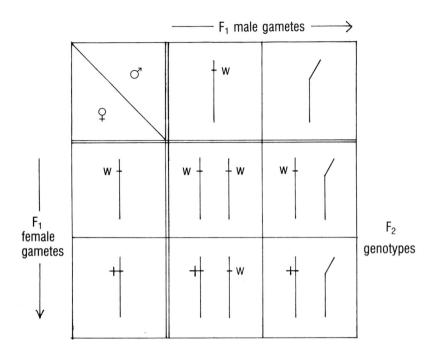

| F₂ phenotype ratios | Females
White eye ; Red eye
1 : 1 | : | Males
White eye ; Red eye
1 : 1 |

ii. Cross a Red Eye Female with a White Eye Male

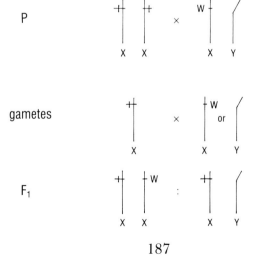

The F_1 generation of this cross consists of red eyed females and red eyed males in a 1:1 ratio.
A male F_1 is then crossed with a female F_1:-

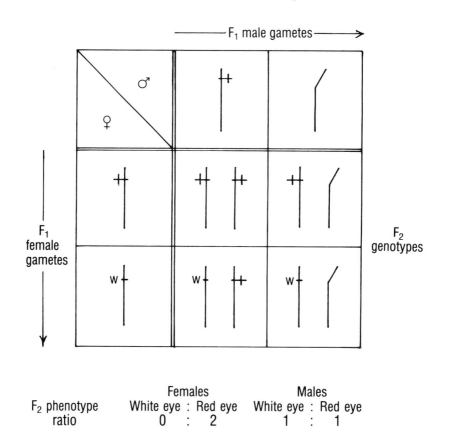

F_2 phenotype ratio

	Females		Males	
	White eye	Red eye	White eye	Red eye
	0	2	1	1

The results obtained from these sex linked, single gene crosses should be compared with the autosomal (the autosomes are all the chromosomes not associated with sex) monohybrid cross in which the crossing of a heterozygous female with a heterozygous male results in a ratio of:-

3 : 1
Dominant   Recessive

for both males and females.

An Example of The Epistatic Interaction of Two Genes to Give Four Different Comb Shapes in Poultry with an F_2 Ratio of 9 : 3 : 3 : 1

The F_2 ratio of comb shapes in poultry demonstrates the way that one gene can modify the expression of another. The modification of the expression of one gene by another is called EPISTASIS.

The F_2 ratio of this genetic study may give rise to the belief that comb shape in poultry is identical to the pattern of inheritance when there is a dihybrid cross. Both give the 9 : 3 : 3 : 1 ratio. However the poultry study examines four different expressions of only one character, the expression of the character (comb shape) being modified by the interactions of the alleles present in two genes.

The two genes that specify comb shape each have alternative alleles:-

Gene 1 alleles :   R or r
Gene 2 alleles :   P or p

The combinations of alleles that produce the four comb shapes are shown below.

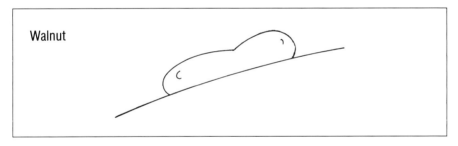

The combination of alleles contributed by the two genes must include R/P to give walnut phenotypes.

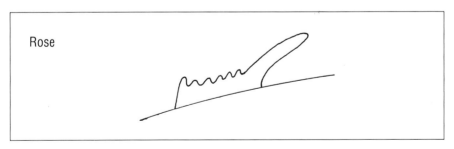

The combination of alleles contributed by the two genes must include R with pp to give rose phenotypes.

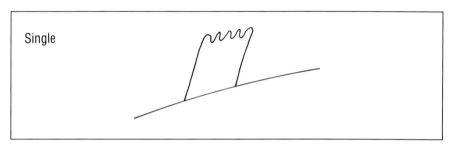

Pea

The combination of alleles contributed by the two genes must include P with rr to give pea phenotypes.

Single

The only combination of alleles that can produce this comb shape is rr/pp.

A pure bred 'rose' parent hen is crossed with a pure bred 'pea' cock.

	♀		♂
P	RR/pp	x	rr/PP
gametes	R/p	x	r/P
F_1		Rr/Pp (all walnut)	

An F_1 male is mated wth an F_1 female:-

♀ \ ♂	R/P	R/p	r/P	r/p
R/P	RR/PP ①	RR/Pp ①	Rr/PP ①	Rr/Pp ①
R/p	RR/pP ①	RR/pp ②	Rr/pP ①	Rr/pp ②
r/P	rR/PP ①	rR/Pp ①	rr/PP ③	rr/Pp ③
r/p	rR/pP ①	rR/pp ②	rr/pP ③	rr/pp ④

① = Walnut ② = Rose ③ = Pea ④ = Single

F₂ phenotype ratios   Walnut : Rose : Pea : Single
                          9   : 3   : 3 :   1

Another Example of The Epistatic Modification to the Expression of Genes – Black, White and Brown Mice in a 4 : 3 : 1 Ratio in the F_1 Generation

The presence or absence of pigments in the hairs of mouse coats depends on the interactions of two epistatic genes. There are two alleles for each of the coat colour genes.

Gene 1 alleles :   C or c
Gene 2 alleles :   B or b

Black coat is produced whenever C/B are present.
Brown coat is produced whenever C/b are present.
White coat is produced whenever c is present, irrespective of the alleles present in the second gene.

A cross is arranged between parents with the following genotypes:-

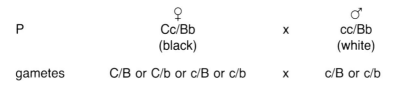

P   ♀   ♂
          Cc/Bb   x   cc/Bb
          (black)   (white)

gametes   C/B or C/b or c/B or c/b   x   c/B or c/b

The prediction for the F$_1$ generation can be made using a table:-

♀ \ ♂	c/B	c/b
C/B	Cc/BB ①	Cc/Bb ①
C/b	Cc/bB ①	Cc/bb ②
c/B	cc/BB ③	cc/Bb ③
c/b	cc/bB ③	cc/bb ③

① = Black   ② = Brown   ③ = White

F$_2$ phenotype ratios   Black : Brown : White
                           3   :   1   :   4

Note: In genetic crosses discussed later in this appendix the oblique stroke is omitted and different letters imply different genes.

The Apparent Epistatic Reversion to The Wild Type in Sweet Peas to Give a 9 : 7 Ratio

Epistasis can be the situation in which one gene prevents the expression of another. For example biochemical pathways have end products that depend upon the correct functioning of enzymes for each biochemical conversion along the pathway. The production of the coloured pigments in the flowers of sweet peas depends on such a biochemical pathway. If the last two biochemical conversions are considered, each conversion requires the correct enzyme:-

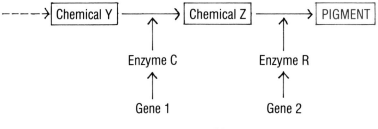

Two alleles exist for each of the genes that control the synthesis of enzyme C and enzyme R.

Gene 1
The alleles are C and c. C has to be present if the pigment is to be made.

Gene 2
The alleles are R and r. R has to be present if the pigment is to be made. Both C and R must be present, therefore, if the pigment is to be made. This situation allows a cross between two apparently recessive (white) plants to give rise to plants in the F_1 and F_2 generations that show the pigment in their flowers.

	♀		♂
P	ccRR	×	CCrr
gametes	cR	×	Cr
F_1		CcRr (all coloured)	

The table shows the genotypes resulting from crossing F_1 plants:-

♀ \ ♂	CR	Cr	cR	cr
CR	CCRR ①	CCRr ①	CcRR ①	CcRr ①
Cr	CCrR ①	CCrr ②	CcrR ①	Ccrr ②
cR	cCRR ①	cCRr ①	ccRR ②	ccRr ②
cr	cCrR ①	cCrr ②	ccrR ②	ccrr ②

① = Coloured ② = White

F_2 phenotype ratio 9 coloured : 7 white

It should be noted that no mutational change was necessary in this cross to produce the 9 : 7 ratio.

The Genetic Consequences of a Mutation That Occurs During Mitosis (a somatic mutation) in Nicotiana tabacum (tobacco)

Tobacco flowers can be carmine, pink or white. Two genes control flower colour, each of the genes being epistatic to the other. The alleles found in the two genes are:-

$$\text{Gene 1} : \text{C c}$$
$$\text{Gene 2} : \text{R r}$$

The phenotypes resulting from the combinations of alleles in the genes are:-

CR   carmine
Cr   pink
cR } white
cr }

A parental cross is arranged so that the F_1 are CcRR:-

	♀		♂
P	CCRR (carmine)	×	ccRR (white)
gametes	CR	×	cR
F_1		CcRR (carmine)	

A mutation occurs in an F_1 plant so that:-

$$\text{CcRR} \rightarrow \text{CcRr}$$

Provided that the mutation occurs early enough in the growth and development of the plant to produce flowers that are all of the genotype CcRr, these mutant flowers can be used for self-pollination to give rise to the F_2 genotypes in the table:-

F₁ mutant   ♀ CcRr   x   ♂ CcRr

gametes   CR or Cr or cR or cr   x   CR or Cr or cR or cr

♀ \ ♂	CR	Cr	cR	cr
CR	CCRR ①	CCRr ①	CcRR ①	CcRr ①
Cr	CCrR ①	CCrr ②	CcrR ①	Ccrr ②
cR	cCRR ①	cCRr ①	ccRR ③	ccRr ③
cr	cCrR ①	cCrr ②	ccrR ③	ccrr ③

① = carmine
② = pink
③ = white

F₂ phenotype ratios   9 carmine : 3 pink : 4 white

The back-cross, using pollen from a mutant F_1, with a double recessive plant gives:-

	♀		♂
Backcross of the male F_1	ccrr	×	CcRr
gametes	cr	×	CR or Cr or cR or cr

♀ \ ♂	CR	Cr	cR	cr
cr	cCrR ①	cCrr ②	ccrR ③	ccrr ③

① = carmine
② = pink
③ = white

F_2 phenotype ratios	1 carmine	:	1 pink	:	2 white

196

An Example of Incomplete Dominance in Snapdragons

In snapdragons there are two alleles for flower colour that exhibit incomplete dominance. The genetics of this species can be studied for red, white or pink flowers as follows. Cross a pure bred, red plant with a pure bred white plant:-

	♀		♂
P	RR	x	WW
gametes	R	x	W
F_1		RW pink	
F_1 gametes	R or W	x	R or W
F_2	RR :	2RW :	WW
	red	pink	white

Another example of incomplete dominance is given in Chapter 16 for sickle cell anaemia in humans.

An Example of Codominance in the Genetics of Human Blood Groups

Codominance occurs when both alleles in a heterozygote are expressed fully and equally in the phenotype. An example is that of the ABO blood groups in humans. The following table relates the blood group alleles to the chemicals present in the blood. The blood groups are defined by the antigens present in the red cells as shown in the table overleaf.

BLOOD GROUP	GENOTYPE	ANTIGENS PRESENT IN RED CELLS	ANTIBODIES IN SERUM
O	$I^O I^O$	None	Anti-A, Anti-B
A	$I^A I^A$ or $I^A I^O$	A	Anti-B
B	$I^B I^B$ or $I^B I^O$	B	Anti-A
AB	$I^A I^B$	A and B	None

I^A and I^B are codominant
I^O is recessive

An Example of a Dominant Inhibitor Allele in Chickens

A dominant allele that inhibits the production of feather pigment in chickens operates in the following way:-

Gene 1
    I is the dominant, inhibitor allele which prevents the pigmentation of feathers.

    i is the recessive, non-inhibitor allele.

Gene 2
    C is the dominant allele for coloured feathers.

    c is the recessive, white feather allele.

The following combinations of alleles are expressed by the following phenotypes:-

    I/C gives white Wyandotte
    I/c gives white Leghorns
    i/C gives coloured chickens
    i/c gives white Leghorns

Crossing a pure bred Wyandotte cock with a pure bred white, double recessive hen gives an F_1 generation that is heterozygous for both genes:-

	♀		♂
P	iicc	x	IICC
gametes	ic	x	IC
F_1		IiCc	

The presence of the dominant, inhibitor allele specifies that all the F_1 hens and cocks are white Wyandottes.

The following genetic cross shows the genotypes and phenotypes that result from a cross between an F_1 hen and an F_1 cock:-

	♀		♂
F_1	IiCc	x	IiCc
gametes	IC or Ic or iC or ic	x	IC or Ic or iC or ic

♀ \ ♂	IC	Ic	iC	ic
IC	IICC ①	IICc ①	IiCC ①	IiCc ①
Ic	IIcC ①	IIcc ①	IicC ①	Iicc ①
iC	iICC ①	iICc ①	iiCC ②	iiCc ②
ic	iIcC ①	iIcc ①	iicC ②	iicc ①

The phenotype ratio in the F₂ generation is:-

①		②
White	:	Coloured
13	:	3

An Example of the Interactions of Super-genes in the Snail Cepaea nemoralis

The phenomena of super-genes and super-alleles were briefly discussed in Chapter 4. In the snail Cepaea nemoralis, shell colour and patterns of banding are controlled by three loci where super-genes are found on the chromosomes. One locus controls the background shell colour, and there are super-alleles for brown, pink and yellow shells. The second super-gene locus determines whether there will be five dark bands superimposed upon the basic shell colour, or whether the shells should be unbanded. The third locus has an epistatic effect on the banding pattern. When the dominant allele for the third locus is present it suppresses four of the five bands that would otherwise be specified by the 5-band super-allele of the second gene locus.

The following scheme shows the interactions between the three loci that control the genetic inheritance of shell colour and banding patterns:-

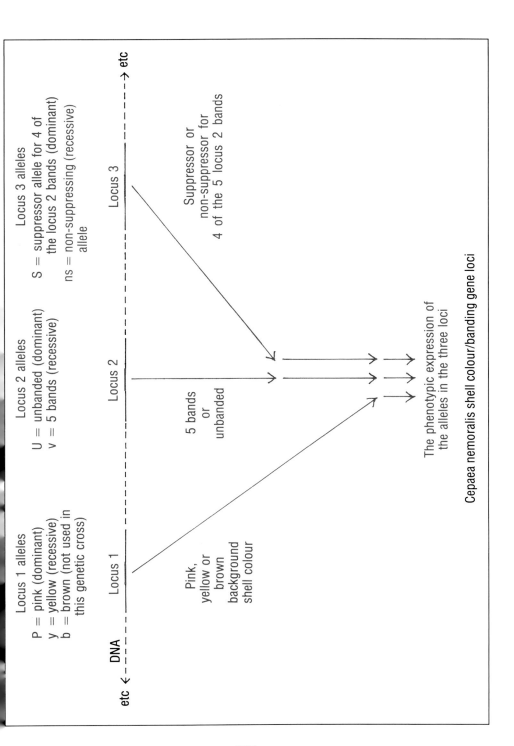

Cross 1

Locus 1 alleles – P is the dominant pink allele.
y is the recessive, yellow allele.

Locus 2 alleles – U is the dominant unbanded allele.
v is the recessive, five bands allele.

A pure bred, pink, 5 bands female is crossed with a pure bred, yellow, unbanded male:-

	♀		♂
P	PP/vv	x	yy/UU
gametes	P/v	x	y/U
F_1		Py/Uv	

All the F_1 snails are pink and unbanded.

F_1 snails are then crossed.

	♀		♂
F_1	Py/Uv	x	Py/Uv

gametes P/U or P/v or y/U or y/v x P/U or P/v or y/u or y/v

So far, the super-alleles act in the same way as the alleles of a dihybrid cross to give an F_2 phenotype ratio:-

9 : 3 : 3 : 1
pink unbanded : pink 5 bands : yellow unbanded : yellow 5 bands

However, the presence of an epistatic third locus causes a deviation from the 9:3:3:1 expectation in some genetic crosses. This is demonstrated in cross 2.

Cross 2

For the sake of simplicity, all the snails used in this cross are pure bred for basic shell colour. Therefore the alleles that are of interest in this case only occur in loci 2 and 3.

A pure bred, 5 bands female which is also pure bred for the recessive, and therefore ineffective third locus allele, is crossed with a pure bred, unbanded male which is heterozygous for the dominant, modifying, third locus allele.

Locus 2 alleles — U is the dominant, unbanded allele.
v is the recessive, 5 bands allele.

Locus 3 alleles — M is the dominant, modifying allele.
— is the recessive, ineffective allele.

	♀		♂
P	vv/– –	x	UU/M –
gametes	v/–	x	U/M or U/–
F_1	Uv/M – 1	:	Uv/– – 1

All these F_1 snails are unbanded.

If the F_1 males and females are allowed to mate at random the following crosses apply:-

	♀		♂
	50% : 50%		50% : 50%
F_1	Uv/M – : Uv/– –	x	Uv/M – : Uv/– –
gametes	U/M or 3U/– or v/M or 3v/–	x	U/M or 3U/– or v/M or 3v/–

Note that identical alleles are identical base sequences in DNA situated in different chromosomes. There are therefore eight possible allele combinations that can be put into F_1 female gametes. Three of the eight are U/– and three are v/–.

The same situation arises in the eight possible combinations that can be put into the F_1 male gametes, three out of eight of which will be U/– and three will be v/–. Hence the number 3 in the 3U/– and 3v/– allele combinations provided in the gametes of the F_1 generation.

If some allele combinations occur three times as frequently as others, the frequency of their fertilization by each of the four allele combinations in gametes of the opposite sex also increases three times.

The F_2 genotypes shown in the table below show that some genotypes occur three times as frequently as others (3 × 1). Yet other genotypes can occur nine times as frequently as some of the others (3 × 3).

♀ \ ♂	U/M	3U/–	v/M	3v/–
U/M	UU/MM ①	3UU/M – ①	Uv/MM ①	3Uv/M – ①
3U/–	3UU/– M ①	9UU/– – ①	3Uv/– M ①	9Uv/– – ①
v/M	vU/MM ①	3vU/M – ①	vv/MM ②	3vv/M – ②
3v/–	3vU/– M ①	9vU/– – ①	3vv/– M ②	9vv/– – ③

F_2 genotypes

The phenotype ratios are:-

         ①   ②   ③

unbanded : 1 band : 5 bands

    48   :   7   :   9

Sex Determination

In many species, there are specialised chromosomes for determining the sex of an individual. It is evident from Drosophila and human genetic studies that genes on the sex chromosomes also control characters other than sex. This simple study shows how the inheritance of the normal number of sex chromosomes occurs (abnormalities of sex chromosomes inheritance are given for humans in Chapter 16) in those diploid species that have specialised sex chromosomes (many fish and flower species have no sex chromosomes – their sexual inheritance is controlled by genes that are widely spread throughout their complement of chromosomes).

    X is the X chromosome
    Y is the Y chromosome

a) Sex Determination in Many Animal Species

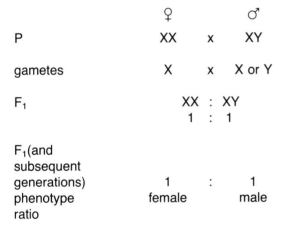

	♀		♂
P	XX	x	XY
gametes	X	x	X or Y
F₁		XX : XY	
		1 : 1	
F₁ (and subsequent generations) phenotype ratio	1 female	:	1 male

This equal ratio of males and females applies, provided the mortality at each stage of the life cycle is the same for both males and females, and provided that the numbers of X and Y gametes from the male achieving fertilisation is the same.

Sex determination depends on the inheritance of the sex chromosome donated by the *male* gamete (as in this example) in many animal species, including humans.

In humans, a gene on the Y chromosome has been identified at the Imperial Cancer Research Laboratories in London which initiates hormone production to give a male. The absence of the gene in XX individuals gives a female – more details of this male determining gene are given in Chapter 16.

b) Sex Determination in Birds

In birds it is the inheritance of the sex chromosome from the *female* gamete that determines the sex of an individual:-

	♀		♂
P	XY	x	XX
gametes	X or Y	x	X
F_1		XY : XX	
		1 : 1	
F_1(and subsequent generations) phenotype ratio	1 female	:	1 male

'Mother' Effect

Chemicals in the cytoplasm of the female gamete can modify the inheritance of an individual in some cases. The way this is done is described for Drosophila melanogaster on pp 458-460. An example in Limnea pereger is given on pp 469-471.

The Differentiation of Cells

Mitotically dividing cells contain identical sets of chromosomes in each individual. It would seem that identical sets of genes in all cells in an individual ought to be expressed in an identical way in every cell. However, the specialisation of cells begins early in embryonic development, and this process of cell differentiation is not yet well understood. An attempt is made in Chapter 14 to describe a few aspects of cell differentiation.

Chapter 5 Questions

1. How does the dominance or recessiveness of alleles operate in diploids? Start with a heterozygous male and a heterozygous female to illustrate how recessive alleles fail to be expressed in many of their offspring.

2. Use a table to illustrate the standard procedure for carrying out dihybrid cross predictions in which there is *no* autosomal linkage.

3. Why is Drosophila melanogaster a species of particular value for carrying out genetic crosses?

4. If Drosophila parents are a homozygous grey-bodied female and a homozygous ebony-bodied male, predict:-
   i. The genotype ratios of the F_1 and F_2 generations.
   ii. The phenotype ratios of the F_1 and F_2 generations.

5. How can an individual showing the dominant character be evaluated as being either homozygous dominant or heterozygous?

6. How is autosomal linkage used for mapping the positions of genes on the chromosomes of Drosophila melanogaster? What is the essential property of male meiosis that allows Drosophila mapping in this way?.

7. Why does the sex of Drosophila melanogaster determine the inheritance of non-sexual characters? Use eye colour as an example.

8. What is epistasis? How does it operate for determining comb shape in poultry? How does it operate for determining coat colour in mice?

9. What genetic inheritance causes a 9 : 7 ratio in the F_2 generation?

10. What process occurring during cell division can lead to a 9 : 3 : 4 ratio of phenotypes in the F_2 generation?

11. What are the genetic reasons for human blood group inheritance?

12. How can a dominant inhibitor allele lead to a 13 : 3 ratio of phenotypes in the F_2 generation?

13. How are basic shell colour and shell banding controlled in Cepaea nemoralis?
14. How is sex determined:-
    i. In man and many animal species?
    ii. In birds?
    iii. In fish and flowers?

6

EXTRA-CHROMOSOMAL INHERITANCE

DNA is found, usually as small closed circles, in several places in cells other than in chromosomes.

Eukaryotic Extra-chromosomal DNA

There are several organelles, collectively classified as plastids, in eukaryotic cells that contain extra-chromosomal DNA:-

(i) Mitochondria of all eukaryotes.
(ii) Chloroplasts in plants.
(iii) Leukoplasts (colourless) in the tubers, endosperms and cotyledons of plants.
(iv) Chromoplasts (pigmented plastids) in plants.
(v) Elaioplasts (oil storing, colourless plastids) in plants, mostly monocotyledons
(vi) Kinetoplasts (mitochondrial extensions) close to the cilia of ciliated protozoa

It is possible, but not yet proven, that fibre producers such as the centrioles of cell division also contain their own DNA.

Some symbiotic associations between bacteria or viruses with protozoa confer on the protozoa characters that are controlled by the DNA of the symbiotic microorganism.

Circular DNA, in the form of small plasmids, are found, in numbers varying from one to about twenty, floating in the cytoplasm of eukaryotic cells. Some cells have no plasmids in them.

Prokaryotic Extra-chromosomal DNA

Bacterial cells usually have several plasmids in addition to their single, circular chromosome. These small DNA circles have proved to be very valuable for genetic engineering (this subject is discussed in Chapter 8).

The inheritance of prokaryotic cells can also be altered by the presence, in their protoplasm, of DNA similar to that found in viruses.

Extra-chromosomal DNA found in Both Prokaryotes and Eukaryotes That is Derived from Mobile Genetic Elements

The subject of mobile genetic elements is quite broad. Chapter 13 discusses a few aspects of these important genetic fragments. It will be seen in Chapter 13 that mobile genetic elements can have RNA instead of DNA as their material of inheritance, in some cases.

A Pointer to Extra-Chromosomal Inheritance

If reciprocal crosses give non-Mendelian results, and if 'sex-linkage', pages 185-188, and 'mother-effect', chapter 14, have both been ruled out, the characters that show non-Mendelian inheritance ratios are likely to be controlled by genes in extra-chromosomal DNA:-

		♀		♂
P		Pure Bred Strain A	x	Pure Bred Strain B
F_1			All Strain A	

However if the reciprocal cross gives:-

	♀		♂
P(reciprocal) cross	Pure Bred Strain B	x	Pure Bred Strain A
F_1		All Strain B	

then extra-chromosomal inheritance is indicated.

The Relationship Between the DNA in Mitochondria and Chromosomal DNA

The size of the mitochondrial circle of DNA is similar to the small, circular plasmids found in bacteria. It contains only about thirty genes.

The genes of mitochondrial DNA have been shown to control the synthesis of:-

Mitochondrial rRNA (different from cytoplasmic rRNA).
Some components (polypeptides) of mitochondrial ribosomes.
Some mitochondrial tRNAs (some of these are different from cytoplasmic tRNAs).
Some inner membrane components of ATPase (the enzymic complex in the inner membrane of mitochondria that assists in the synthesis of ATP through aerobic respiration)
Some components of the cytochromes in the inner mitochondrial membrane.
Some genes for drug resistance

However, very many of the biochemical reactions in mitochondria depend upon products from chromosomal genes.

In mitochondria the DNA is not associated with histones and at least one of the codons in mitochondria of the yeast Saccharomyces cerevisiae codes for a different amino-acid by comparison to the cytoplasmic mRNA codon for the same amino-acid. UGA codes for tryptophan in mitochondria, whereas this codon is a 'termination' codon for the cytoplasmic mRNA.

The ribosomes of mitochondria include rRNA molecules with base sequences different from cytoplasmic, ribosomal rRNAs. The sedimentation constant for mitochondrial ribosomes is also different from cytoplasmic ribosomes.

The Study of Mitochondrial Inheritance in Yeasts

Yeasts lend themselves to studies of mitochondrial inheritance because they can continue to grow using anaerobic respiration if they inherit mutant genes for aerobic respiration in their mitochondria.

Mitochondria can fuse together to allow the recombination of mitochondrial genes and, together with the study of deletions and restriction enzyme analysis (see Genetic Engineering, Chapter 8), this has led to the mapping of a few mitochondrial genes.

The rough relationship between the positions of some mitochondrial genes in Saccharomyces cerevisiae are:-

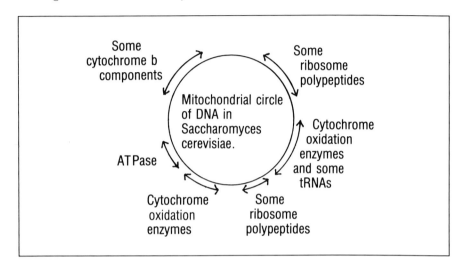

There is some (inconclusive) evidence that mitochondria were once primitive independent organisms during an era after the evolution of the photosynthetic process had begun to supply oxygen to the world's atmosphere. They may have established symbiotic relationships with anaerobic cells that gradually evolved into the endosymbiotic relationship which now allows eukaryotic cells to use mitochondria for aerobic respiration.

There needs to be considerable further study of the endosymbiotic relationship between mitochondria and eukaryotic cells before any definite conclusions can be drawn about the beginnings of their association. The molecular and structural evidence may no longer exist, because of evolutionary changes during the long period since their original association. It is *possible* that mitochondria arose *within* cells by advantageous mutations followed by genetic drift and natural selection after oxygen was produced in considerable quantities by photosynthesis to be added to air or dissolved in water.

Mitochondrial ribosomes are sensitive to the same drugs as bacterial ribosomes, whereas eukaryotic cytoplasmic ribosomes have a different response to drugs.

A Pointer to Mitochondrial Inheritance

Mitochondrial inheritance is indicated if the cytoplasmic inheritance factor seems to be donated by the gamete that contains the larger number of mitochondria. Often, but not always, the female gamete donates the larger number of mitochondria to the zygote. An example in Neurospora crassa is given in Chapter 7, pages 223-224

The Relationship Between the DNA in Chloroplasts and Chromosomal DNA

In the same way that mitochondria depend to a considerable extent upon the products controlled by chromosomal DNA, so too do chloroplasts partly depend upon chromosomal products as well as having their own DNA. Chloroplast DNA codes for a proportion of the polypeptides needed for their function as photosynthesisers. The form of chloroplast DNA is circular and small and, in many plant species, there are several circles of DNA, each of which may be identical. Between 10 and 60 small, circular DNA fragments have been observed in the chloroplasts of those plant species studied so far.

The genes of chloroplast DNA control the synthesis of:-

Chloroplast rRNA.

Some chloroplast ribosome polypeptides.

Some chloroplast tRNAs (some are different from cytoplasmic tRNAs).

Some hydrophobic polypeptides needed for the conversion of light energy into chemical energy in the membranous thylakoids of chloroplast grana.

Ribulose-1-5-diphosphate carboxylase which fixes CO_2 into the Calvin CO_2 (dark reaction) cycle of photosynthesis.

Most biochemistry books discuss the processes of photosynthesis and show precisely where the processes occur in the cells of plants.

It is thought that chloroplast DNA codes for about 100 different polypeptides, most of which still have to be identified.

The two components of ribosomes in chloroplasts have sedimentation constants close to those of bacterial ribosomes.

The amino-acids in chloroplasts are less specific in their attachments to tRNAs than the very specific cytoplasmic amino-acid/tRNA associations for polypeptide synthesis.

The control of the synthesis of many chloroplast components has been shown to come from chromosomal DNA:-

The replication and transcription of the chloroplast DNA.

Some ribosome polypeptides.

Some tRNAs.

Some mRNAs.

Soluble enzymes.

One constituent of the enzymes needed for CO_2 fixation.

Some enzymes needed for the conversion of light energy into chemical energy.

Some constituents of the chloroplast double membrane.

Some examples of inheritance controlled by chloroplast genes are given in Chapter 7.

Investigations are in hand to try to discover the evolutionary association between chloroplasts and the eukaryotic plant cells found in the contemporary world, but not much is known as yet.

Other Plastid Inheritance in Eukaryotes

Not much is known about the inheritance controlled by the DNA in plastids other than mitochondria and chloroplasts.

Cytoplasmic Plasmids in Eukaryotes

The plasmids in yeast are suitable for genetic engineering.

Inheritance Controlled by Endosymbiosis In a Single Cell Eukaryote – Paramecium aurelia 'Kappa' Particles

Many well known 'cell organelles' such as chloroplasts and mitochondria are classified as organelles because of the regularity of their occurrence in cells, even if they originated as endosymbionts. So long as endosymbionts only occur in a small proportion of cells their role in controlling inheritance in their host species must be regarded rather differently from the inheritance role of cell organelles.

An example of endosymbiotic inheritance is that of the 'Kappa' particles that are found in the protozoan Paramecium aurelia. The number of 'Kappa' particles per protozoan varies from nil (common) to about a hundred. There are two types of 'Kappa' particles:-

NON-BRIGHT and BRIGHT

Most 'Kappa' particles are 'non-brights', but these have the ability to mutate spontaneously into 'brights' during Kappa particle replication.

The 'brights' confer upon those paramecia that contain them the ability to kill paramecia without 'brights' inside them. However 'brights' paramecia cannot kill other 'brights'. 'Brights' Kappa particles cannot replicate, but 'non-brights' can.

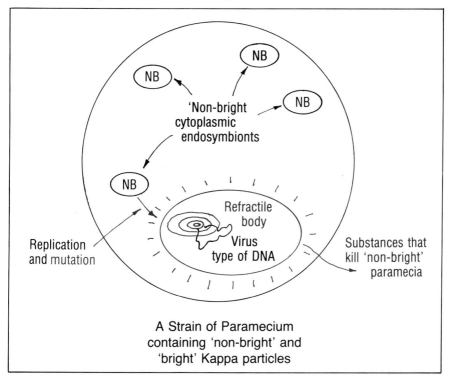

A Strain of Paramecium containing 'non-bright' and 'bright' Kappa particles

'Brights' kill 'non-brights' by paralysis, vacuolation and other distortions. A colony of Paramecium aurelia can be cured of killer 'brights' if the rate of 'non-brights' produced by paramecium replication is much faster than the production of new 'brights'.

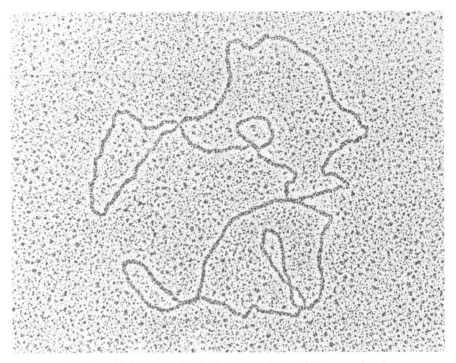

Mitochondrial DNA molecule from an oöcyte of *Xenopus laevis*.
From *The Cell* by D.W. Fawcett. Photograph by Igor David.
With permission of W. B. Saunders Co.

Plasmids in Prokaryotic Cells

The term 'plasmids' refers to all fragments of non-viral DNA, other than the chromosome. In prokaryotic cells the plasmids occur as small, circular fragments of DNA floating in the protoplasm. There are usually a few plasmids in bacterial cells, and sometimes there are several.

Plasmid genes control several inherited phenotypic characters of bacteria. The Sex (F) Factor, the Hfr (high recombination frequency factor), R (resistance) Factor, and the Col Factor (in killer bacteria) are described in Chapter 7.

When they occur in bacteria the F and R plasmids are made of DNA having only about 2% of the length of chromosomal DNA. Other plasmids are usually smaller than the F and R plasmids.

Plasmid replication and the number of plasmids per bacterium are partly under the control of chromosomal genes and partly under the control of genes in the plasmids themselves.

The Usefulness of Plasmids in Prokaryotes

Plasmids in bacteria have proved to be very useful for genetic engineering. This topic is discussed in Chapter 8.

Chapter 6 Questions

1. Where is DNA found apart from in the chromosomes:-
   i. In prokaryotic cells?
   ii. In eukaryotic cells?
2. How is extra-chromosomal inheritance usually detected?
3. How is mitochondrial inheritance usually detected?
4. What characters are known to be controlled by mitochondrial DNA?
5. What products are known to be under the control of genes in chloroplast DNA?
6. What is the inheritance of Paramecium aurelia and how is it expressed?

7

SOME EXAMPLES OF INHERITANCE DETERMINED BY EXTRA-CHROMOSOMAL DNA

The extra-chromosomal DNA particles discussed in Chapter 6 have yielded some specific genetic information about the ways they control inheritance. Several experimental studies have shown that extra-chromosomal inheritance can be divided into the following broad categories:-

1. Extra-chromosomal Inheritance in Prokaryotic cells

Prokaryotic extra-chromosomal inheritance can be sub-divided into:-

a) Characters controlled by genes contained in mobile, viral or bacterial DNA that invades prokaryotic cells and alters their characters. Examples of mobile, bacterial, extra-chromosomal DNA are given in this chapter, pages 219-221. Chapter 13 discusses mobile viral DNA.
b) Characters controlled by genes in those plasmids that are non-mobile.

2. Extra-chromosomal Inheritance in Eukaryotic Cells

Eukaryotic extra-chromosomal inheritance can be sub-divided into:-

a) Characters controlled by genes in the DNA of organelles such as mitochondria and chloroplasts. Examples are given in this chapter on pages 223-227.
b) Characters controlled by genes in plasmids. For example the plasmids of yeast can be used for genetic engineering.
c) Characters controlled by viral, mobile, cancer-inducing genes. Examples are given on pages 489-491 of chapter 15.
d) Characters controlled by viral, mobile genes that have an evolutionary or disease bearing significance.

Categories 1 and 2 are now discussed with specific examples.

1. Extra-chromosomal Inheritance in Prokaryotic Cells

A. Mobile, Bacterial, Extra-chromosomal Inheritance

Three well known examples of genes found in prokaryotic plasmids are the Sex (F) Factor, the R (resistance) Factor and the Col Factor in the bacterium species Escherichia coli.

i. The Sex (F) Factor in E. coli

a) The Conversion of a Recipient Bacterium into a Donor Using the Sex (F) Factor

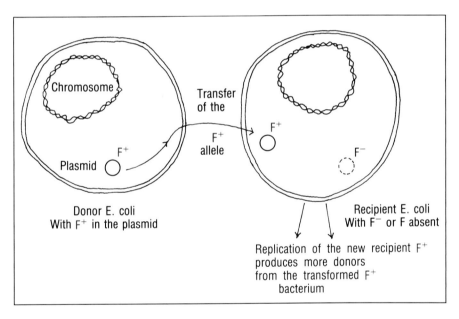

F^+ bacteria are those that can donate their chromosomes into recipient F^- (or F absent) bacteria. There are several genetic strains of E. Coli, each exhibiting a variety of characters, and this transformation behaviour, when F^+ is passed into a new F^+ individual, ensures variety among those bacteria that are donors in any locality. This in turn leads to a wider variety of genotypes and phenotypes upon which genetic drift and natural selection can operate.

b) The Creation of High Frequency Recombinant (H fr) Bacteria at Times when F^+ acts as an Episome

Those E. coli containing the F^+ sex factor sometimes take the plasmid DNA into their chromosomes. Whenever this happens the donor bacterium has an increased ability to transfer its chromosome into a recipient bacterium.

A gene acting in such a way that it is sometimes used, when in the plasmid, to produce one character, and sometimes incorporated into the chromosome to produce a different character, is called an EPISOME.

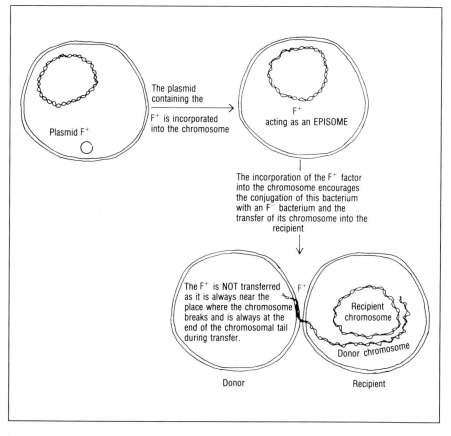

Crossings-over between the chromosomes of the donor and recipient bacteria create recombination, so the Hfr episome F^+ assists in the creation of recombinants. Again this increases variety among E. coli upon which genetic drift and natural selection operate.

ii. The R (resistance) Factor

The presence of the R (resistance) factor in a bacterial plasmid increases the resistance of such bacteria to antibiotics. The R factor may consist of two different groups of genes, one group determining anti-biotic resistance and the second encouraging the drug resistance plasmid to move into the bacteria that come into contact with R (resistance) factor E. coli.

Species other than E. coli can accept the R factor plasmid. It should be noted that the R factor does NOT become incorporated into the bacterial chromosome.

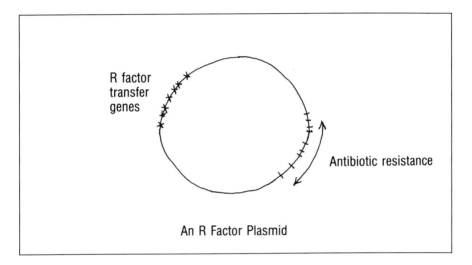

An R Factor Plasmid

iii. The Col Factor

The Col Factor is a plasmid gene found in some bacterium species (E. coli is one of the species) that gives a small proportion of bacterium populations the ability to make the poison COLICIN. Those individuals that can make the poison are killed by it, but other individuals of the same species are resistant to the poison. However bacteria of other species are also usually killed by colicin, so the suicide of a Col factor bacterium removes local competition for the essentials of life, thereby benefiting other individuals of the same species.

Transfer of the Col factor plasmid DNA is common, but it may need the assistance of the F^+ gene, acting as an episome, to create high frequency recombination.

B. Inheritance Controlled by Prokaryotic, Non-Mobile, Plasmid Genes

Some plasmid genes, are contained in the non-mobile plasmids.

2. Extra-chromosomal Inheritance in Eukaryotes

If sex linkage and mother effect (page 469) have both been ruled out, yet there are differences in the phenotypic expression of a character between reciprocal crosses, there is likely to be extra-chromosomal DNA that determines the character.

	♀		♂
P	Strain A	x	Strain B
F_1		Phenotype 1	

But:-

	♀		♂
P	Strain B	x	Strain A
F_1		Phenotype 2	

Extra-chromosomal inheritance in eukaryotes is known to control some characters in a variety of species:-

   i. Anti-biotic resistance in protozoa.
   ii. The inability to respire anaerobically in yeast.
   iii. Male sterility in flowering plants.
   iv. The production of chlorophyll in the leaves and stems of some plant species.

a) An Example of Extra-chromosomal Inheritance in Variegated Plants

Variegation in some plant species, such as Antirrhinum majus, is caused by the presence of chloroplasts in some of the leaves and stems while in others the chloroplasts are absent. Some of the genes in seeds produced from those flowers growing on white stems are different from the genes in the seeds taken from the green stem flowers. There are no sex chromosomes in plants so sex linkage is ruled out, even though reciprocal crosses give different F_1 phenotypes:-

	♀		♂
P	Green	x	White
F_1		All Green	

But:-

	♀		♂
P	White	x	Green
F_1		All White	

In this example it seems that the extra-chromosomal inheritance contribruted by the female gamete controls the presence or absence of chloroplasts. Often the number of organelles in eggs is much greater than in male gametes and, if an organelle (plastid) gene controls a character, it is the female gamete that controls the expression of the gene. This is not invariable. There are known examples in which it is the male gamete's plastid gene that controls the expression of a character.

b) An Example of Inheritance Controlled By a Mitochondrial Gene – Respiration in the Fungus Neurospora crassa

Several genes that control respiration and drug resistance have been shown to be in the DNA of mitochondria.

Many species of fungus can multiply by both sexual and asexual methods. Sexual eruptions occur out of the asexual phase of fungal species to produce male and female gametes. In the fungus Neurospora crassa the female gamete contains a much larger quantity of cytoplasm than the male gamete.

It has been shown, by crossing male and female gametes of known genotypes, that the inheritance of respiratory defects depends on inheritance from the female gamete. The characters of the F_1 generations that result from reciprocal crosses demonstrate this:-

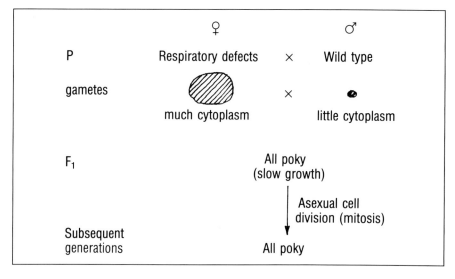

By contrast the reciprocal cross proceeds as follows:-

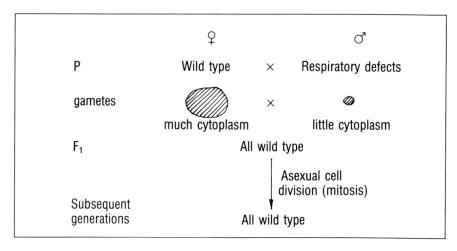

c) Examples of Chloroplast Inheritance

i. The Interactions of Chloroplasts and the Nucleus in Higher Plants

Two species of closely related higher plants are crossed (their offspring are infertile) to demonstrate inheritance controlled by chloroplast genes interacting with nuclear genes:-

	♀		♂
P	Oenothera muricata	x	Oenothera hookerii
F_1		All green (with some yellow sectors)	

By contrast the reciprocal cross gives:-

	♀		♂
P	Oenothera hookerii	x	Oenothera muricata
F_1		All yellow (and they die)	

The explanation of these results is that some nuclear genes from Oenothera muricata prevent the normal chloroplast development in hookerii. If chloroplasts are taken from yellow parts of the F_1 plants and put into Oenothera hookerii cells they develop into green chloroplasts, even after a prolonged existence in a cell under the control of unfavourable muricata genes.

ii. Another Example of Chloroplast Inheritance – Streptomycin Resistance in the Unicellular alga Chlamydomonas reinhardi

A summary of the products of chloroplast genes, together with their relationships with some products of nuclear genes in providing the structures and enzymes of chloroplasts, were given in Chapter 6, pages 213-214. It is with the intention of providing a specific example of chloroplast inheritance that a character unrelated to chloroplast structure – resistance to the antibiotic streptomycin – is shown here.

The alga Chlamydomonas reinhardi has, in its vegetative cells, a haploid nucleus, about 20 mitochondria (the mitochondria are not involved with this study, it is believed) and only one chloroplast. The life cycle of the species has both a sexual phase and an asexual, vegetative phase. In the processes of producing gametes two nuclear types, of opposite vegetative mating types are needed – mt^+ and mt^-. These two types are usually indistinguishable from each other by visual inspection.

Nuclear inheritance of mt^+ and mt^- proceeds in the conventional way in the production of four gametes. However chloroplast inheritance is *always* that contributed by the vegetative mt^+ nucleus. The chloroplast genes donated by the cell with the mt^- are thought to be broken down in any mt^+ cell with which the mt^- nucleus fuses. If α^+ represents streptomycin resistance and α^- represents susceptibility to streptomycin, the following diagram shows how the inheritance of resistance to streptomycin occurs in C. reinhardi by way of the chloroplasts:-

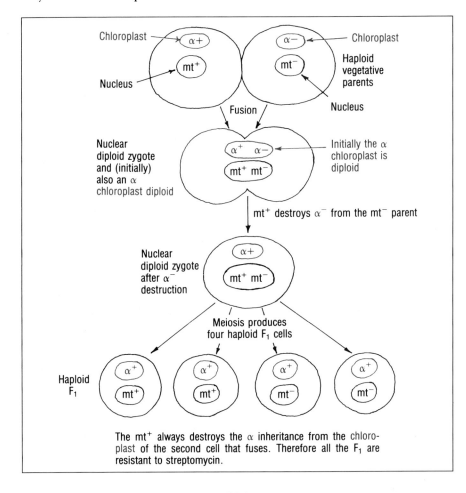

By contrast when the nuclear and chloroplast genes are associated differently in the vegetative parental cells:-

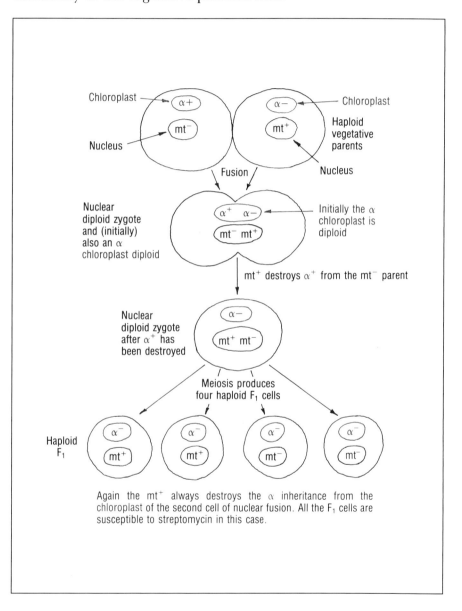

Chapter 7 Questions

1. How does the plasmid sex (F) factor change a bacterium from being a recipient for DNA into a donor?
2. How does the F^+ sex factor operate as an episome to create high frequency recombination (Hfr) in E. coli?
3. How is a 'col factor' suicide of benefit to a bacterium species such as E.coli?
4. What result from reciprocal crosses would suggest extra-chromosomal inheritance in eukaryotes?
5. From which gamete is extra-chromosomal inheritance of greater influence as a rule?
6. Give examples of mitochondrial and chloroplast inheritance.

8

GENETIC ENGINEERING

The distinction needs to be made between two important ways by which man alters the inheritance of living things.

Breeders have been interfering with nature for many centuries by selecting useful phenotypic characters in a large number of plant and animal species. By doing so they have refined living things to be as close as possible to genetic homozygosity, so that they will breed 'true', according to man's needs. Within the constraints imposed by the damaging effects of inbreeding, caused when damaging, recessive alleles become homozygous, the breeds and varieties created by man's artificial selection are those that give rise to new generations that are predictable and which breed true.

In addition to these traditional breeding skills, the last few years have produced one of the most potent and brilliant advances in man's ability to manipulate genes – GENETIC ENGINEERING. Quite a large number of scientists, mostly working in Europe and in the United States of America, have made contributions to new discoveries and to new techniques in the synthesis of genes and for introducing genes into bacterial plasmids. Enzymes were discovered in the late 1960s that can reverse the normal processes of DNA transcription into mRNA. It became possible to isolate particular mRNA molecules and these were put into solution. The separate constituents of DNA – four bases, deoxyribose and phosphate – were then added to the solution in quite large numbers. In the presence of the enzyme reverse transcriptidase, the sequence of bases in the mRNA was transcribed into the sequence of pairing bases in DNA. In other words, it became possible to synthesise the gene which encodes the particular mRNA molecules in solution.

The history of discovery in the field of genetic engineering is very well described in a textbook from Germany, published by VCH. The English translation of the book is mentioned in the bibliography for this chapter.

The brief account on genetic engineering given in this chapter includes the description of a method for inserting a synthesised gene into a bacterial plasmid. Recently it has become possible to manipulate synthesised genes in higher organisms, especially in yeasts. With the advance of knowledge it may eventually become possible to manipulate genes in all species safely, including genes in humans. This gives rise to the very real possibility of removing defective genes and replacing them with wild type alleles in human germ cells, thereby eliminating family defects but permitting desirable family traits to be passed on. It may also become possible to identify 'best sets' of human alleles that can be collected together and inserted into human DNA by genetic engineering.

The new science of genetic engineering raises several moral and legal issues. It will certainly be necessary to restrain scientists in their ability to 'play God'. The codes of practice governed by the loose language of the Hippocratic Oath make the clinical decisions of doctors very difficult to reach in some cases. Striking the correct balance between the enforcement of parliamentary legislation to constrain irresponsible or unnecessary experimentation, and the practical need to remove serious genetic defects from families and the human species will not be easy.

There is also the possibility of the artificial synthesis of cells according to the specifications of scientists. This has enormously broad implications for the future of mankind, both for good and for evil. In the same way that the application of atomic energy has led to its uses for good – generating electricity and combating cancer – and for evil – making nuclear weapons – so, too, could genetic engineering lead some scientists towards the genetic improvement of life on this planet and others towards making genetic horrors. Once knowledge becomes available to mankind, it is very difficult to exercise world-wide control of the technological and biological uses to which it is put. Genetic engineering offers to doctors, veterinary surgeons and breeders a step further forward than the methods they used before its discovery. So long as it can be shown that it gives a success rate at least as good as other forms of treatment to living things, with as few undesirable side-effects, then scientists have a good case for continuing their efforts to improve the genetic stock of humans and of other species.

Much of the detail of this chapter describes the method for introducing synthesised genes into the plasmids of the bacterium Escherichia coli. It was very difficult persuading micro-organisms to accept 'foreign' genes into their DNA. It was also very difficult to find techniques for making genes in large numbers 'in vitro' so that they can be stored and used whenever needed. It was not until the 1970s that the techniques described later in this chapter gave rise to genetic engineering. In the last decade it has become possible to:-

i. Isolate prokaryotic genes in complete purity.
ii. Synthesise eukaryotic genes by using their specific mRNA and the reverse transcriptidase enzyme. These synthesised genes of eukaryotic species have their introns removed before being inserted into bacterial plasmids.
iii. Isolate genes from both prokaryotic and eukaryotic species. These can be studied in gene libraries in the absence of the living organisms from which the genes were taken or transcribed.
iv. Accumulate gene libraries from which a large number of different, desirable genes can be quickly and easily obtained.
v. Introduce desirable genes into micro-organisms. The genes can have their origins in any species and can be introduced into the DNA of bacteria, yeasts or bacteriophage viruses. When the micro-organisms multiply they can translate the introduced genes into desirable products.
vi. Set up commercial businesses based on the products derived from genetic engineering.

Practical Uses for Genetic Engineering

(i) Medical or Veterinary

(a) Enzyme Production for Individuals with Defective, Mutant Genes for Enzymes

The NBS Biological Wall Chart *'Inborn Errors of Metabolism'* * shows that mutant genes exist for most of the essential biochemical conversions in human cells. The chart shows that a high proportion of essential metabolic conversions in humans have defective alleles spread through the human population. Fortunately, only a small proportion of the human population inherit the double recessive

---

* Obtainable from: New Brunswick Scientific,
                    NBS Biologicals,
                    Edison House,
                    163 Dixons Hill Road,
                    North Mymms,
                    Hatfield, Herts, AL9 7JE

condition for these defective alleles. In rare instances individuals do inherit the double recessive mutant alleles, and this can be lethal. If 'wild type' genes for the essential enzymes needed for cell metabolism are introduced into bacteria these micro-organisms can make the enzymes deficient in a double recessive mutant individual, and the enzymes can be 'fed' into a patient.

(b) Hormone Production for Individuals with Defective Mutant Hormone Genes

Insulin, human interferon and human growth hormone can all be made by bacteria, into which the purified genes for each of these hormones have been introduced by genetic engineering.

(c) Ante-natal Diagnoses of Genetic Defects in Human Embryos

Consideration can be given by doctors and parents for deliberate abortion of foetuses that show serious genetic defects. An analysis of the DNA taken from foetal cells, floating in the fluid around the foetus, can reveal genetic disorders such as sickle-cell anaemia or more serious inherited diseases.

(d) To Increase Disease Resistance (other than by making anti-biotics)

Some pathogenic bacteria have to cling to human or animal tissues in order to multiply and cause disease. If a non-pathogenic strain of the bacterium (or a suitable sugar in diet) is introduced into the patient, the non-pathogen (or the sugar) can substitute for the pathogenic bacteria, so preventing large numbers of the pathogens from clinging to the human or animal tissues. Infectious diarrhoea in animals can be treated in this way. Suitable strains of non-pathogenic bacteria can be made by 'genetic engineering'.

(e) Antibiotics Production

The actinomycete fungi produce several thousand antibiotics. The biochemical synthesis of the antibiotics often requires a dozen (or more) genes. Some of the antibiotic genes are found on the plasmids in these fungi. The recombination of desirable antibiotic genes in plasmids by genetic engineering may lead to more effective anti-biotics, produced by these fungi in large and economic quantities.

(ii) Detoxification

Man's need to use insecticides and herbicides has led to undesirable longer term consequences caused by the persistence of these toxic chemicals. Some plasmid genes allow their hosts to use toxic substances as their source of carbon, the entire set of degradation genes being in some cases in the plasmids. In others the plasmid degradation genes rapidly produce a chemical from the toxin that can be dealt with by the normal degradation processes of cells. Genetically engineered bacteria may be able to detoxify regions contaminated by persistent toxins.

(iii) Animal Feed Synthesis

Methanol is a constituent of 'natural gas', derived from fossil fuels. The carbon in methanol can be removed from air enriched with methanol and converted into cellular, carbon-containing molecules. The bacterium species Methylophilus methylophilus (ASI) mops up the carbon in methane quite efficiently. However if ammonia in the air is the source of cheap nitrogen this species is rather inefficient at converting ammonia into glutamate (one of the amino-acids), the means of introducing nitrogen into the metabolism of the cell. Luckily E. coli have an efficient ammonia conversion biosynthesis and the genes for this biosynthesis have been introduced into M. methylophilus by genetic engineering to allow that species to convert methane and ammonia into glutamate. Glutamate can be persuaded to move outwards through the bacterial membranes by altering the biotin concentration, which alters the fatty-acid constitution of the bacterial membrane, so allowing the extrusion of glutamate in large quantities into the broth in which the bacteria multiply exponentially. Glutamate is used for animal feed production.

(iv) To Improve Disease Resistance in Crops

Plasmids play a part in disease resistance. Plant breeders are continuously faced by the problem of a build-up of disease whenever one strain of a crop is used for more than one season. Improved disease resistance by genetic engineering of the plasmids in commercial crops may lessen the need to keep changing the strains grown from one year to the next.

(v) To Loosen or Remove The Specificity of Desirable Symbiotic Associations

By introducing 'foreign' genes into Rhizobium species of bacteria it is possible to make those bacteria that were once host specific for a legume take on the ability to infect and nodulate a large number of legume species, so increasing nitrogen fixation, thereby increasing crop yields.

Plasmid and Viral DNA Used for Genetic Engineering

(a) Plasmid DNA

Any extra-chromosomal DNA is a plasmid. Plasmids are found in all types of bacteria, in yeasts, in fungi and in some mammals. Their DNA codes for a wide range of characters:-

i) Adhesion by animal pathogenic bacteria to the gut lining. An example is diarrhoea-causing bacteria.

ii) Virulence.

iii) Biosynthesis of antibiotics.

iv) Resistance to antibiotics.

v) Restriction enzyme (page 236) synthesis.

vi) The ability to transfer chromosomal and plasmid DNA from one microorganism (or cell) to another.

vii) The degradation of carbon compounds (that are sometimes toxic).

viii) The control of nodulation in legumes.

Genetic engineering sometimes makes use of the plasmid genes themselves. However, much of genetic engineering makes use of another property of plasmids – the ability to accept 'foreign' genes into their DNA and to translate the 'foreign' genes into biochemical products or other inherited characters.

Plasmids occur as small circular DNA fragments. The cells of bacteria and yeast usually contain one or more of them.

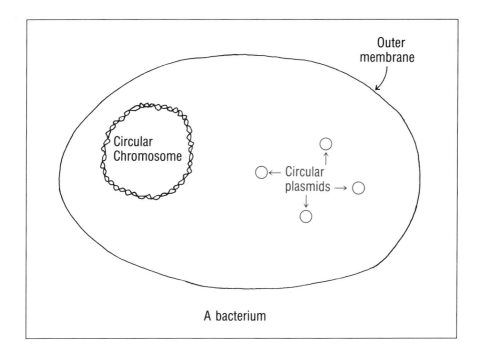

A bacterium

Bacterial Plasmids

A bacterium species commonly used for genetic engineering is Escherichia coli. Three of the plasmids found in this species – Col E1, PMB 9 and PCR1 – are suitable for accepting 'foreign' genes.

Fungal Plasmids

The yeast, Saccharomyces cerevisiae, contains the plasmid Scp 1, which is suitable for genetic engineering.

(b) Viral DNA

Bacteriophage lambda, and the viruses SV40 and Polyoma have been used successfully.

The Properties of Plasmids That Make Them Suitable for Genetic Engineering

Some plasmids have a position on their DNA that responds to enzymes in such a way that 'foreign' DNA can be inserted into the circular plasmid. The identification of the positions and the necessary enzymes for inserting 'foreign' DNA were crucial steps forward in making genetic engineering possible.

The place where 'foreign' DNA can be inserted into a plasmid is called a 'RESTRICTION SITE'.

The enzymes used for DNA insertion are called 'RESTRICTION ENZYMES'.

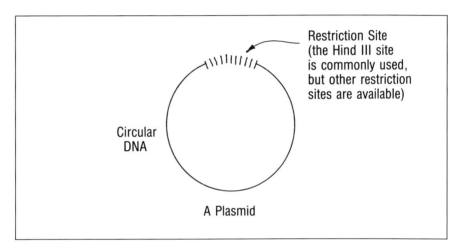

A Plasmid

The Synthesis of a 'Desirable' Gene and the method by which it is put into a Bacterial Plasmid

Chapter 4 described eukaryotic genes as having 'redundant' loops in them called 'introns'. In order that eukaryotic genes can be stored in 'gene banks' for insertion into prokaryotic bacteria the introns have to be removed, otherwise the bacteria would include the codes in the introns in translation and make incorrect polypeptides. The following rather complex procedure is therefore necessary to make it possible to store genes in 'gene banks' and to use the stored genes to make desirable products from bacteria.

The Synthesis of Complementary DNA (cDNA) for Eukaryotic Genes

When genes are inserted into bacteria by genetic engineering, the bacteria transcribe each inserted gene from a continuous sequence of bases in the DNA. By contrast, the genes in eukaryotic species have base sequences that are interrupted by intron loops. It would therefore cause the wrong polypeptide to be made by bacteria if a eukaryotic gene were introduced directly into a bacterial plasmid. There would also be considerable problems in purifying eukaryotic DNA.

It is, however, possible to synthesise the sequence of bases in the exons of an eukaryotic gene using the rather complicated procedure described later in this chapter. Briefly, this is done by extracting and isolating the mRNA that was transcribed by the desired, eukaryotic gene. Enzymes and chemicals are then used to reverse the normal transcription process. This is done *in vitro* and the DNA base sequence pairs up with the mRNA base sequence. The DNA single strand made in this way is then used as the template for a second, pairing nucleotide chain to complete the DNA structure of a synthesised gene. The synthesised gene can then be inserted into a bacterial plasmid. Transcription and translation by the bacterium of this genetically engineered, inserted gene make the desired product specified by the eukaryotic gene.

The Storage of Genes in Gene Banks

Having purified mRNAs and desired genes it is economic to store them in a form and in conditions that do not allow them to fragment or mutate. It is also convenient to keep only a small gene stock from which it is possible to produce very large numbers of desired genes. This can be done by inserting the cDNA genes into the plasmids of bacteria by genetic engineering. The bacteria multiply very quickly so that it is possible to produce desired products from them rapidly and economically.

Gene banks can be used to store:-

 i) mRNA appropriate to the desired gene.
 ii) cDNA appropriate to the desired gene.
 iii) Genetically engineered bacteria clones containing the desired gene in a plasmid.

It is now appropriate to describe the procedure for inserting an eukaryotic gene into a bacterium.

Stage 1 – The Synthesis of Single Stranded cDNA using a Stored mRNA as the Template

The synthesis of single stranded cDNA, in which occurs the same order of bases as those in one of the DNA strands of *the coding part* of the desirable gene, depends on an enzyme called Avian Myeoblastosis Virus (AMV) transcriptidase. This enzyme, in the presence of the mRNA for the gene and segments of nucleotides, can use the mRNA as a template for single stranded cDNA synthesis:-

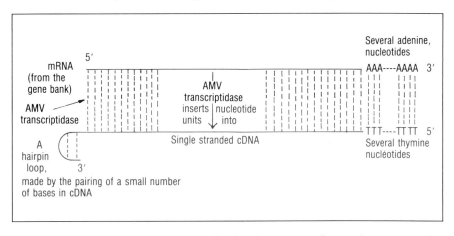

The single strand of cDNA can be broken away from the mRNA by alkali hydrolysis. In the presence of the enzyme E. coli DNA polymerase or AMV reverse transcriptidase the pairing second strand of cDNA can be made if nucleotides are provided:-

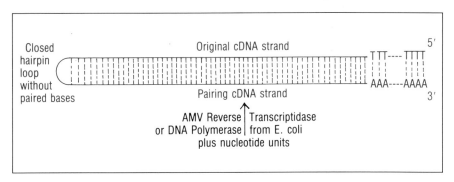

The closed hairpin loop can be broken open using an enzyme called S1 nuclease, which comes from the fungus Asperigillus oryzae (high salt concentration and low temperature are needed), and this enzyme operating in suitable conditions does not break the DNA outside the loop.

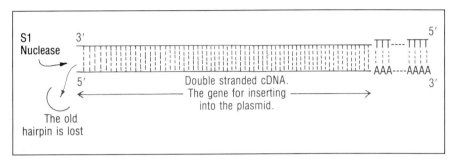

Stage 2 – The Introduction of Double-Stranded cDNA into Microorganisms (E. coli in this case)

Having made double-stranded cDNA it is necessary to persuade micro-organisms to take it in through their surrounding membrane. The cDNA molecules are made *in vitro* in very large numbers so that, even if, as usually happens, a high proportion of cDNA molecules fail to penetrate the membrane of a bacterium, there are sufficient numbers of bacteria that do accept the cDNA. Those bacteria that have taken in the cDNA have to be identified and these can be produced in enormous numbers later.

Bacteria take up cDNA through their outer membranes more easily if calcium chloride is added to the broth in which the bacteria multiply. Bacteria are said to be 'COMPETENT' when they are in a condition to take in cDNA through their membranes.

Stage 3 – The Transformation of a Microorganism by cDNA

Transformation (taking on a new genotype and phenotypic character) of microorganisms such as E. coli can be brought about by inserting the double-stranded cDNA into one of its plasmids. However it is necessary to prepare both the cDNA and the restriction (insertion) site on the plasmid (often the Hind III site) so that enzymes can insert the cDNA into the plasmid. To do this, bases at the ends of the cDNA molecule must pair up with the bases at the broken ends of the plasmid DNA.

The following scheme shows how double stranded cDNA is modified so that it can enter the Hind III restriction site on the plasmid:-

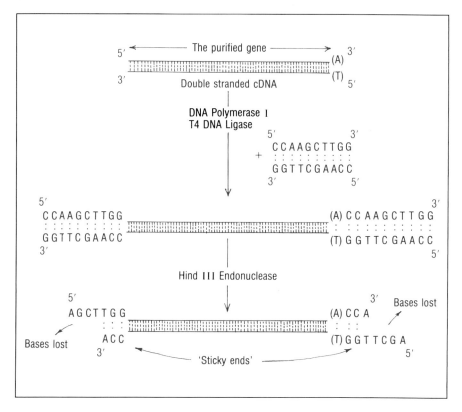

The addition of the 'sticky-ends' to the purified gene makes it possible for the gene to enter the Hind III site of the plasmid.

The plasmid must first be prepared at one of its restriction sites – the Hind III site. A RESTRICTION ENZYME is used in suitable conditions to break the double strand of the plasmid DNA to make plasmid bases available to pair up with the 'sticky ends' of the transforming gene:-

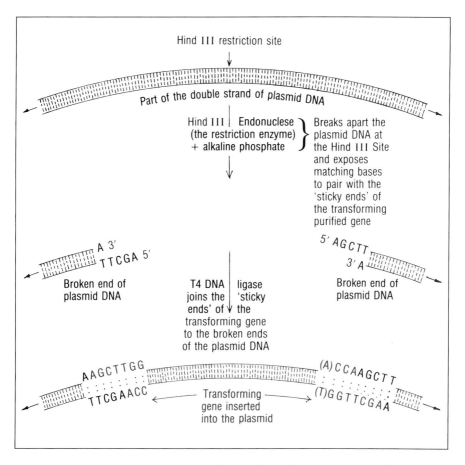

A variety of eukaryotic or prokaryotic wild-type desirable mutant genes can be inserted into several plasmid sites in this way. When the bacteria reproduce in large numbers the plasmids, with the transforming cDNA in them, are replicated and pass into new bacterial cells. Then the translation of the transforming genes by millions of bacteria into chemicals (polypeptides) gives economic yields in several cases.

In recent genetic engineering it has been possible to find restriction enzymes and restriction sites that allow the insertion of transforming genes without the need for making 'sticky ends'.

Stage 4 – The Identification and Isolation of Bacteria In which Desired Genetically Engineered Transformation Has Occurred – SCREENING

The aim of genetic engineering is to introduce a desired foreign cDNA into a bacterial plasmid and for the bacteria to reproduce, so creating a very large number of clones (bacteria of identical, desirable genetic inheritance – in this case all having the transforming cDNA in them).

The introduction of cDNA into bacterial plasmids is rarely achieved for only one type of cDNA at a time. Bacteria with the 'wrong' cDNA will therefore grow side by side with bacteria containing the 'desired' cDNA.

All these bacteria are stored as frozen colonies on millipore filters placed on agar. It is then necessary to identify those colonies of bacteria that contain the desired cDNA and to ensure that they produce the desired industrial or medical chemical. The identification procedure is called SCREENING. It is carried out in two stages.

(i) Primary Screening

Primary screening identifies those colonies of bacteria that *might* contain the desired cDNA in their plasmids. Radioactive cDNA containing ^{32}P of the desired base sequence is introduced into a colony of bacteria that are multiplying on a millipore filter. Some of the competent bacteria will take up the desired cDNA in its entirety, but others will only take up part of the cDNA, after it has partly broken into fragments. Any radioactivity caused by cDNA transformed bacteria, when placed on an X-ray plate, shows where there *may* be bacterial clones containing the entire desired cDNA.

Colonies containing radioactive bacteria are plated out at low densities and wherever radioactivity occurs in the colonies there *may* be the desired cDNA clones, uncontaminated by bacteria of other strains.

However, before the bacteria are grown in large quantities for industrial, environmental or medicinal purposes, it is necessary to check that the desired cDNA has indeed been incorporated into the plasmids of the cloned bacteria.

(ii) Secondary Screening

Because there may have been the fragmentation of the desired cDNA before or during the uptake into a bacterial plasmid it is necessary to carry out secondary screening. This is to ensure that the bacterial plasmid has been transformed by the *complete*, desired cDNA. It also checks that the desired chemical is produced by the bacteria. There are various ways of carrying out these checks.

(a) Secondary Screening of the cDNA

(i) Analysis of the base sequence of purified cDNA taken from restriction sites in the transformed bacterial colony.

(ii) Quantitative measurement of the length of the super-coil of a plasmid. This length is checked against the super-coil length when the complete cDNA is known to be inserted at the plasmid restriction site.

(iii) Restriction enzyme analysis when the base sequence of the plasmid restriction site is known.

(b) Chemical Product Analysis

(i) Electrophoresis distinguishes between polypeptide molecules by making use of their electrical properties to separate them. Their distance moved across absorbent paper can be used to make a rough guess at the polypeptide produced.

(ii) Techniques are available for making a complete amino-acid sequence analysis of the products of genetic engineering.

Some Uncertainties about Genetic Engineering

(a) The Multiple Effects of Genes

Because some genes have multiple effects, accelerate or decelerate the products of other genes or overlap each other, at one time producing one product, at other times a second product from the same length of DNA, it is not always possible to predict the consequences of genetic engineering. It is not impossible that dangerous micro-organisms could be made accidentally.

(b) Increasing the Proportion of Inviable Alleles

Genetic engineering encourages the perpetuation of inviable recessive alleles when it is used to make metabolic enzymes for mutant individuals. Such individuals then produce offspring and increase the proportion of inviable mutant alleles in the population. It may, at some time in the future, be possible to manipulate genes in the gametes of mutant individuals in a way that corrects the damaging mutation, so allowing mutant parents to have normal children.

Eukaryotic Genetic Engineering

In the same way that the genetic engineering of bacteria presented considerable difficulties to scientists, so too has the genetic engineering of eukaryotic species encountered several obstacles to progress. One of the major problems has been the difficulty of verifying the successful incorporation of desirable genes into the replicating parts of an eukaryotic cell's genome. Screening has become possible by using drug resistant and/or mutant marker genes which have close physical attachments to desirable genes on transforming DNA. In the case of drug resistant genes, screening identifies those cells that grow in the presence of the drug. The implication is that the desirable gene is also in the transformed cell. In the case of mutant genes, special techniques are available using cell fusion for identifying cells containing desirable genes. This technique is described later (pages 247-251). Further screening is then needed to make a positive identification of the biochemical products (polypeptides) translated from the desirable genes. Desirable genes and drug resistant genes can be obtained either from living cells or they can be synthesised *in vitro*. At the present state of knowledge the genetic engineering of eukaryotic cells is confined to those genes that produce functional or structural polypeptides. Although some of these polypeptides are hormones which control overall characters such as growth, in general gross features in multicellular organisms cannot be controlled by genetic engineering in eukaryotes.

In order to take in transforming DNA, eukaryotic cells must be COMPETENT. They will then only be of long term use if transformed cells are taken from 'established cell cultures' and even then not many cells will show permanently transformed characters. Cells in culture can be divided into those that can only divide about 20 times – PRIMARY CELL CULTURES – and those that can divide indefinitely – ESTABLISHED CELL CULTURES.

A further serious problem for the genetic engineering of eukaryotic cells is that of gene regulation. Regulatory genes (an example was given in chapter 4, pages 159-160) usually have to lie adjacent to the gene they regulate. Deregulation, caused by the separation of the transforming gene from its controlling genes, can have serious consequences (an example of gene transfer which results in human cancer is given in Chapter 15, page 489). There is much research needed before desirable genes can be incorporated into a transformed cell in the 'correct' part of the genome together with their regulatory genes.

In some cases the successful transfer of transforming DNA into an eukaryotic cell is not followed by the replication of the transforming DNA fragment at cell division. In this case the transformation of a cell is of no long term value.

Yet another problem for eukaryotic genetic engineering is that the polypeptide produced by the transcription of a gene in a bacterium is different from the polypeptide produced from the same gene in eukaryotes. The simple sequence of amino-acids is made more complex in eukaryotes by the addition of sugar residues to a proportion of the amino-acids. Thus the success of a colony of bacterial clones in producing a desirable character, (usually an end product of a biochemical pathway), is no guarantee that the cloned bacterial gene will operate correctly if its product transforms a eukaryotic cell. The screening of the polypeptide product synthesised by transformed eukaryotic cells therefore has to be carried out.

Colchicine has proved to be a very useful chemical for genetic engineering. It arrests the processes of mitotic cell division when chromosomes are shortest and thickest (see Chapter 9, page 271). Techniques are available for breaking down the nuclear membrane and outer membrane of cells to release the short, thick chromosomes. Each short, thick chromosome has a different molecular mass and can therefore be separated and identified by differential centrifugation. Chromosomes known to be the carriers of desirable genes can therefore be separated and identified by differential centrifugation. When the chromosomes containing transforming genes are put into a cell culture medium, a very small proportion of cells ingest the transforming gene on its chromosome by phagocytosis (the outer cell membrane is modified to surround a small volume of bathing medium), and genetic transformation can occur if the ingested chromosome reaches the nucleus of the recipient cell.

Genetic engineering in eukaryotic cells has been of great value in mapping the positions of genes on the chromosomes of those diploid species in which crossings-over occur in both sexes during meiosis. Genetic mapping has been relatively easy in haploid species such as bacteria and in diploid species such as Drosophila melanogaster in which crossings-over at meiosis only occur in one sex. In the eukaryotic cells of most diploid species genetic mapping has been much more difficult, but recent advances in eukaryotic genetic engineering has allowed the assignment of genes to particular chromosomes. It has also allowed the distance apart of genes on chromosomes to be estimated. Genetic mapping is discussed further in Chapter 16 in relation to human genes.

A complication that occurs very frequently in those cells that do take in potentially transforming DNA is that there may have to be several cell divisions before a transforming gene is incorporated into the translated genome of the recipient cell. Nearly all the DNA ingested by competent, recipient cells is fragmented either before or after ingestion. Only one fragment from each chromosome is likely to have a centromere (page 273). The location in a cell of those DNA fragments without a centromere but which contain desirable, potentially transforming genes is therefore uncontrolled. Many of the cells that arise from a cell that has ingested DNA into it never receive the transforming gene. Together with the unpredictability of the locus where a transforming gene fragment is incorporated into a recipient cell's genome, the absence of the transforming gene makes the success rate for transformation by genetic engineering very low. This is true even when calcium phosphate is added to transforming DNA and when dimethylsulphoxide is added to recipient cells. Nonetheless considerable progress is being made in the genetic engineering of higher organisms and a review of some success is now given.

A Summary of Methods Used to Insert Desirable Genes into Eukaryotic Cells

(i) Cell Fusion followed by Genetic Transformation

Eukaryotic cell fusion is encouraged by:-

a) The presence of a few species of virus (Sendai virus is an example). It is necessary to remove the pathogenic properties of such viruses by irradiation before they are safe to use.

b) Chemicals such as polyethylene glycol.

Cell fusion causes extra DNA to be present in the fused cell. If some of this extra DNA is included in the replicating part of an established cell's genome, genes in the extra DNA may be expressed in cells arising from the fused cell. Any of the extra DNA that does not become part of the fused cell's replicating DNA is shed at each cell division. When fusing cells both come from the same species it is usual that chromosomes are shed during cell division until the number reduces to the approximate number of the diploid cells of the species. When fusion occurs between cells taken from different species it is usual that the chromosomes from one of the species are almost all shed, while those of the second species are largely or entirely retained. It is this feature of cell fusion that allows the genetic mapping of genes by contransformation frequencies. An example is given on pages 555-556 of human/mouse cell fusion.

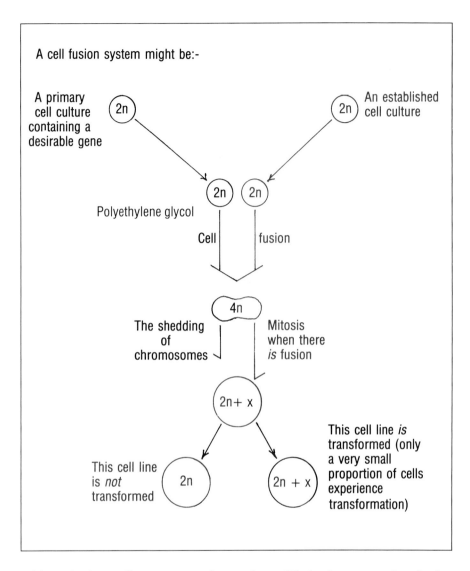

This technique allows cotransformation of linked genes to imply the distance apart of genes on a transforming chromosome or on a chromosome fragment.

Various techniques have been developed for the detection of fused cells that are likely to contain desirable genes. The simplest schematic presentation of how this can be done is shown on the next page.

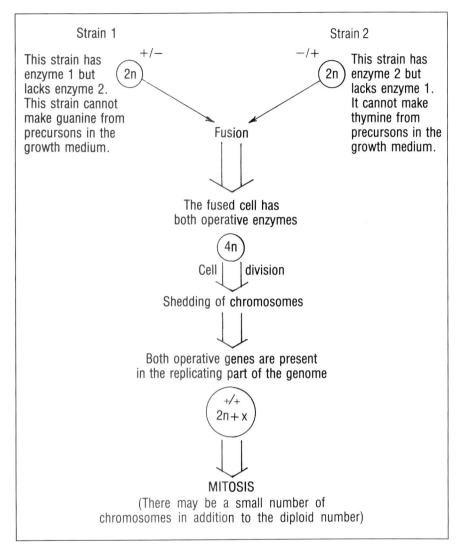

A specific example of how mutant cells can be used to identify fused cells is one in which there are two biochemical pathways for DNA synthesis – the normal pathway and an alternative pathway called the scavenger pathway.

The normal pathway for DNA synthesis can be deliberately inhibited by adding the drug aminopterin to the medium containing cells. The scavenger pathway uses enzymes to synthesise the constituent molecules of DNA. Mutant cells are available in which guanine cannot be used for DNA synthesis using precursor molecules. Other mutant cells which cannot place thymine in DNA are also known and these cannot make thymine from precursor molecules.

A cross between the two mutant strains of cells is arranged on a medium containing the precursor molecules needed by the scavenger pathway for DNA synthesis. Neither of the mutant strains can grow on their own. However fusion on the medium containing the scavenger precursor molecules hypoxanthine, aminopterin and thymidine (HAT medium) does result in growth. The following scheme illustrates this:-

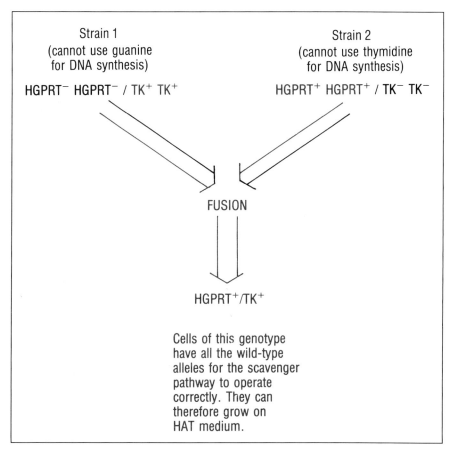

The rare occasions in which there is cell fusion can be identified by this means. Having identified those cells that have fused using the HAT/mutant identification process, there are then slightly different consequences in those fusions involving cells of the same species and those involving cells of different species.

Same Species Fusions

There is only the occasional loss of chromosomes so the fused cells retain 4n chromosomes for very many generations of cells. The expression of genes in fused nuclei depends on the dominant/recessive relationships of the four alleles present for each gene.

Different Species Fusions

When there is the fusion of cells taken from different species the chromosomes from one species are retained at the expense of those from the second species, which are largely shed. This has permitted considerable advances in the genetic mapping of genes. Fused cell lines exhibit transformation for some of the genes that are expressed at the biochemical level in cells. Examination of fused cells after several mitotic divisions can identify which chromosome(s) (or parts of chromosome(s)) have caused genetic transformation. This allows a number of genes to be ascribed to particular chromosomes. In the case of human/mouse fused cells it has been possible to find out upon which chromosomes some human genes are located. This is because human chromosomes are shed and mouse chromosomes are retained. Each fused human (dominant alleles) / mouse (recessive mutant) cell retains different human chromosomes. The progeny cells of each cell line are examined for the expression of human genes, and the human chromosome upon which each expressed human gene is found can be identified. Further genetic mapping to find out how far apart genes are on the same chromosome can then be carried out by cotransformation frequencies (page 556).

(ii) The Microinjection of DNA into the Nuclei of Eukaryotic Cells

Desirable genes in DNA that is either synthesised in vitro or extracted from living cells can be injected into the nucleus of an eukaryotic cell. Cells at the earliest stages of embryonic development can have DNA injected into them in this way. The embryo cells must be separated from each other at the four cell stage before the microinjection of a large number of DNA fragments is made into one of them. If the first embryo cell shows no evidence of being transformed, the other three embryo cells can be injected in turn in the hope that one of the separated cells will be transformed by the injected DNA in the required way. The pronuclei of fertilised oöcytes can be genetically engineered in a similar way. Wild-type globin genes have been successfully injected into the replicating parts of mutant rabbit oöcyte DNA using this technique. In mice, growth hormone genes have been injected into the nuclei of embryo cells to increase the overall size of those mice that develop from cells transformed in this way.

(iii) The Use of Plasmid Equivalents Derived from Viruses

The Simian Sarcoma Virus 40 (SSV 40) has a circular genome similar in size to a plasmid. The DNA of this virus species can be genetically engineered so that it loses its damaging properties and so that it includes 'foreign' desirable genes in its genome. Eukaryotic cells that are invaded by SSV 40 DNA can be divided into:-

a) Permissive Host Cells

The viral DNA enters the nucleus of the host cell. Viral genes specify the production of those polypeptides necessary for viral DNA replication and in consequence an enormous number of viral particles are produced in the host cell. This leads to the rupture (lysis) and death of the host permissive cells. Monkey cells can be permissive for SSV 40.

b) Non-Permissive Host Cells

In other species such as rats and mice the biochemistry of cells does not support the biosynthesis of viral particles. However the SSV 40 virus DNA does sometimes become incorporated in the replicating part of the host cell's DNA. This can lead to cancer, proving that viruses either carry cancerous genes or that they can deregulate host cell genes so that they become cancerous. The genetics of cancer is discussed more fully in Chapter 15.

The genetic engineering of the SSV 40 genome using restriction enzymes is similar to that of bacteria. However there are always undesirable consequences of introducing the genetically engineered SSV 40 genome into eukaryotic cells, even if the cells are non-permissive and cancer is not a consequence. A further problem arises because the desirable gene introduced into the viral genome is not always correctly translated in transformed, non-permissive eukaryotic cells. Permissive cells die, even if there is a short-lived correct expression of the transforming gene. This technique is therefore only suitable for finding out the biochemical products and phenotypic expression of genes introduced into eukaryotic cells. It is not suitable for altering the long term genotype of established cell lines.

(iv) The Insertion of Whole Cells into Animal Embryos

If cells from a different genetic line are inserted into a developing animal embryo of the same species at the blastula stage, the genes in the inserted cells may be expressed in those parts of the adult animal that develop from the inserted cells. If an inserted cell develops later into the gonads, the genotype of the inserted cell may be transmitted in the gonad derived gametes into future generations. This technique is not suitable at present for use in humans because there is not yet sufficient certainty in this type of genetic engineering to make it safe.

Having discussed some general aspects of eukaryotic genetic engineering it is now appropriate to divide eukaryotic cells into animals, fungi and plants and to examine each of these in turn.

The Genetic Engineering of Animals Cells

The main thrust of animal cell genetic engineering has concentrated on mammal cells in the hope that techniques will eventually be found for the genetic engineering of defective human genes. The removal of inherited diseases in both humans and animals by means of genetic engineering in germ cells and embryos is a real possibility. Genetic engineering will also lead to the improvement of animal stocks with better commercial characters.

The glib writing of a commentator does no justice to the scientists who identified the techniques now used for the genetic engineering of animal cells. Therefore no complaint is intended by the statement that the unreliability of genetic engineering in its present form makes it unsafe to attempt the genetic engineering of human gametes, human zygotes or human embryo cells. It should be emphasised that uncontrolled genetic engineering has enormous potential for damage as well as for improving the genetic stock. Very close supervision by experienced scientists in this field is therefore an absolute requirement.

The following brief list outlines some of the advances made in the genetic engineering of mammal cells:-

(i) The assignment of genes to particular chromosomes.
(ii) The determination of how far apart genes are on the same chromosomes.
(iii) The screening of embryos for genetic defects.
(iv) The transformation of animal cells using 'foreign' DNA from a variety of sources. In some cases germ cells have been transformed; in others somatic cells include transforming DNA in the replicating part of their genome.
(v) The discovery of some of the genetic causes of cancer.
(vi) The ability to combat some types of cancer cells.
(vii) The insertion of 'foreign' cells into embryos in such a way that the 'foreign' cells pass on their genes into the gonads of the developing animal. Genes in the 'foreign' cells can be passed on into new generations which arise from the gametes produced by the adult that develops from the genetically engineered embryo.
(viii) The introduction of specific antibody genes into fused cells. This aspect of genetic engineering has considerable potential for the medical profession and is now considered in a little more detail.

Cell Fusion for Producing Animal Antibodies

Antibodies are produced by animal immune systems in response to the invasion of their bodies by particles from some alien source (a brief description of the genetics of antibody production is given in Chapter 14, pages 464-469). In the past one method (in addition to vaccination) of obtaining antibodies to fight human diseases was by injecting the disease into large animals. The antibodies produced by the immune system of infected animals were later extracted from the animal by bleeding. However there have always been problems of isolating pure colonies of specific antibodies and separating them from other antibodies in the animals' blood. Another disadvantage of this method was the limited quantity of a specific antibody that could be extracted from each animal. In recent years genetic engineering has both increased the production of antibodies and has allowed specific antibody production without contamination by other antibodies.

Cancer cells produce antibodies very efficiently. Mouse cancer cells can be taken from a cell culture and fused with individual, normal cells taken from the spleen of mice. Each spleen cell in the living animal has its genetic system modified for the production of only one antibody. After fusion there is, to start with, the production of both cancer cell antibodies and the specific antibody produced by the individual spleen cell that fused. Mutant cancer cells occasionally arise which fail to produce any antibodies at all. Fusions of such mutant cancer cells with individual spleen cells give rise to a small proportion of fused cells that:-

(i) Divide continuously with cancerous growth in culture medium.
(ii) Produce the antibody specific to the spleen cell.

Fused cells capable of producing one specific antibody can be isolated and grown in perpetuity in culture for medicinal antibody production. Screening for the desirable antibody gene system is done using the HGPRT and HAT medium. Further screening is carried out by identifying the specific antibody required by doctors for use in patients. A common problem encountered by the medical profession is that the chromosome carrying the antibody genes is lost quite frequently during cell division. Genetic markers on the chromosome must be used to ensure the presence of the antibody gene. This is done very simply by growing the cells on a medium lacking particular vital chemicals specified by those marker genes on the same chromosome as the antibody gene. Pure antibody cultures can be used for the following medicinal or veterinary purposes:-

(i) To associate with specific antigens that are found on the surfaces of cancer cells. This allows antibodies labelled with radioactive elements to identify the location of cancerous tumours or individual cancer cells in animals.
(ii) The recognition of cancer cells by specific antibodies produced by mutant, genetically engineered cancer cells. This allows cancer combating drugs to be carried by these antibodies to antigens that are only found on the surfaces of cancer cells. The drugs fight all cancer cells in the body *in situ*, so giving some hope that many of those cancers that spread through the body (see Chapter 15) can be destroyed. It is possible that some of the antibodies which recognise specific cancer antigens can kill cancer cells. Such antibodies, produced by the genetic engineering of pure antibody culture, may reduce the doses or even the use of those drugs which fight cancer, but which also produce undesirable side effects.

The Genetic Engineering of Fungal Cells

When the hyphae of two different strains of a fungus grow so that they touch each other they can fuse their cells locally to form cells containing chromosomes from both strains. This process doubles the number of chromosomes and increases the number of plasmids in the fused cell. The nuclei of fused cells must also fuse together. The fused fungal cell then divides mitotically. Chromosomes and plasmids are shed as it does so. The chromosome and plasmid number is gradually reduced to the haploid number found in the hyphae cells of each strain. However the genome of the cell line that grows from the fused cell is a mixture of DNA from each fungal strain. The selection of desirable combinations of phenotypic expressions has considerable potential for improving the antibiotic properties of fungi.

The Genetic Engineering of Plants

When attempts were first made to engineer plant cells genetically, the cellulose cell wall presented formidable barriers to the translocations of DNA into plant cells. Techniques are now available for removing plant cell walls without damaging the 'living' parts of the cell. In many ways genetically engineered plants arise much more readily from single cells because plant cells divide vegetatively more readily than animal cells. Now that the problems created by cellulose cell walls are being resolved, progress in plant cell genetic engineering is likely to be rapid. Among the successes of plant genetic engineering are:-

(i) The Production of Homozygous Genes in Plant Cells with the Desirable Allele in Both Chromosomes of Diploid Plant Species

This is achieved by persuading a haploid pollen precursor cell in the anther to undergo chromosome replication without the migration or separation of the chromosomes at cell division. Two identical sets of chromosomes are produced in this way in a single cell. If the cell can be further persuaded to divide mitotically to produce clone cells in culture, individual cells can develop into mature plants with the required characters.

(ii) The Fusing Together of Plant Cells After Their Walls Have Been Removed by Cellulases and Pectinases

The fusion of plant cells can be done with cells taken from the same species. Fusing cells can also be taken from different species. The recombination of desirable alleles in a new plant line can be achieved by this method. During mitotic cell divisions there is the shedding of chromosomes, as in animal fused cells. In plants there is the occasional recombination (a random and uncontrollable event) of DNA from both of the fusing cells.

(iii) The Transformation of Plant Cells Using Plasmid DNA

Plasmids often carry disease resistance genes. The potential for improving commercial crop disease resistance is therefore considerable.

Plant cells can be grown in cultures where they often aggregate in calluses. Sometimes (in the presence of cellulase) plant cells can multiply to produce cells that are not held together by cell walls. The rate of cell division in plant cell cultures can be accelerated by suitable additions of plant hormones. The odd thing is that plants arising from totipotent, separate cells, having a single cell from which they all arose by mitotic cell division, show considerable variation in phenotypic expression. For example disease resistance varies considerably between mature plants derived ultimately from a single cell. The explanation is, perhaps, that there is the uneven distribution of plasmids at cell division, so conferring different disease resistant inheritance to different cell lines.

The selection of plant cells containing desirable genes can be achieved in much the same way as for animal cells. The chromosomes are engineered in such a way that a known mutation exists in the same chromosome or plasmid as the desirable gene. At present it is only possible to screen for characters that are expressed at the cellular level. Osmotic properties, disease resistance, herbicide resistance and the end-products of biochemical pathways are detectable at the cellular level. The selection of broader characters that will be expressed in mature plants is not possible at present.

One very useful discovery has been the existence of a bacterial plasmid that induces rapid plant cell mitotic divisions in conditions and places in plants which would not normally encourage large-scale plant cell division. The large plasmid has been called 'tumour inducing plasmid', and this is abbreviated to Ti. It is found in the bacterium Agrobacterium tumifaciens. When the Ti strain of

A. tumifaciens infects a plant, a tumour forms (usually called a gall) and the cells taken from the tumour have very good totipotent qualities of great value for genetic engineering. Totipotency is explained in chapter 14, page 444.

Ti induced galls are produced when the Ti plasmid becomes part of the genome of the plant cells it invades. Another useful property of the Ti plasmid is that it can be genetically engineered to include genes that specify desirable properties. The potential for the genetic engineering of plants using the Ti plasmid to produce both totipotent plant cells and to introduce desirable genes into plants is very great. However, it must not be forgotten that the problems of inserting a desirable gene, together with its control genes, apply to plant cells as well as to animal cells. A further drawback is that Ti does not transform monocotyledons. Nonetheless there is much scope for the genetic engineering of plants for the benefit of crop breeders.

Genetic Engineering and AIDS

The AIDS viruses, collectively called Human Immuno-deficiency Viruses, HIVs, are retroviruses. Retroviruses have a genome made of single-stranded RNA. They create double-stranded DNA analogues of themselves after they invade a host cell. They are able to do this because each retrovirus particle contains the enzyme reverse-transcriptidase, and also because their DNA anologues can generate the synthesis of reverse transcriptidase when they become incorporated into the host cell's genome. A general summary of the ways the inheritance materials of viruses is expressed is given in Chapter 13.

Several types of human cells can be infected by HIV. However, it is in certain specialised white blood cells that HIV has its worst effects.

The white blood cells that are most vulnerable to HIV are T4 cells. This is because T4 cells have the misfortune to have a polypeptide on the outside of their plasma membrane that acts as a receptor for HIV viruses. When HIVs stick to these protein receptors, T4 cells ingest the viral particles. Reverse transcriptidase then does its work and the viral DNA analogue is quite often included into the T4 cells' genomes.

HIV then lies dormant until the arrival of a disease. Some of the T4 cells multiply in response to an invading disease, these being the T4 cells with a specific antigen on the outside of their outer plasma membrane which recognises the disease. When T4 cells multiply, mRNAs corresponding to the bases that encode viral polypeptides are made at the same time. Many HIV viral genomes are made at this stage. These viral genomes persuade small parts of the outer cell membrane of T4 cells to envelop them. They also take with them some cellular polypeptides which they cannot make for themselves. T4 cells are killed when their outer plasma membranes are lost to HIV. The process, of course, damages some important functions of the human body's defences against disease.

T cells in general can either kill viruses directly by binding with the antigens on viral surfaces, and then ingesting the virus in a destructive way, or they can attach to an invading virus antigen and carry the viral particles to other white cells of the immune system for viral destruction.

T cells can kill virally infected human cells. They also make polypeptides that allow other white blood cells to act more efficiently in their immunising effects. Interferon is one very effective defence polypeptide which is produced by T cells.

White B-cells in blood are switched on for their immunisation function by T cells. B-cells have several very important functions in the immune system of mammals such as humans. The most important function of B-cells is that they produce antibodies. There are several important specialisations of antibodies:-

> i) They bind to antigens on the outsides of viruses that are not already attached to, or inside, cells. This can prevent viruses from attacking human body cells.
> ii) They allow phagocytes to bind to the virus/antibody complex, the antibody having sites on its surface which conform with receptor sites on the surfaces of phagocytes. Phagocytes are a type of white blood cell.
> iii) They break cell membranes to which viruses are attached.
> iv) They recognise the antigens in viruses, attach to them, and white blood cells then attack the virus/antibody complexes and kill the virus by breaking its membrane with enzymes.
> v) They can activate other enzymes produced by cells in blood. These enzymes can in turn destroy the membranes of infected cells.

In brief, therefore, it has to be stated that the cumulative effects of killing T cells is very serious indeed. With HIV it is the killing of T4 cells in particular that is exceptionally damaging.

T4 Cells and The Spread of HIV in a Human Individual

Whenever T4 cells lose their outer plasma membranes to thieving HIV viruses, a large number of membrane-bound viral particles invade new T4 cells, so gradually reducing the ability of humans to be immune to invading disease. T4 cells divide in response to each new invading disease, so spreading the HIV virus particles to more and more cells of the immune system. Eventually, so many of the immune system cells are infected by HIV that the symptoms of immunodeficiency begin. So begins a steady downward spiral, with premature death the inevitable end for the AIDS victim.

Lines of Research into Cures for AIDS

Modified HIV viruses, with reduced powers of invading T4 cells, are at present too dangerous to use as a vaccine. Nonetheless, an AIDS vaccine has been used, without official approval, in efforts to stop an AIDS epidemic in Zaire.

Most scientists in the search for a cure for, or for the prevention of AIDS have concentrated on the use of chemicals or polypeptides that attack the virus at different stages of its life cycle. The chemicals and polypeptides used so far have been chosen for one of the following functions:-

 (i) To prevent the HIV from binding to the outer membranes of human cells, especially T4 cells.
 (ii) To prevent reverse transcription.
 (iii) To prevent new HIV RNA particles forming when human immune system cells, especially T4 cells, divide.

A Potential Role for Genetic Engineering in Fighting AIDS

Professor N. Letvin of Harvard Medical School in the United States of America has indentified, in 1991, a human polypeptide, CD4, that promotes the production of antibodies. It is known that HIV binds to CD4. Antibodies then attach to the CD4/HIV combination. Phagocytes can now ingest and destroy the antibody/CD4/HIV complex, because phagocytes bind antibodies to specific sites on the outsides of their outer plasma membranes. HIV is killed in this way.

It has yet to be demonstrated that an immune defence system, similar to that provided by vaccination, can be produced in mammals in response to innoculation with the CD4 polypeptide. If it is, then this polypeptide may prove to be a safer way to provide immunity against HIV and AIDS than the use of HIV particles of lower virulence than the viruses of the present epidemic. One serious problem here is that there is more than one HIV viral strain, and there may already be very many. Whether or not one polypeptide, allied to phagocytes, can deal with all the strains of HIV is very much open to question.

If all the HIV strains can be tracked down, it may be possible to produce several different polypeptides in considerable quantities by the genetic engineering of bacteria or yeasts. Such polypeptides could be used as substitutes for live virus vaccinations. Commercial companies are already producing CD4 in case it proves to be an effective preventer of AIDS.

A simplified scheme showing the mode of action of CD4, using arbitrary shapes for illustrative purposes, is:-

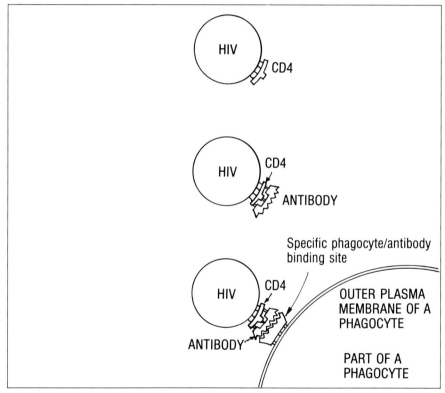

Phagocytes can extend their outer plasma membrane to envelop the whole complex bound to them. The hope is that they will be able to do so before HIV infects the vital T4 cells.

Some further examples of how mobile genetic elements, such as viruses, can be used for genetic engineering are given in Chapter 13.

Several PBR 322 plasmids taken from E. coli can be seen in this electron micrograph. This plasmid is often the one chosen for the insertion of desirable genes by genetic engineering.

Magnification x 31,000

*Photograph by Dr. Gopal Murti.
With permission from The Science Photo Library, London.*

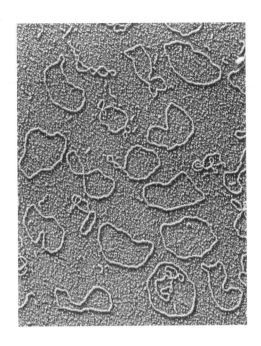

Scanning electron micrograph of a human T-lymphocyte (a type of white blood cell). Following infection by HIV virus, the surface microvilli (the long projections in the picture) are replaced by small, rounded protrusions.

Magnification x 360

*Photograph by NIBSC.
With permission from The Science Photo Library, London.*

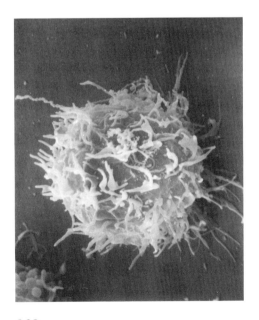

Chapter 8 Questions

1. What ethical anxieties arise from the ability of scientists to manufacture and manipulate genes?
2. What can be done using contemporary techniques of genetic engineering? What should not be attempted?
3. Give some examples of characters controlled by plasmid DNA.
4. What makes plasmids suitable for genetic engineering?
5. What is a restriction site?
6. How is complementary-DNA (cDNA) made?
7. How is cDNA inserted into a bacterial plasmid?
8. How is screening carried out after genetically engineered transformation has occurred?
9. What are the dangers of genetic engineering?
10. Some complications of genetic engineering are peculiar to eukaryotes. What are they?
11. How are genes inserted into eukaryotic cells?
12. How is cotransformation used to map the distances apart of eukaryotic genes on their chromosome?
13. Of what usefulness is preferential gene shedding after the fusion of cells?
14. What is the scavenger pathway for DNA synthesis?
15. What is $HGPRT^+$ and what is $HGPRT^-$? What are TK^+ and TK^-?
16. What genetic differences are there between the fusions of cells taken from the same species and the fusions of cells taken from different species?
17. Can the microinjection of DNA into cells be used instead of cell fusion for the genetic engineering of eukaryotic cells?
18. Can viral DNA be used for genetic engineering instead of plasmid DNA?
19. What can be the consequences of inserting a whole cell into an embryo?

20. What can be done by genetic engineering to alter the characters of fungi?
21. What can be done by genetic engineering to alter the characters of animals?
22. What can be done by genetic engineering to alter the characters of plants?
23. What problems are posed for genetic engineers by control genes and by upstream and downstream genetic systems?
24. What possibilities are there for curing AIDS?

9

THE BEHAVIOUR OF CHROMOSOMES DURING CELL DIVISION IN DIPLOIDS – MITOSIS AND MEIOSIS

Before the structure of DNA was known, and before the electron microscope allowed very high magnification of chromosomes, scientists had to rely upon the chemical staining of chromosomes and were restricted by the low magnification imposed by the optics of the light microscope. Discovery was therefore limited by the stained appearance of the chromosomes in dividing cells at certain stages of cell division. Scholars communicated to each other in the classical languages of Greece or Rome in those days of early research work on chromosomes, and Greek was chosen to describe the several stages of mitosis and meiosis. The traditional terms, used to describe each stage of cell division are given, together with appropriate photographs, later in this chapter. It is the intention of this chapter to reduce the processes of mitosis, and meiosis to their significant points, in the aftermath of modern discovery. The discussion is limited to cell division in diploids.

MITOSIS IN DIPLOIDS

Although cases in Drosophila melanogaster are known of crossings-over between homologous pairs of chromosomes during mitosis, the normal behaviour of homologous chromosomes during mitosis is for them to lie quite far apart from each other in the nucleus. They are also independent of each other after the nuclear membrane has broken down during cell division. The replication of each chromosome of an homologous pair therefore takes place independently of the second chromosome of the pair.

In cells that are not preparing to divide, the chromosomes are long and thin, with a low degree of coiling or shortening. The number of chemical groups per unit area that respond to stains when the chromosomes are elongated is low, so they are not easily visible in non-dividing cells, even when stains are added.

The following sequence of events takes place in those cells that undergo division by mitosis.

(i) **The Replication of Chromosomes (sometimes, and incorrectly, called 'The Resting Phase' because stains are not visible)**

The replication of chromosomes occurs in cells when the chromosomes are long and thin. Local areas of coiling, called heteropycnotic bodies, may respond during this stage to chemical stains, but it is unlikely that the elongated replicating chromosomes are visible to anything other than the electron microscope at this stage.

DNA replication requires several enzymes. Among the important ones are:-

(a) Endonuclease – this enzyme is an essential initiator of replication. It hydrolyses DNA by introducing double strand breaks, and allows the two strands of DNA to unwind.

(b) DNA polymerase

(c) Polynucleotide ligate

These two enzymes synthesise new pairing nucleotide chains and the ligate stitches nucleotide units together to build the pairing chains.

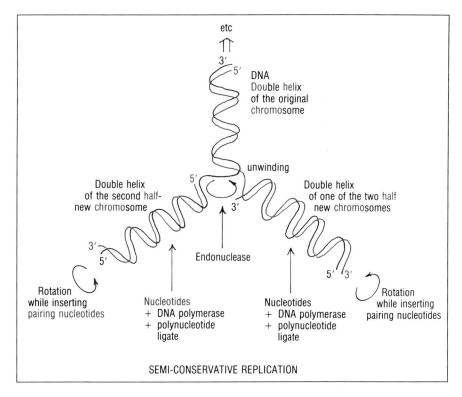

Base pairing along each new chain occurs, in the presence of suitable building units and enzymes. The insertion of bases along both new polypeptide chains always starts at the 5′ end of each new chain and proceeds towards the 3′ end. The building of a new pairing nucleotide chain is straightforward along the leading chain template. This is because the 5′ end of the new nucleotide chain can start to be inserted as soon as the 3′ end on the original chromosome presents unpaired bases. The diagram on page 268 shows the direction of insertion of bases along the leading chain template.

An electron micrograph showing the fork in DNA where DNA replication is taking place is shown on page 289.

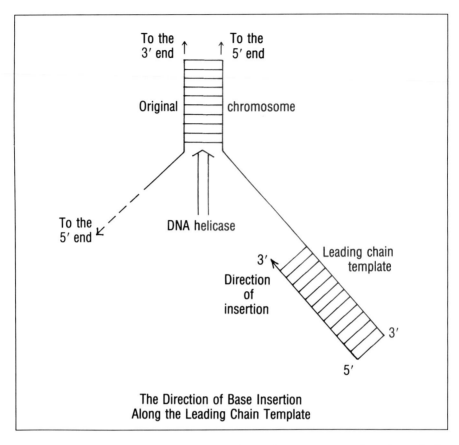

The Direction of Base Insertion
Along the Leading Chain Template

However at the 5′ end of the original chromosome's lagging chain it is not possible to start inserting pairing bases. On the lagging chain it is therefore necessary for a considerable length of the chain to become unpaired before any new, pairing bases can be inserted, starting at the 5′ end of the new nucleotide chain. The process is illustrated on the facing page.

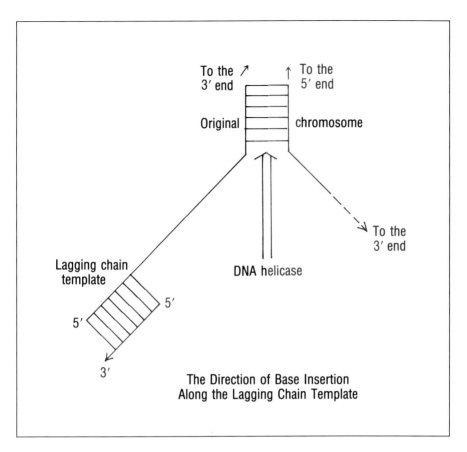

As the bases of the original chromosome become more and more unzipped, the insertion of new, pairing nucleotides takes place further along the lagging chain. The diagram on the next page shows the ways that pairing bases are inserted along both the leading and lagging chains when a greater length of the original chromosome than was shown in either of the last two diagrams has produced longer leading and lagging chains.

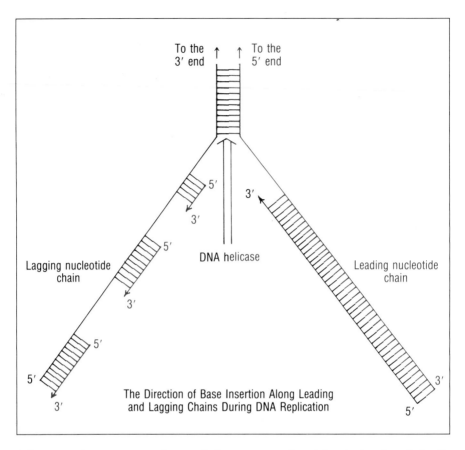

The Direction of Base Insertion Along Leading and Lagging Chains During DNA Replication

The gaps between sections of the new nucleotide chain that is built onto the lagging chain are filled in later, again from the 5' end to the 3' end of each additional section. In this way the lagging chain is fully replicated, more or less at the same time as the replication of the leading chain.

The name given to the process of producing new chromosomes is 'the SEMI-CONSERVATIVE replication' of DNA. The term 'semi-conservative' is appropriate as only one of the two nucleotide chains from the original chromosome is conserved in each of the two half-new chromosomes. Matching nucleotide chains, appropriate to each of the original chains, are synthesised and pair up to give two half-new chromosomes (often called chromatids) each of which is usually identical to the original chromosome. The replication of each chromosome in the mitotic cell division of somatic cells provides each new cell with chromosomes identical to those in the zygote (the fertilised egg), from which they ultimately arise. In some species an enzyme exists that can cut out parts of DNA damaged during replication and replace them with 'correct' base sequences.

The Shortening and Coiling of Chromosomes – PROPHASE

The shortening and coiling of the chromosomes, following DNA replication, allow them to become more and more visible when stains are present.

During this part of mitosis the nucleolus (or nucleoli) becomes smaller.

Replicated chromosomes appear to be joined at a point where they are thinnest, and the contrast between the thick and the thin parts become more and more accentuated.

If *one pair* of replicated, homologous chromosomes were to be examined in the nucleus just before the nuclear membrane breaks down, they would have the following appearance:-

A pair of replicated, homologous chromosomes, lying quite far apart in the nucleus

The Migration of the Replicated Chromosomes To Where They Line Up on The Equator of Mitotically Dividing Cells – THE MITOTIC DANCE

By the time the replicated chromosomes are short and thick, when they are also intensively coiled, the nuclear membrane breaks down to allow the replicated chromosomes to move more widely in the cell. Their movements, called 'The Mitotic Dance', take them to the equator where they line up and attach themselves to a newly formed spindle made of polypeptide fibres.

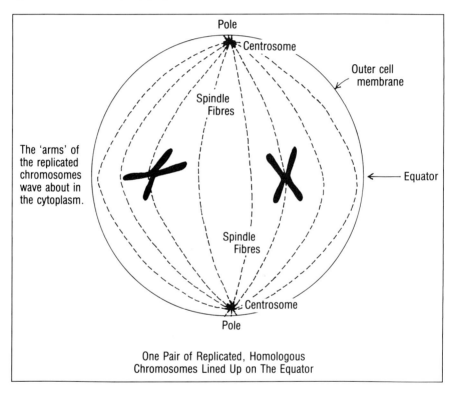

One Pair of Replicated, Homologous Chromosomes Lined Up on The Equator

A centrosome forms at each pole of the cell, anchored to the outer cell membranes by fibres. Spindle fibres extend towards the equator from each centrosome and the replicated chromosomes line up on the equator at the end of their 'mitotic dance'.

Attaching Each Chromosome of a Replicate to a Spindle Fibre So That It Will Be Taken To The Opposite End of the Cell, Away From Its Sister Replicate – METAPHASE

Every chromosome has a place within it that can attach to a spindle fibre. The name given to the specialised 'attachment part' of a chromosome is the 'centromere'.

In the case of replicated chromosomes each sister replicate has a centromere, at the same position along the length of each of the chromosomes:-

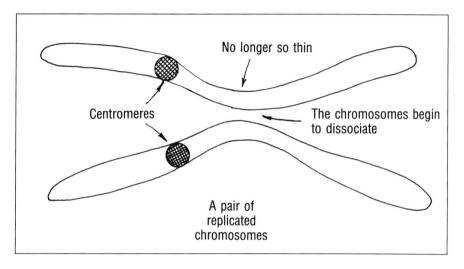

When the replicated chromosomes line up on the equator, the centromere of each sister chromosome becomes attached to a spindle fibre in opposite sides of the cell. At the same time there are forces of attachment that hold the 'tails' of the chromosomes as seen overleaf:-

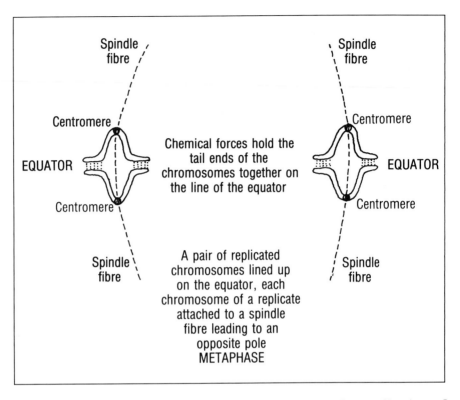

It should be noted that the cell has been 4n since the replication of the chromosomes.

Moving the Diploid (2n) Number of Chromosomes Into Each Half of The Cell – ANAPHASE

The spindle fibres can contract. Held fast by the anchored centrosomes at the poles of the cell, the fibres pull each sister chromosome to opposite sides of the cell.

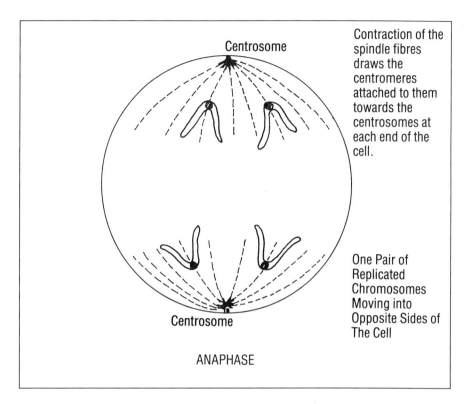

The forces holding the tails of the sister replicates together break down and 2n chromosomes move into each half of the cell. The set of chromosomes in each half of the cell is identical to the set in the original cell.

The Formation of A Dividing Membrane To Create Two Cells from One Cell

There is the replication of cell components, such as the organelles, at the same time as chromosome replication occurs. An organisation exists to ensure that each half of the dividing cell receives all the components it needs. A membrane then forms along the line of the equator:

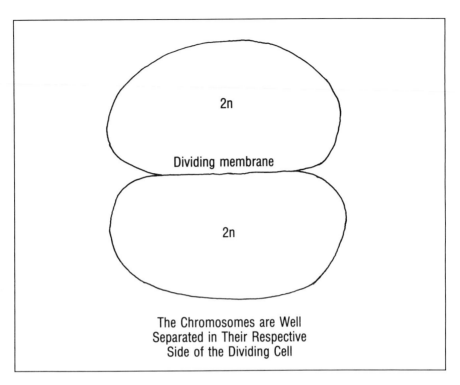

At about the same time as the two sets of chromosomes approach their respective centrosomes a nuclear membrane envelops the chromosomes, the nucleoli in the nuclei enlarge, the spindle disappears and mitosis is complete.

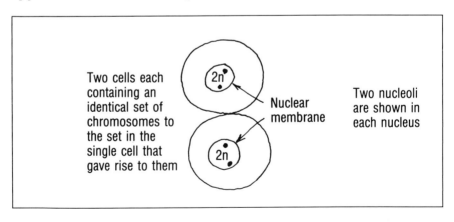

The chromosomes are long and thin in the two new nuclei, so they no longer respond to stains.

The Specialisation of Cells Arising From Mitotic Cell Division

One of the astonishing facts of nature is that a single fertilised germ cell can divide mitotically in an organised way to give rise to a very large number of cells, each having identical chromosomes. Yet each cell can take on a specialised form and function. The specialisation of mitotically dividing cells is called 'differentiation' and is one of the subjects of Chapter 14.

MEIOSIS IN DIPLOIDS

Sexual reproduction depends upon the formation of germ cells (GAMETES) by specialised male and female body parts. Fertilisation and mitotic cell division follows the penetration of eggs by male gametes in suitable environments to encourage the growth and development of a new individual of a sexual species.

The processes of the type of cell division that produces gametes, MEIOSIS, ensure that the fertilised eggs, from which a new generation develops, have the same number of chromosomes in them as the somatic cells in the previous generation. MEIOSIS therefore has to reduce the number of chromosomes in eggs, pollen and sperms to a half of the chromosome diploid number in somatic cells.

A simple scheme showing the production of gametes, fertilisation and the development of a new individual is shown on page 278. The first 2n cell produced by fertilisation is called the ZYGOTE.

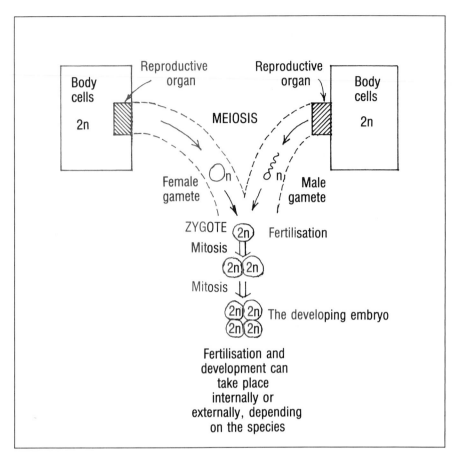

The processes for providing gametes is similar, but not identical, for animal, plant and fungal diploid species. The following diagrams show that there are slightly different processes for making male and female gametes from the specialised cells in testes, anthers, ovaries and ovules.

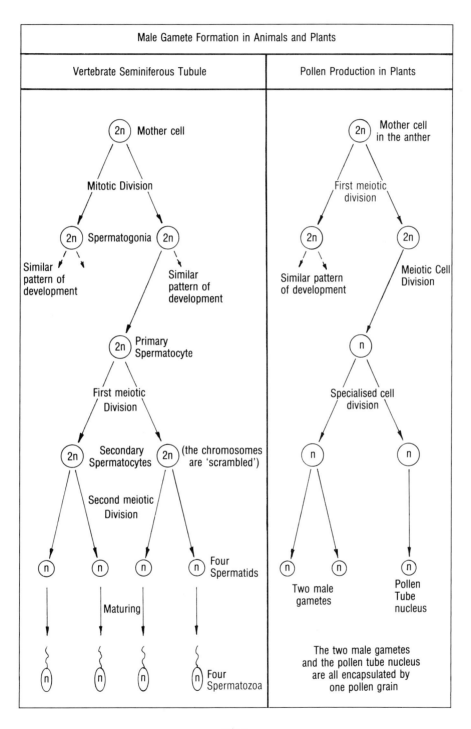

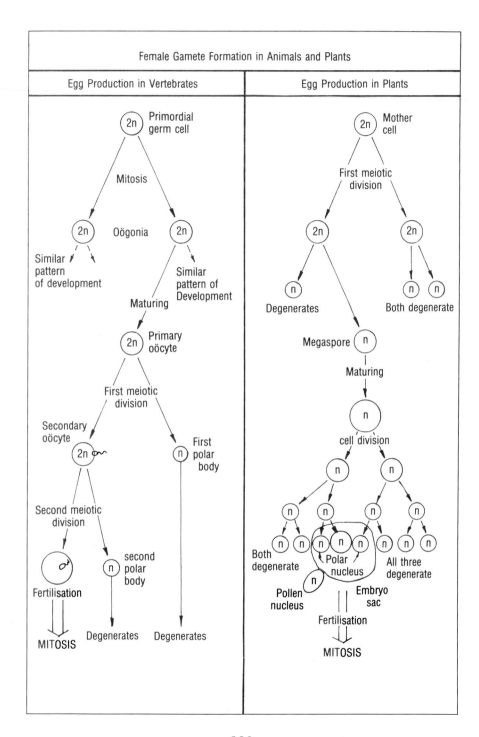

Meiotic cell division

There are *two* cell divisions in meiosis.
In the first meiotic cell division there are:-

(i) The replication of chromosomes.
(ii) The 'scrambling' of genes by crossings-over.
(iii) The random orientation of the 'scrambled' chromosomes into each half of the dividing cell.
(iv) The production of two diploid cells, each of which contains unique chromosomes and therefore unique diploid sets of chromosomes.

In the second meiotic division there is no further replication of DNA, but there is the random migration of unique chromosomes into each half of the dividing cells to provide each gamete with a unique HAPLOID set of chromosomes.

The First Meiotic Cell Division

Whereas replicating, homologous chromosomes lie widely separated during the early stages of mitosis, the replication of DNA in meiosis occurs with the homologous pairs of chromosomes lying flush with each other:

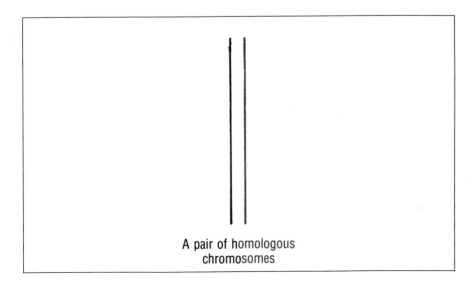

A pair of homologous chromosomes

Replication of the chromosomes occurs to give rise to:-

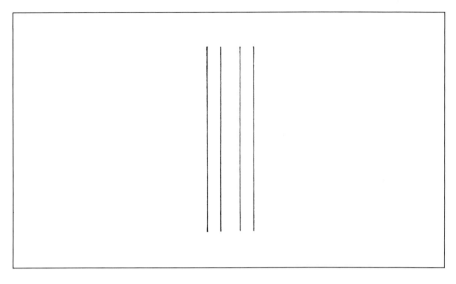

The replicating chromosomes then alter their shape, after shortening and thickening, so that they appear to touch each other:-

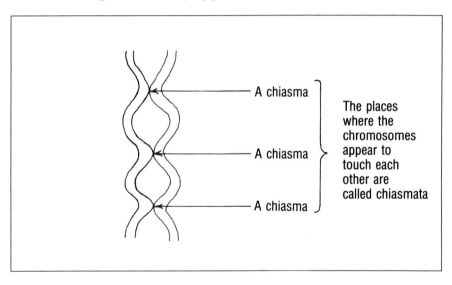

It has been established that, at the chiasmata, each chromosome can break and the broken ends can join up with the broken ends of a chromosome from the other replicate. This process is called 'crossing-over'.

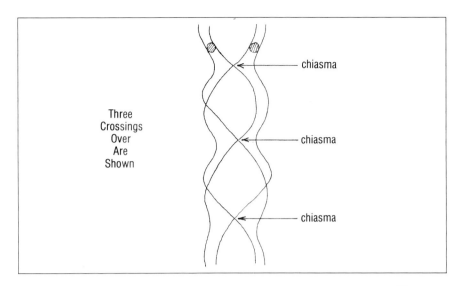

When the replicated chromosomes move to the equator to line up, before the cell divides, they do so as 'scrambled' chromosomes, each of which is a composite of more than one of the original chromosomes:-

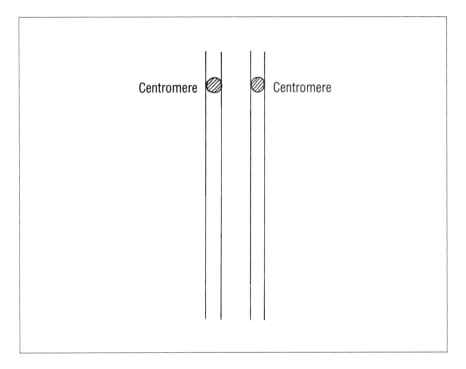

An apparatus, similar to the spindle of mitosis, emerges after the nuclear membrane has broken down. The 'scrambled' chromosomes move to the equator where their centromeres attach to spindle fibres. Forces of attachment hold their tails together:-

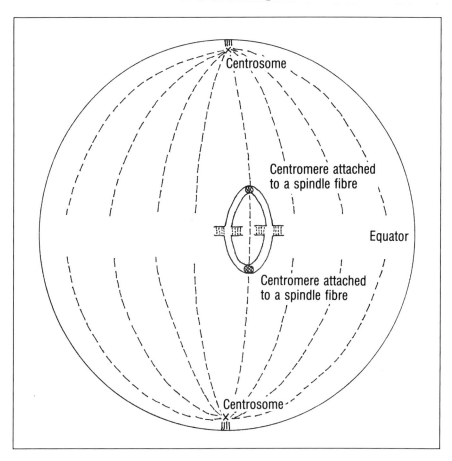

The diagram above shows the 'scrambled' replicates of only one pair of homologous chromosomes. There would also be the 'scrambled' replicates of all the homologous pairs of chromosomes lined up on the equator at this phase of the first meiotic cell division.

When the spindle contracts to pull equal numbers of chromosomes into each half of the cell, the chromosome complement of each of the two cells produced at the end of the first meiotic cell division will be:-

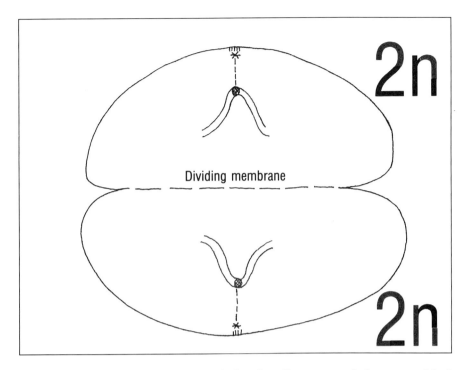

The orientation, north or south in the diagram, of the scrambled chromosomes is random. There will therefore be considerable genetic differences between each of the two cells produced by the first meiotic division.

At this stage both cells are still diploid.

The Second Meiotic Divsion

There is no wastage of genetic material in male gamete production such as there is in the decay of female polar bodies. It is therefore convenient to consider the second meiotic cell division in male gamete production.

In the second meiotic cell division there is NO further replication of DNA. It is, therefore, during this division that the number of chromosomes is reduced from the diploid number (2n) to the haploid number (n). In each of the two cells produced by the first meiotic division a new spindle forms in such a way that single, 'scrambled' chromosomes can move into opposite sides of each cell.

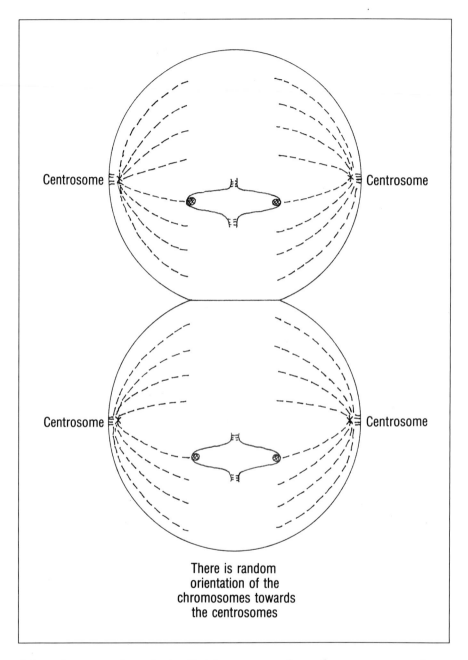

After the second meiotic division, to produce four gametes, there will be four haploid cells containing unique 'scrambled' complements of chromosomes:-

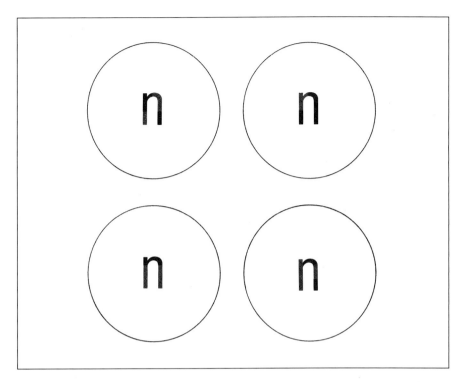

The random orientation of the chromosomes during the second meiotic cell division places a unique complement of genes in each gamete. This genetic variation in gametes allows the phenotypic expression of the alleles in the offspring of a single pair of mating adults to vary considerably. Natural selection operates on the variation shown in the offspring of each mating pair of adults, and the existence of differences between individuals is an important contribution for allowing evolutionary changes.

Chapter 10 discusses the relationship between variation caused by mutation and variation caused by the 'scrambling' of chromosomes during meiosis.

The processes of meiosis in female gamete production have to concentrate large quantities of chemicals into eggs so that nutrition and differentiating chemicals are packed into each egg. After fertilisation the nutrients and differentiating chemicals in the egg have to sustain and direct mitotic cell division until the developing embryo can gain nourishment from some source outside the egg. At each meiotic division in egg production there is therefore the need to concentrate as many desirable chemicals as possible into one of the two products and to allow the other (the polar body) to be infertile and wasted. Nevertheless the scrambling of chromosomes and their random orientation makes each egg unique in its genetic composition, so assisting the crossings-over and random chromosome orientation of male gamete production in producing the unique genetic inheritance of offspring resulting from sexual reproduction. Crossings-over do not occur in every case of gamete production. There is *NO* crossing-over in *male* Drosophila spp.

The scrambling of genes during meiosis explains the differences in the phenotypes observed by Charles Darwin. It is the reason for differences between brothers and sisters. Natural selection then operates in such a way that, in general, only those individuals best suited to the demands of their environment survive. Natural selection tends to reduce the variation among individuals, hence reducing the proportion of genes for which there are alternative or several alleles. It is estimated that the proportion of genes for which more than one allele exists may be as low as 5% in each species. Nonetheless meiotic gene scrambling is the major contemporary mechanism for creating variation among diploid individuals of a species.

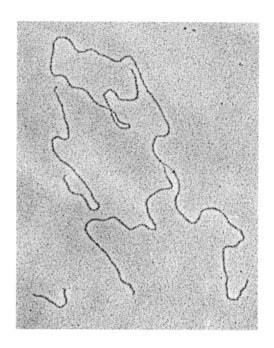

Transmission electron micrograph showing the fork in DNA where DNA replication is taking place. The replicating DNA in the picture is taken from Euplotes spp.

Magnification x 30,100

*Photograph by Dr. Gopal Murti.
With permission from The Science Photo Library, London.*

Mitosis Photographs

The sequence of photographs on the five pages that follow show some of the important phases of mitosis in Lilium regale. The photographs were taken by John McLeish and Brian Snoad and are reproduced with permission of MacMillan & Co. Ltd.

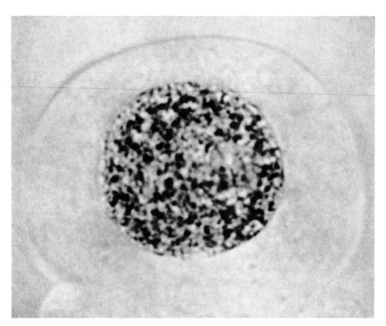

Interphase in a root-tip cell. The chromosomes are not individually distinguishable. Magnification × 2040

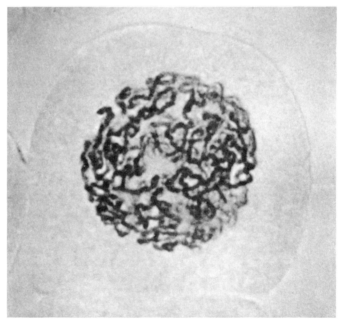

Prophase. The chromosome threads are now quite distinct. Magnification × 2040

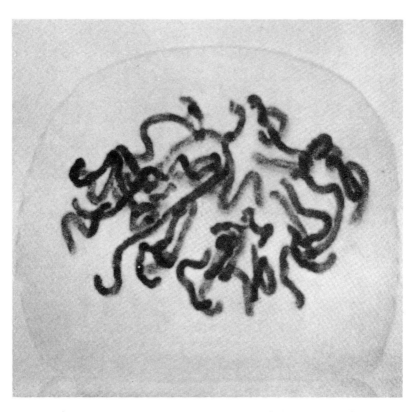

Prophase. The chromosomes are reaching their maximum degree of contraction and the nuclear membrane is breaking down.
Magnification × 2400.

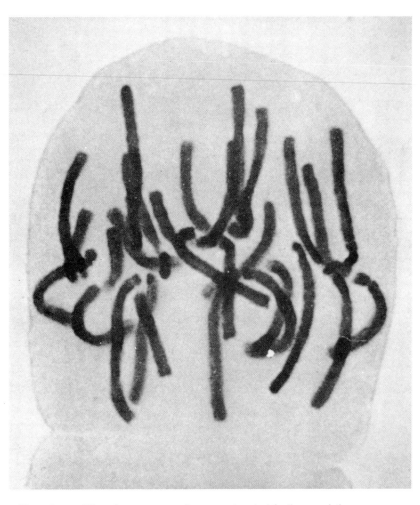

Metaphase. The chromosomes have contracted further and they are now arranged upon the equator of the spindle. The spindle is not visible when the Feulgen staining method is used.
Magnification × 2400

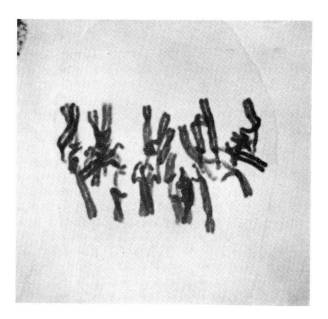

Anaphase. The chromosomes are still arranged on the spindle equator but their centromeres have divided and are beginning to move apart.
Magnification x 1800

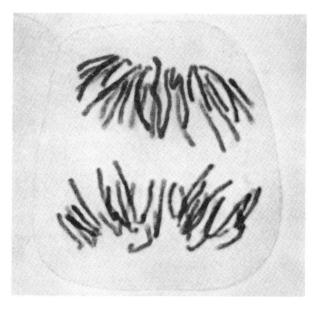

Anaphase. The daughter chromosomes have separated and they are moving to opposite poles of the spindle. Magnification x 1800

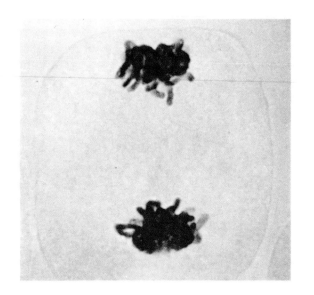

Telophase.
Chromosomes associate closely in two widely separated regions of the original cell.

Magnification x 1700

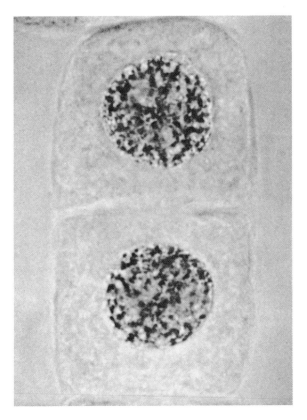

Interphase.
Two new nuclei have now been formed which are completely separated by the new cell membrane.

Magnification x 1780

Meiosis Photographs

The simplified introduction to meiosis on pages 277-288 now needs to be considered in two further stages. The first of these is the visual evidence obtained from stained chromosomes. The second stage is to describe the cellular events at each phase of meiotic cell division, against the background of a photograph. It is extremely difficult, and at the biochemical level impossible, to capture all the events of meiosis in one sequence of photographs. The captions to some of the photographs of meiosis therefore fill in some details that cannot be seen. It is hoped that the visual evidence of the photographs themselves, coupled with further information in their captions, will allow a mind's eye picture beyond that of the stained chromosomes themselves. The captions indicate that which cannot actually be seen.

Greek/English Terms

The following summary of the English translations of Greek descriptions of the appearances of chromosomes may help to unravel some of the stages of meiosis as they are described in the conventional terms used by geneticists.

MEIOSIS
GREEK/ENGLISH TRANSLATIONS

FIRST MEIOTIC DIVISION

Chroma (colour) soma (body)
Cyto (Kytos = cell) plasm (plasma = moulded thing)
Mitos (thread); osis (production)
Zygote (zygon = yolk)
Meiosis (reduction)
Inter (between) phase (phasis = appearance)
Chroma (colour) tid (eidos = likeness)
Centro (kentron = centre) mere (meros = part)
Pro (before) phase
Centro (centre) some (soma = body)
Meta (beyond) phase
Ana (back) phase
Telo (telos = end) phase
Diploid (diploos = double; eidos = likeness)
Haploid (haplos = single; eidos = likeness)

SECOND MEIOTIC DIVISION

Leptotene (lepton = thin; taenia = ribbon)
Zygotene (zygon = yolk; taenia = ribbon)
Bivalent (bis = two; valens = made strong)
Pachytene (pachys = thick)
Diplotene (diploos = double)
Diakinesis (dia = through; kinesis = movement)

Stages of meiosis in the plant *Crepis capillaris (n = 3)*

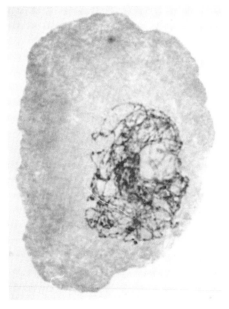

a

FIRST MEIOTIC DIVISION

a) *Leptotene*
Following DNA replication during so-called interphase, unpaired elongated chromosomes can be seen. Not shown is that one end of each chromosome is anchored by a telomere to a specific site on the inner side of the nuclear membrane. Such specific sitings of telomeres allow the pairing of homologous chromosomes.

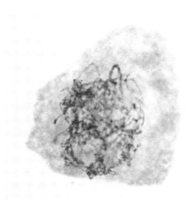

b) *Early Zygotene*
Each chromosome lies flush with its homologue. There is not much shortening or thickening of the chromosomes. Synapsis/pairing occurs. The photo shows that some regions are synapsed while others are still unsynapsed.

c) *Pachytene*
Each chromosome that replicated during interphase has now begun to shorten and thicken. The 4n cell has also completed histone synthesis (not visible) by this stage. Also not shown is that, during pachytene, a polypeptide structure called the synaptoneural complex develops and runs along the length of each homologous pair of chromosomes. Localised electron-dense areas, which act as recombination nodules, occur from place to place in the synaptoneural complex.

d) *Diplotene*
Pairs of chromosomes now exist as bivalents. The bivalents begin to separate from their binding polypeptides, except at the chiasmata. The chiasmata positions may have been determined by the electron-dense regions of the binding polypeptides. Chiasmata vary in frequency in different parts of every chromosome. In general, however, the longer the chromosome the more chiasmata there are likely to be.

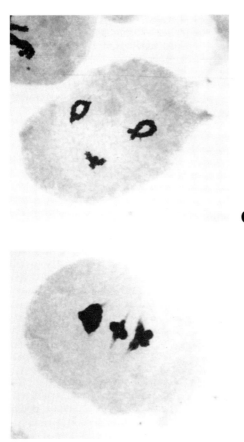

e) *Diakinesis*
The chromosomes are now very short and thick. The chiasmata are clearly visible in the photo.

f) *Metaphase I*
The nuclear membrane disappears before metaphase I. The spindle fibres then become visible during metaphase I. Bivalents migrate to the equator where the tails of the chromosomes that point towards opposite poles are held together. Each centromere, holding a pair of chromatids, attaches to a spindle fibre on either side of the equator.

g) *Anaphase I*
Homologous pairs of 'scrambled' chromosomes are pulled towards opposite poles in their respective halves of the cell.

h) *Telophase I*
A new nuclear membrane surrounds the chromosomes in each half of the cell. A dividing membrane separates each half of the cell. Each new cell contains pairs of chromatids, each of which is unique following crossings-over.

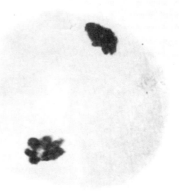

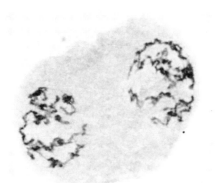

i) *Interkinesis*
There may be a short period of time between the first and second meiotic divisions; or the period of time between the two divisions may be more prolonged. During interkinesis there is the elongation of chromosomes.

SECOND MEIOTIC DIVISION

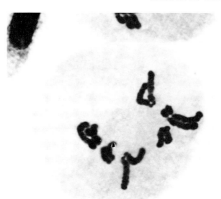

j) *Metaphase II*
The chromosomes migrate to the equator, this time without any further replication of DNA.

k) *Early Anaphase II*
l) *Late Anaphase II*
Individual chromatids move to opposite sides of the cell containing them, towards their respective poles.

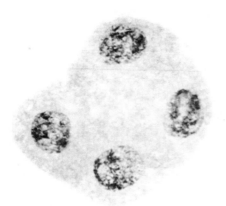

m m) *Telophase II*
The nuclear membrane of each haploid pollen cell forms. Each cell is also surrounded by its own outer plasma membrane. Two meiotic cell divisions, following only one replication of DNA, have now created haploid, genetically scrambled, unique pollen cells.

Meiosis photographs by Dr. G.H. Jones, taken from Genetics. *by Fincham, J.R.S., Butterworth, with permission.*

This ends the meiotic sequence in *Crepis capillaris*.

An excellent photograph showing crossings-over in an animal species is taken from *Locusta migratoria*.

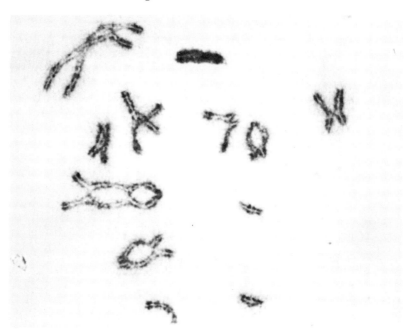

A particularly clearly analysable diplotene from locust (*Locusta migratoria* spermatocyte). Note the structure of the chiasmata with only two of the four chromatids exchanging at any one chiasma.

Photograph by courtesy of Dr G.H. Jones.

The Phenotypic Expression of Genetic Recombination Caused by Crossings-Over in the Ascomycete Fungus, Sordaria brevicollis

The higher Ascomycetes, such as the genus Sordaria, produce many asci, in which the spores are found, in their sexual, fruiting bodies. The production of an ascus arises out of the fusion of two haploid cells, one from each parent fungus.

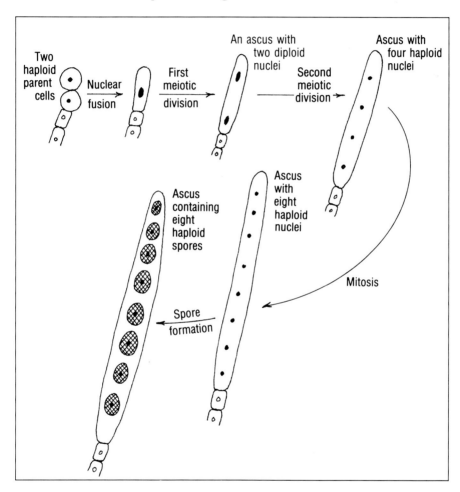

Wild-type Sordaria produce black ascospores. If there is no crossing-over at the first meiotic division the fusion of two pure bred, haploid, parent cells produces the following sequence of events in the formation of an ascus containing spores of different appearances:-

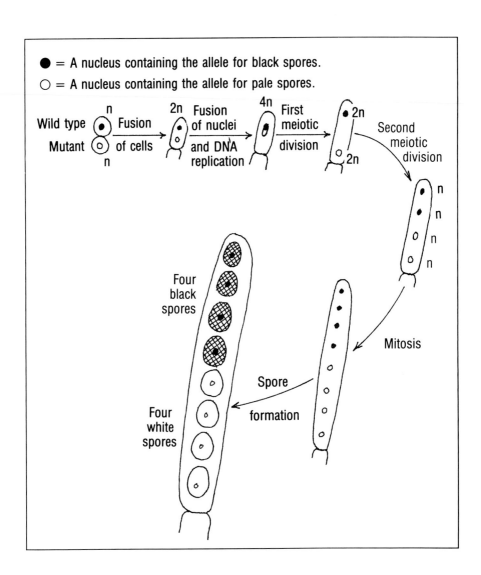

First Meiotic Cell Division

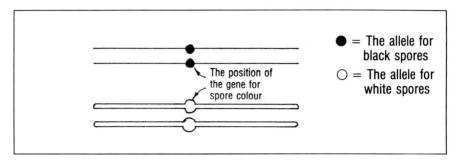

It makes a difference where, in relation to the spore colour gene and the centromere, the crossing-over occurs.

1. If crossing-over occurs on the far side of the spore colour gene from the centromere:-

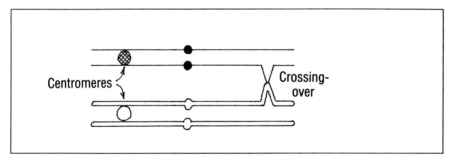

The resulting chromosomes are:-

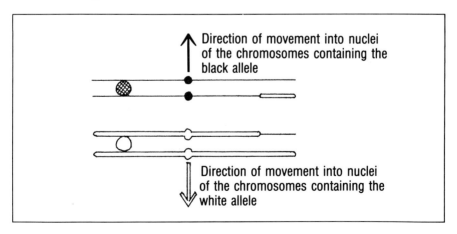

If replicating chromosomes move in the directions shown, after crossing-over is complete, they produce the following disposition of spores in the asci after the second meiotic division and mitosis have been completed:-

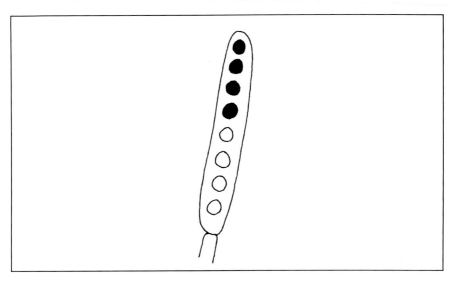

However if the orientation of the chromosomes, after crossing-over is complete, is:-

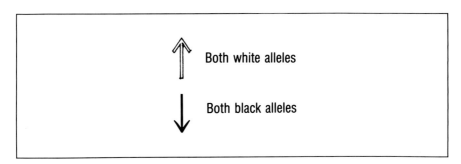

they produce the following arrangement of spores in the ascus:-

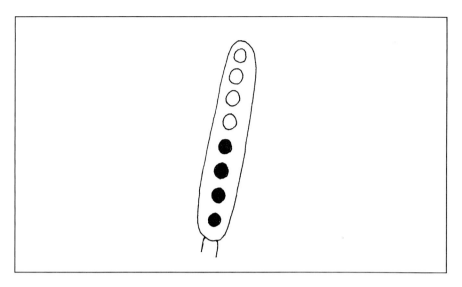

2. If a crossing-over occurs between the spore colour gene and the centromere:-

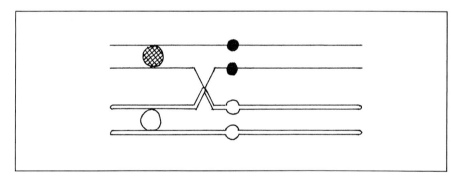

the resulting chromosomes produced from a crossing-over such as this are shown in the diagram at the top of page 306. The inclusion of dark and pale arrows with question marks indicates that there is random orientation of the black and white alleles upwards or downwards from each crossed-over chromosome replicate.

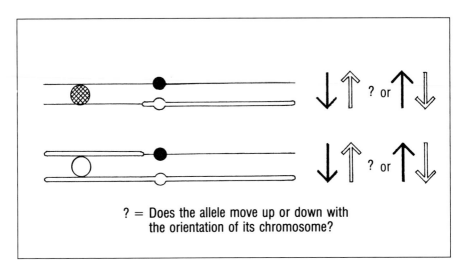

The pair of chromosomes in the upper position of the diagram above occupy the top two haploid nuclei after the second meiotic division, and the lower pair of chromosomes move into the lower two haploid nuclei. The orientation up or down of the scrambled, upper and lower pairs of chromosomes determines the pattern of spore colours in the asci. The ways in which different spore patterns can come about by different orientations of chromosomes, in which crossing-over occurs between the spore colour gene and the centromere, are shown in the diagrams on pages 307 and 308.

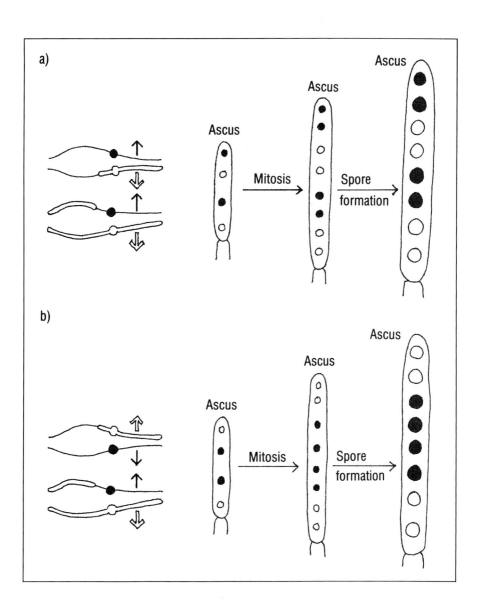

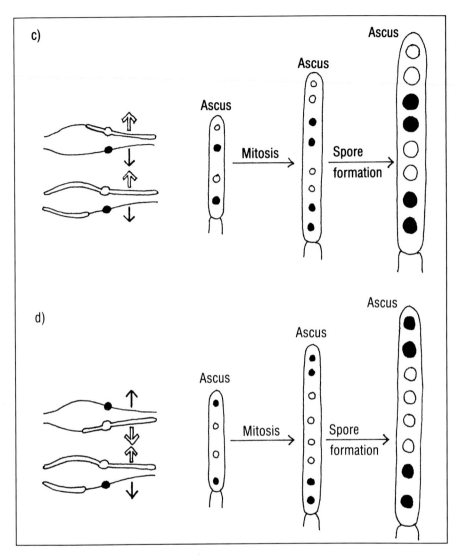

Exceptional Spore Patterns Caused by Mutations

Exceptions to these spore patterns are occasionally observed because there are mutations from 'pale allele' to 'black allele', or from 'black allele' to 'pale allele'. If the mutation occurs in one of the meiotic cell divisions the mutation will ultimately affect two spores in an ascus:-

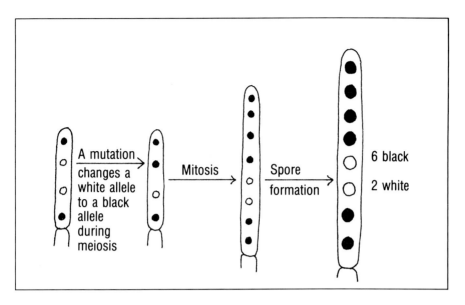

If the mutation occurs in the process of mitosis, only the colour of one spore will be altered:-

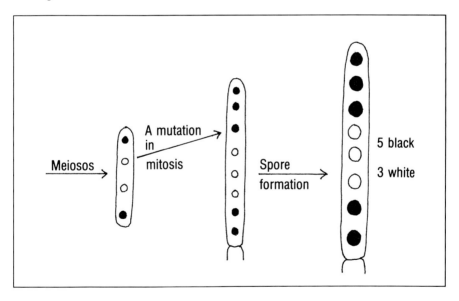

The reproduction system for making spores in Ascomycetes has been an excellent source of genetic information, therefore, and has greatly assisted our understanding of recombination by crossing-over at meiosis.

Comparisons Between Mitosis and Meiosis

(a) Similarities

(i) The replication of DNA takes place in both processes.
(ii) There are structural similarities of spindle, centromeres, and centrosomes in both the mitotic and the meiotic apparatus.
(iii) Chromosomes line up on the equator before cell division in both processes.
(iv) Although the gametes receive the haploid number of chromosomes, whereas the cells produced by mitosis contain the diploid number, there are nonetheless equal numbers of chromosomes that are drawn into each half of dividing cells in both mitosis and meiosis, mitosis providing 2n to each cell and meiosis providing n to each cell.

(b) Differences

(i) Four cells are produced from one cell in meiosis, whereas two cells are produced from one cell in mitosis.
(ii) There is one cell division in mitosis, whereas there are two cell divisions in meiosis.
(iii) Homologous chromosomes replicate when lying flush with each other during the first cell division of meiosis, whereas the replication of homologous chromosomes in mitosis takes place with homologous chromosomes quite widely separated from each other.
(iv) Crossings-over occur as a common occurrence in meiosis (exceptions to this general rule are known – for example there is no crossing-over in male Drosphila spp.). Crossings-over are very rare in mitosis.
(v) The random migration of 'scrambled' chromosomes during the two cell divisions of meiosis ensure that there are considerable differences between the alleles found in the genes of the set of chromosomes that randomly orientate into each gamete (provided that a proportion of the parental genes have alternative alleles). In mitosis, by contrast, each new cell usually has exactly the same alleles in its chromosomes as the cell from which it arose.
(vi) Cells produced by mitosis contain the diploid number of chromosomes, whereas gametes have the haploid number.

(vii) Eggs (usually large cells) contain cytoplasmic chemicals of great importance for cell differentiation, and these chemicals are distributed in eggs to modify the expression of genes in the cells that are produced by mitotic divisions originating in the single cell that was the fertilised egg. Sperms and pollen do not have this unequal distribution of chemicals for influencing subsequent mitotic divisions. By contrast there is often not such a far reaching distribution of chemicals in somatic cells, after the early mitotic cell divisions are completed, even if specialisation by differentiation has to occur from mitotic cell division. Differentiation, hormones and ageing modify gene expression in somatic cells. Chapter 14 discusses these three factors.

(viii) Cells produced by mitosis have the capacity to produce other cells by mitotic cell division in developing organisms. Many somatic cells retain the ability to divide, even in the mature organism, in response to damage or environmental stress. In general, cells produced by meiosis can only give rise to further cell division if a male gamete fertilises a female gamete. Whenever there is no fertilisation gametes degenerate and die.

Chapter 9 Questions

1. What are the essential processes of MITOSIS?
2. What are the essential processes of MEIOSIS?
3. Compare mitosis and meiosis.
4. Describe male gamete production in animals and plants.
5. Describe female gamete production in animals and plants.
6. Compare male and female gamete production.
7. What are the implications of:-
   a) mitotic crossings-over
   b) meiotic crossings-over
   for evolution? Does it matter at what stage of the life-cycle and in which organ mitotic crossings-over occur?
8. Describe the meiotic implications of spore colour patterns in the fungus Sordaria brevicollis.

10

MUTATIONS

It is the intention of this chapter to describe some characteristics of mutations themselves. The discussion about 'chance' acting on the transmission of the mutations and about 'the environment' acting on the phenotypic expression of mutations is left for Chapter 12. The chance factor is called 'genetic drift' and environmental factors are collectively called 'natural selection'. Both topics are discussed at some length later in relation to population studies.

To a considerable extent, at least, evolution has ultimately depended on mutations. Mutations are quite varied in their forms and some examples are given in this chapter.

It must be made clear that nearly all mutations are damaging. A high proportion of them are lethal. Nearly all the rest confer serious disadvantages to individuals in which a somatic mutation occurs, or to those individuals unfortunate enough to inherit mutations through the gametes from which they develop. It must also be made clear that, in sexual species, variation among individuals of a species depends to a very great extent upon crossings-over at meiosis. The scrambling of those genes with alternative alleles is the major cause of variation among individuals of a sexual species.

Mutations can occur in some parts of the genome without having much effect, if any, on the phenotypic expression in the mutant individual or in its offspring. Two types of examples illustrate this point. The first is that of multiple genes. Multiple genes for controlling just one phenotypic character are fairly common. A mutation in one of the multiple genes will have little, if any, effect on the phenotypic expression of the other unmutated, multiple genes. The second is that of 'redundant' DNA. 'Redundant' DNA in the introns of higher organisms can accommodate some degree of genetic change without altering the translation of the exons in the gene. Quite a high proportion of the genome is probably made of introns in many species of higher organisms. Genetic variation in introns, caused by mutations, has the potential to throw up base sequences that can evolve into useful genes without altering existing gene expression.

A further consideration for geneticists when studying the evolutionary consequences of mutations is that a great variety of base sequences can be produced in a small segment of DNA by routine translocations of DNA. For example mobile genes control the synthesis of an enormous variety of antibodies in mammals. The ways in which a small segment of DNA can control the production of a large number of polypeptides using mobile genes is described in Chapter 14, pages 464-469. The translocation of mobile genes does sometimes produce new base sequences by mutations, but much of the variation among the antibody polypeptides has other causes.

There are many damaging alleles in populations of each species but they usually occur as rare, recessive alleles that are only expressed in that very small proportion of a population which inherits both damaging alleles for the gene. This does not cause the inviability of sexual species, so recessive, damaging, mutant genes can persist for very many generations. Some examples of damaging, persistent alleles in the human species are given in Chapter 16, including a persistent, dominant, damaging allele.

Broad Categories of Mutation

There are four broad categories of mutation

(i). The Addition or Deletion of a Small Number of Adjacent Bases in Genes

The consequences of adding $3n + 1$, $3n + 2$ or $3n$ bases into a gene or of deleting them were discussed in Chapter 1 – the working out of the genetic code using frameshift mutations. The addition or deletion of $3n + 1$ or $3n + 2$ adjacent bases causes the introduction of 'nonsense' codons. However the addition or deletion of $3n$ DNA bases causes the addition or deletion of only one amino-acid to or from a polypeptide. Provided that the amino-acid change occurs at some position in the polypeptide that does not alter the functional region of the molecule, then the addition or deletion of $3n$ bases can be viable. Those $3n$ mutations that do alter the functional region of a polypeptide are likely to be harmful.

(ii) The Substitution of 3n + 1, 3n + 2 or 3n Bases in a Gene

Base substitutions are discussed in some detail in Chapter 11. In most cases the substitution of 3n + 1, 3n + 2 or 3n bases leads to alterations in the corresponding amino-acids that are put into polypeptides. However it is also possible that base substitutions leave the corresponding amino-acids unaltered. This is because the genetic code is degenerate.

(iii). Large Alterations to the Structures of Chromosomes

Using chemical stains it is possible to identify large alterations in chromosome structure. Some of these large structural changes have led to heritable chromosome variation within a species. In a few cases large alterations to chromosomes have led to speciation.

(iv). Alterations to the Number of Chromosomes

Examples are given later in this chapter of some species in which changes to the number of chromosomes have occurred without too much damage to individuals or to the species. In some cases, change to the number of chromosomes has conferred advantages to individuals. It is likely that the fusion of two chromosomes in an individual female ape caused the separation of anthropoids from other apes.

Successful Mutations

The numbers of chromosomes, the positions of genes on the chromosomes and the alleles for each gene in every species are those that have survived the tests of natural selection. The same statement applies to the numbers of plasmids, the positions of genes on them and the alleles for each plasmid gene. The expression of the codes in DNA in living things have hitherto worked well enough to permit success in reaching maturity and in producing offspring. Mutations therefore very often modify successful genes and are usually harmful. Rarely mutations confer phenotypic viability and the passing on of the mutation to subsequent generations. Of course the degree of selective advantage, neutrality or selective disadvantage for viable mutations is different in each individual case.

Whenever dominance and recessiveness apply to a gene in which there is a new mutation there will be no phenotypic distinction between homozygous dominants and heterozygotes. A recessive mutation will not always be expressed in the early generations after its occurrence and may spread quite widely in a population before being expressed as a phenotypic character in double recessives.

It is also true that some alleles in genes depend on the environment for their expression. For example the chlorophyll allele in green plants can only be expressed in the presence of light. Potentially beneficial mutations may be lost if the individual in which the mutation occurs is subjected to local, unusual and damaging environmental factors, to which the majority of the population are not normally exposed.

Although successful mutations are only a small proportion of all the mutations that occur, they have given rise to alternative or to a number of alleles for a proportion of genes in each species. The proportion of genes which have different alleles is quite small in most species, if not in all species. Having homozygosity for the majority of genes ensures the stability of basic form for a species. Having a small proportion of genes for which the species has alternative, or several alleles permits some variation among individuals of a species, within the constraints imposed by the invariable genes. In a broad sense the multiple effects of the so called invariable genes dictate the 'species', and those genes with alternative or several alleles dictate variation among individuals of the species. Chapter 12 discusses 'species genes' in more detail.

Mutation Rate

Because different species have different methods of reproduction (asexual; sexual) and different time intervals for producing a new generation (bacteria very short; Drosophila spp. short; man quite long), it is worth comparing the mutation rates of some known mutable genes in different species.

Most of the studies so far have chosen those genes that mutate spontaneously most frequently. The upper limit of mutation rates is therefore studied and no figures are available for average mutation rates. A guess is that about one single base mutation occurs per gene locus per million cells produced by either mitosis or meiosis. Larger mutations caused by breakage and reconstruction may be more frequent. Not much reliability can be placed on the 'one in a million' hypothesis and this figure may be greatly over-estimated.

The upper limits for any locus found so far give:-

Species	Mutations per million cells
Drosophila melanogaster	120
Zea mays	106
Man	190
Mice	44

In sexual species females may have a slightly lower rate of spontaneous mutations than males.

When repeating genes, introns, and non-functional exon regions are taken into account, in which mutations may have little or no effect on gene expression, the true mutation rate is likely to be higher than the suggested figures in the table.

The Breeding System and the Rate of Incorporating Mutations into Subsequent Generations

(a) Length of Life Cycle

The mutation rate per locus in germ cells is similar in man and Drosophila melanogaster, measured per generation. When measured per unit of time therefore, the mutation rate per gene locus in Drosophila is 800 times greater than in man. This may explain why crossings-over only occur in female Drosophila, so reducing the combined effects of mutations and crossings-over in producing too much variation within the species. Bacteria have an approximate mutation rate per generation only 0.0001 that of multicellular species.

There is an added advantage when crossings-over only occur in one of the sexes of a species such as Drosophila melanogaster. 'Best sets' of genes that have stood the tests of natural selection can be passed unaltered from generation to generation in the gametes of the sex in which no crossings-over occurs.

(b) The Breeding System
(i) Asexual

Viable mutations will automatically be passed on to all cells arising from the mutant cell.

(ii) Sexual

Viable mutations will only pass into subsequent generations if the mutation occurs early enough in the mitotic development of an organism to pass into the gametes it produces, or if the mutation occurs in the production of gametes themselves by meiosis. There also has to occur the fertilisation process using a gamete containing a mutation, and this will not necessarily occur. The chances of a mutation passing into the new generations of sexual species will also depend on whether it is inbreeding or outbreeding.

The Types of Genes that Are Changed by Mutations

Natural selection acts on the phenotypic expression of:-

(i) Sharply defined, simple Mendelian characters which depend upon the alleles of single genes for their expression. Dominance, incomplete dominance and recessiveness apply.

(ii) Those characters that depend on super-genes for their expression. Again dominance, incomplete dominance, and recessiveness apply.

(iii) Those characters that depend on the complex inter-actions of many genes in which the characters expressed are usually a blend of the parental phenotypes.

It should be noted that a mutation in a single gene can have complex, multiple effects on other genes. In this case a favourable mutation is one in which the several consequences have advantages that outweigh the disadvantages. When this happens the mutation will spread and may become the norm, either as a single allele or as a new super-allele, taking on the role of the wild-type allele or wild-type super-allele.

Some genes have been identified, mostly in the heterochromatin near the extreme ends of chromosomes, which control the rate of spontaneous mutations in all the chromosomes. The allele 'hi' on chromosome 2 in Drosophila melanogaster increases the mutation rate in all the chromosomes. When the allele is homozygous the mutation rate is increased ten times. Mutation inducing genes are an important factor in controlling the adaptability (creating variations in phenotypes) of some species.

Reverse Mutations

Mutations can revert to the wild-type alleles. This is thought to be only about $\frac{1}{10}$ as likely as the:-

$$\text{wild-type} \rightarrow \text{mutant}$$

mutation. This makes the average figure for the reverse mutation about one cell in ten million. Chapter 11 discusses some aspects of point mutations and reverse mutations.

Repair of Mutations

Many mutations in the materials of inheritance are caused by heat energy generated in the cells themselves. However, in the case of DNA, repairs very greatly reduce the number of stable mutations. DNA polymerase inserts pairing bases whenever single-stranded DNA is produced after damage to base sequences. DNA ligase then joins the ends of the repaired section to the appropriate locus in the undamaged DNA. Provided there are no deletions, additions or substitutions of bases during the repair process, repairs to damaged DNA can prevent permanent mutations, so avoiding evolutionary consequences of mutations.

 DNA glycolases (at least 20 of these enzymes have been found so far) recognise 'wrong' bases and catalyse their removal. DNA polymerase and DNA ligase can then insert the 'correct' base sequence in place of the removed section.

The Natural Selection of Phenotypes in Which Mutations Occur

Natural selection operates on the phenotypic expressions of the alleles in populations of every species. There may also occur:-

(i) Stability

(ii) Change

in the proportion of alleles present.

In an unchanging environment there is likely to be the stabilisation of both the phenotype and genotype of each species. However the physical, chemical and biological environments rarely stay unchanged for long, and each species must have a complement of alleles, created by mutations and spread through the population, that allows some individuals of each generation to survive and reproduce in a changing environment. If the species cannot produce individuals that can survive, the species becomes extinct. Over the whole period of life on this planet there have been many more species that have become extinct than have survived. Extinction is, therefore, the rule rather than the exception.

Whenever there is allopatric (in two separate geographical regions) or sympatric (in the same geographical region – having different breeding seasons for example) isolation of two or more groups of individuals from one species, and the groups are later subjected to different environmental pressures, there will be the natural selection of diverging phenotypes. The complement of alleles in each group will also diverge, partly because different genes will disappear in different groups but also because different, rare, favourable mutations will give rise to different, viable alleles and super-alleles for genes and super-genes in each isolated group. There will also be differences in the large, viable, structural changes to chromosomes in the isolated groups.

Some examples of viable mutations that have been observed in a variety of species are given in the following pages.

Naturally Occurring Mutations in Chromosomes

Before the evolution of the photosynthetic apparatus in living things, there was no protective ozone layer against ultra-violet radiation from the sun. Primitive forms of life, early in the evolution of life on earth, were subjected to this powerful agent for mutation, and the properties of the material of inheritance in living things at that time must have included:-

(i) The ability to mutate frequently enough to give rise to variation among individuals in a changing biological as well as a changing physical and chemical environment.

(ii) Enough stability to ensure that some new cells retained the genes needed to direct essential biochemical processes.

During the long period of evolution, there have been differences in environmental causes of mutations – ultra-violet radiation, chemicals, radioactivity, wetting and drying, heat and cold. Cells have also had the capacity for making 'mistakes' in the manufacture of new chromosomes for new cells, irrespective of environmental factors. There are some mistakes that arise spontaneously in cells which can be viable. Others are almost certain to be inviable.

Mutations can be broadly divided into those that involve only one gene, those that involve structural changes to parts of chromosomes larger than one gene, and those that alter the number of chromsomes. Some examples are given on pages 323-330 of structural alterations to chromosomes. Alterations to the number of chromosomes are discussed, with examples, from pages 330-334. Local mutations involving small numbers of bases are discussed in Chapters 1, 11 and 15. Before looking at any specific examples of large mutations it is necessary to mention some general requirements for the success of structural mutations in chromosomes.

The General Requirements for Viability after The Positional Transfer of Parts of Chromosomes Larger Than One Gene

(i) The Need for Chromosome Equality

Unless chromosomes, resulting from the positional transfer of parts of chromosomes that are larger than one gene, can migrate to opposite poles of a dividing cell in an equal way, there are usually problems of viability in individuals having aberrations of equality in their chromosomes. In a sample of 750 spontaneously aborted human babies, for example, nearly half were found to have inviable chromosome aberrations of inequality.

(ii) The Desirability of Having Only One Centromere in Each Chromosome

Chromosome changes that result in the removal of the centromere from a chromosome, or the addition of an extra centromere cause problems at cell division.

The spindle cannot move those chromosomes without a centromere, so mutant chromosomes of this sort may end up on the wrong side of the new dividing membrane at the first or any subsequent cell division after the mutation occurs.

Chromosomes with two centromeres may have each centromere attached to different spindle threads, one on either side of the equator. When the contraction of the spindle fibres tries to pull each centromere to opposite ends of the dividing cell the chromosome may break in a way that causes the unequal movement of chromosome fragments to each side of the new dividing membrane.

A further problem, associated with the attachment of chromosomes to the spindle, is that there is a specific spindle site for attachment of the centromere of each chromosome. The inclusion of two centromeres in a chromosome is therefore likely to cause the breaking of the chromosome, as each centromere will attach to spindle fibres on either side of the equator.

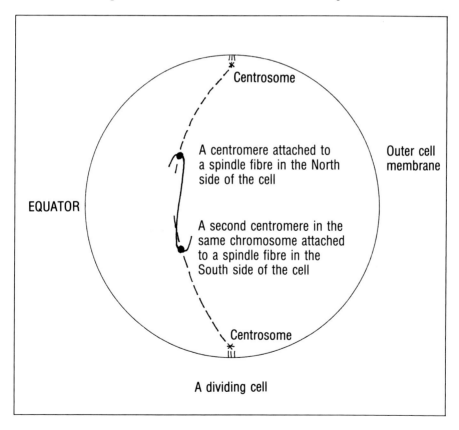

A dividing cell

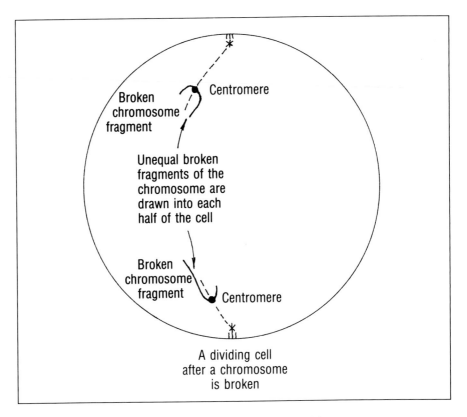

A dividing cell after a chromosome is broken

Categories of Viable Structural Changes in Chromosomes

1. Alterations to the amount of DNA in a chromosome

Additions (most common) or deletions of part of a chromosome can sometimes be viable. Several of the autosomes (non-sex chromosomes) in man, and also the Y chromosome in man, show variable amounts of DNA and the amounts are heritable.

The occurrence of 'redundant' DNA as a high proportion of the genome of many eukaryotic species must have occurred by additions. Additions also give the possibility of new genes, especially when the addition duplicates existing genes. Additions can give additional variety among individuals of a species. They can also lead to different degrees of complexity in different species by adding to the genome new DNA from which new viable genes arise.

The consequences of deletions, by contrast, must in general be harmful. They can be viable if the deletion removes a proportion of repeated genes.

A postulated evolutionary consequence of gene duplication is given on page 367, where the divergence of blood globin genes is illustrated.

2. Alterations to Chromosomes, either within a Single Chromosome or when there is The Exchange of Parts between Chromosomes

The alterations can be divided into:-

(a) Those Changes that Leave the Overall Form of the Chromosomes Unaltered at all Stages of Cell Division

No change in the morphological form of chromosomes at any stage in cell division implies that there is NO change in the position of the centromere in the chromosomes affected by the change. Changes of this sort can be divided into two categories:-

(i) The Inversion of a Segment of a Chromosome Which Leaves the Centromere Unaffected – A Paracentric Inversion

The diagrams show a small part of the double strand of a single chromosome, straightened out to simplify the diagrams:-

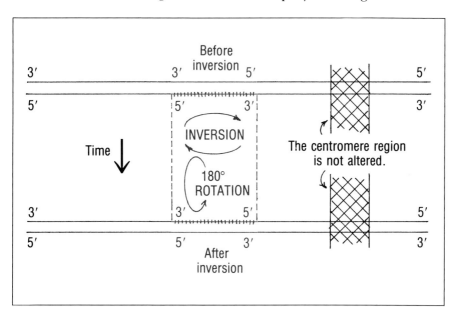

Rotation is necessary when there is inversion, so that the 3' and 5' ends fit with the DNA which is not inverted.

Many inversions have been observed in the genus Drosophila. The reason why inversions work at meiosis is that a loop is formed so that if crossings-over occur the exchange of genes between chromosomes involves the same genes on both chromosomes:-

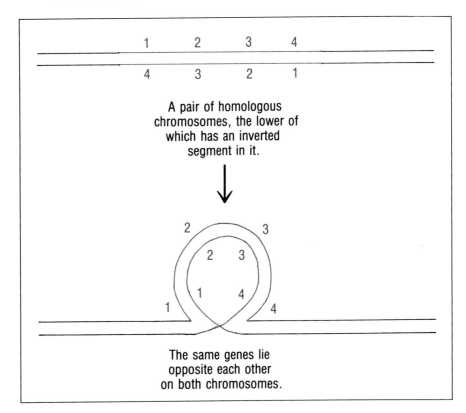

A pair of homologous chromosomes, the lower of which has an inverted segment in it.

The same genes lie opposite each other on both chromosomes.

The loop allows the same genes to lie opposite each other at meiosis, despite the fact that an inversion has taken place. The consequences of not having this arrangement would be deletions of whole genes or of parts of genes when there are crossings-over in this region of the homologous chromosomes:-

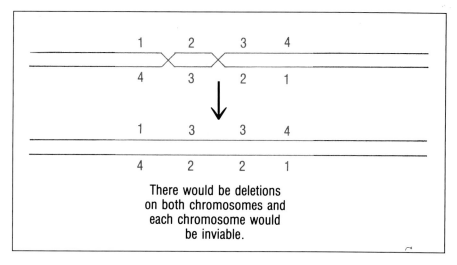

There would be deletions on both chromosomes and each chromosome would be inviable.

(ii) Equal Reciprocal Translocations Between a Pair of Homologous Chromosomes

Although all the genes are still present during mitotic cell divisions, following equal reciprocal translocations, there is no real difference in effect between a translocation of this sort and crossing-over at meiosis. It is therefore difficult to identify such translocations in natural populations.

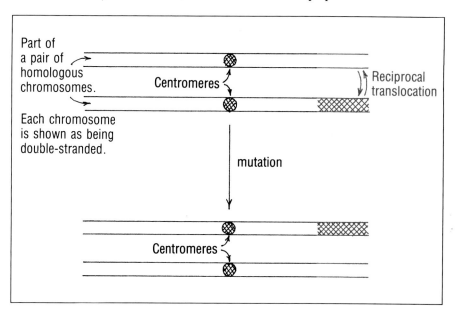

(b) Those Alterations to Chromosomes that DO Alter the Positions of the Centromere but which Can, even so, Result in Viable Chromosomes

(i) Unequal Reciprocal Translocations

Although changes which alter the positions of the centromere in a chromosome cause difficulties of equal migration of the affected chromosomes at cell division, such changes have been observed in several species of plants. Six genera of Onograceae from North America exhibit unequal translocations. However such translocations are known in only very few animal species – one example is found in scorpions; another is in the American cockroach.

An example of unequal reciprocal translocations is:-

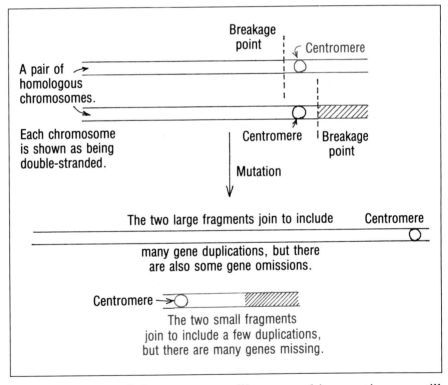

Because all the genes are still present this mutation can still work in the diploid in which it occurs. However there are problems at meiosis if such a mutation is found in the gamete mother cells. Any gamete containing deletions will be inviable.

If unequal reciprocal translocation results in one of the chromosomes being very small it may degenerate, even if it contains a centromere, so reducing the chromosome number in a gamete. This is also likely to cause inviable gametes.

(ii) Pericentric Inversion in Which the Inversion Changes the Position of the Centromere in only One of a Pair of Homologous Chromosomes

An example showing the result of a pericentric inversion is:-

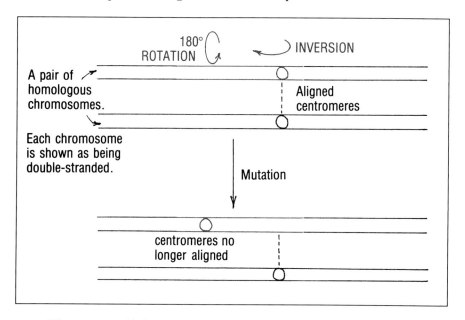

The correct lining up on the equator of pairs of chromosomes on their spindle fibres at cell division is made more difficult. However both chromosomes retain all their genes and may be viable.

Pericentric inversions have been observed in the black rats of South East Asia (Rattus rattus) and in some grasshopper species (Keyacris scurra).

(iii) Centric Shift

The result of a centric shift is similar to that of a pericentric inversion:-

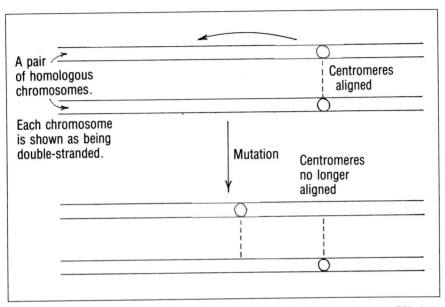

All the genes remain but there may again be problems of lining up on the spindle at the equator during cell division.

(c) Breakages Within the Centromere Itself
(i) Centric Fission

Fission implies breakage and separation of chromosomes. In order that they can subsequently move in association with the spindle fibres at cell division, each broken chromosome fragment must have a part of the original centromere in it.

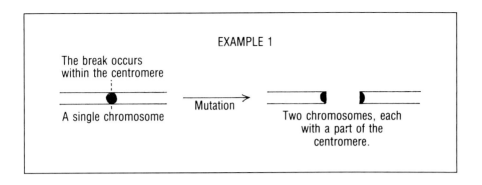

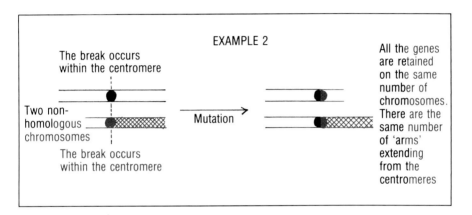

The common European shrew Sorex araneus shows variation in the dipoid numbers of chromosomes arising from centric fissions, in the mountain races of the species.

(ii) Centric Fusion

If a break occurs in the centromere of two chromosomes, each from a different homologous pair of chromosomes, two broken fragments may join and add part of a centromere to the fragment it joins:-

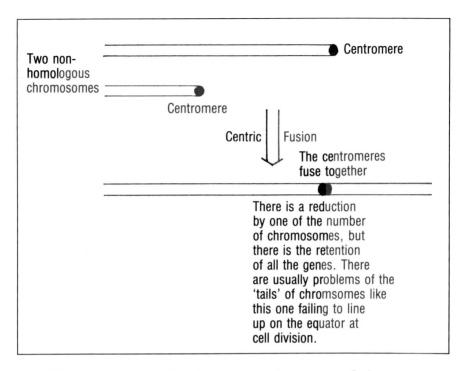

The black rats of South East Asia have sets of chromosomes that differ from those of the black rats of Oceania (42 in SE Asia compared with 38 in Oceania). The sizes of chromosomes in the Oceania black rats imply that fusion of the 4 and 7 chromosomes has occurred in some of them (Australia, New Zealand, New Guinea) while the fusion of the 11 and 12 chromosomes has occurred in the black rats of Ceylon.

3. Alterations to the Number of Chromosomes

(i) Aneuploidy

    (a) The gain of a whole chromosome (polysomy) or

    (b) The loss of a whole chromosome (monosomy)

can occasionally be viable. The loss of a chromosome rarely allows the viability of diploids – in the plant Clarkia ameona in North America 13 chromosomes are viable in some plants instead of the more normal 14 (n = 7). In polyploids aneuploidy is more common than it is in diploids. The plants Claytonia virginica of eastern North America have several species and races exhibiting aneuploidy.

(ii) Polyploidy

In general the addition of an extra *set* of chromosomes to give 3n, 4n etc. by incorrect migration of chromosomes at cell division in sexually reproducing species gives rise to inter-sexes and sterility. Polyploidy is therefore most common in plants and in animal species, such as fish and amphibians, in which no sex chromosomes exist. Sex is determined by a combination of genes distributed throughout their autosomes. Occasionally exceptions to this general rule of having no sex chromosomes are found, and plants having strong male determining Y chromosomes, such as Meladrium, Acnida and others can sometimes outbreed even when they are tetraploid.

Polyploidy leads to the formation of intermediate forms between strains of diploids from which tetraploids were formed. The grass Aegilops umbellata has formed seven distinct tetraploids, all based on a diploid set of chromosomes. The tetraploids were derived by crossing three diploid grasses (Aegilops candata, A. corrosa and A. speltoids) with the naturally occurring Aegilops umbellata. They are all Mediterranean region grass weeds.

Because of problems at meiosis when there is an uneven number of sets (3n, 5n) of chromosomes, some plant and animal triploids have been forced to adopt the spontaneous generation of new individuals. This is called parthenogenesis in animals and apomixis in plants. The Arctic species of blackfly (Gymnopais spp.) and whiptail lizards (Cnemidophorus spp.) from the Americas have triploid hybrids which produce offspring parthenogenically. Most European dandelions (Taraxacum officinale) are triploid, obligate apomicts.

(iii) Polyploidy in Grasses and Wheat

The following evolutionary pathway shows how hybridisation and polyploidy have altered the chromosome numbers in those grass hybrids that gave rise to modern wheats.

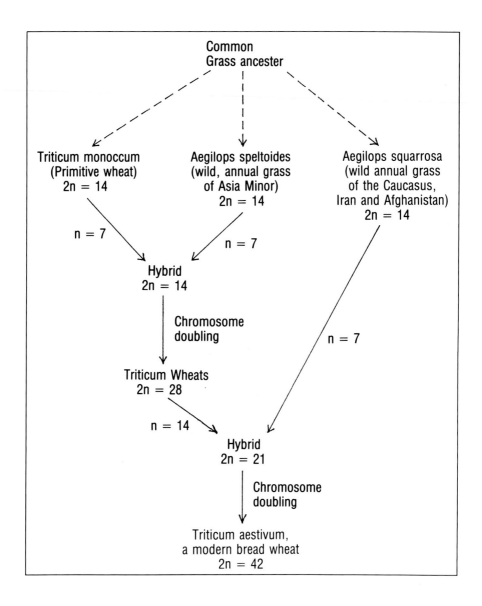

The 21 pairs of chromosomes in Triticum aestivum can be divided into seven groups, each group having three pairs. It may be that there is one pair of chromosomes in each group that comes from each of the three ancestral species. Each pair of chromosomes in any one group controls similar characters, indicating that their genetic function in each of the ancestral grass species was to control similar characters. Presumably the 2n=28 Triticum wheats had groups of two pairs of chromosomes and outcrossing with Aegilops squarrosa introduced a third pair of chromosomes to each group. Although similar to each other, the pairing of chromosomes in one group is specifically between pairs originating from the same ancestral species. A mutant gene has arisen in the 2n=42 modern wheats in chromosome pair V. The mutation prevents chromosomes from different ancestral species from pairing at meiosis. The mutation therefore gave genetic stability and improved fertility by ensuring that genetically equivalent chromosomes from the same ancestral species pair up at meiosis.

Cereal breeders are faced with two main problems when they try to introduce desirable qualities shown by wild grasses such as disease resistance or resistance to cold. The first is that there is no pairing of chromosomes between the wheat and wild grass chromosomes because of chromosome V. The second is that there are genetically unbalanced germ cells produced by the hybrids caused by unequal chromosome orientations when drawn into opposite sides of germ cells by the spindle fibres during meiosis.

When chromosome V is removed from the wheat, the chromosomes from wheat and grass can pair. Sterility of improved hybrids has been removed by using colchicine to double the chromosome number. This chemical causes chromosome replication without cell division in hybrid seedlings and converts sterile plants into fertile crops with improved qualities.

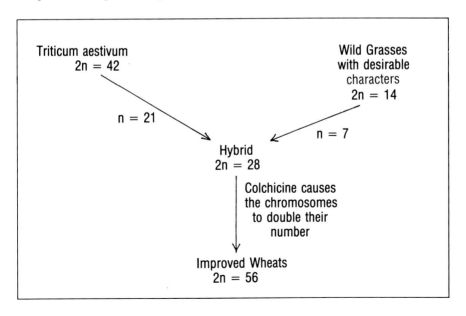

(iv) Polyploidy in Valeriana spp.

The European plant Valeriana officinalis, which has a haploid number of 7, $2n=14$, has evolved an octoploid in Britain. Polyploidy on the edge of the range of a species is not uncommon. The octoploid is classified as an 8n species in which $8n=56$. Valeriana sambucifolia is a diploid species native to Britain in which $2n=56$.

By having the same number of chromosomes the octoploids can outbreed with the native diploids in Britain to produce hybrids that are less fertile and that compete less well than either parental species. The hybrids grow in areas that are not inhabited by either ancestral species, so avoiding competition between ancestral species and the hybrid.

(v) Polyploids in Spartina spp.

Spartina alterniflora, $2n=70$, from North America was introduced into England. It crossed with the native Spartina stricta, $2n=56$, to produce male sterile hybrids Spartina townsendii. The hybrids, $2n=63$, mostly reproduce vegetatively. The hybrid doubled its number of chromosomes, $2n=126$, to produce a species that is genetically isolated from both parental strains. The hybrid is an undesirable water weed that chokes harbours.

4. Small Supernumerary Chromosome Segments

Small supernumerary chromosome segments have been observed in both animals (the meadow grasshopper – Chorthippus parallelus) and in plants (the bluebell – Scilla sibirica). These are found at the ends of, or within the chromosomes and vary considerably in frequency within the species in which they are found.

Evolutionary Changes Caused by Mutations and Other Genetic Factors

The examples of this chapter and those that follow in Chapter 11 show that the chemical and physical properties of DNA allow chromosomes and plasmids to mutate in several different ways. Further genetic variation in many sexual species is generated by stitching together new combinations of alleles in chromosomes by crossings-over during meiosis. Yet more inherited variation is created by mobile genes when they rearrange the linear associations of alleles, or parts of alleles.

In order that there can be evolutionary consequences of genetic changes, it is generally necessary that groups of individuals belonging to one species be isolated from each other. The prevention of cross-breeding between isolated groups of individuals may allow each group to experience different mutations, followed by different consequences of genetic drift and natural selection. In such cases there is likely to be phenotypic and genotypic divergence of each isolated group. If there is sufficient divergence to prevent inter-breeding between groups, even when the isolating barriers are removed, then more than one species has arisen out of a single species.

The combined effects of the phenotypic and genetic variations caused by mutations, crossings-over, mobile genes, isolating barriers between groups of individuals, genetic drift and the natural selection of phenotypes by the environment are the essential constituents of evolution.

Chapter 10 Questions

1. What are the ultimate reasons for differences between individuals of a species?
2. What genetic reasons allow crossings-over at meiosis to have a greater effect than mutations in causing variation among individuals of a species?
3. When do mutations have little or no effect?
4. Are mutations directional in conferring design advantages to individuals in their environment, or are mutations haphazard?
5. Why do damaging alleles persist in populations?
6. Are mutations generally beneficial?
7. What are the broad categories of mutation?
8. What are the properties of successful mutations?
9. Is there any reliable information about the mutation rate in dividing cells?
10. Does the type of breeding system alter the rate of introducing a successful (or non-harmful) mutation into a species?
11. What types of gene can be changed by mutation?
12. Do reverse mutations occur?
13. What can happen to allele frequencies in populations?
14. What environmental factors can increase the mutation rate?
15. What are the general requirements that allow the viability of large alterations to chromosomes?
16. What general categories of viable, structural changes to chromosomes have been identified?
17. When are aneuploidy and polyploidy viable?
18. How has man altered the numbers of chromosomes in producing modern bread wheats and improved wheats from primitive wheats and common grass ancestors?
19. Why is polyploidy more common on the edge of the range of a species than in the main body of the habitat?

11

EVOLUTIONARY STUDIES COMPARING POLYPEPTIDES THAT ARE WIDELY SPREAD AMONG MANY SPECIES

After six years of chemical analysis at Cambridge University, F. Sanger had by 1955 worked out the sequences of amino-acids in the two polypeptide chains of insulin, as well as the way in which the two chains are held together by sulphur bridges.

At the same time that the search was on for the structure and mode of action of DNA, there emerged techniques for working out the orders of amino-acids in polypeptides. This allowed the orders of amino-acids to be compared with the sequences of bases in DNA and in RNA. The painstaking work of Sanger, who struggled for so long with the complexities of insulin structure caused by sulphur bridges, was of immense value later for cracking the genetic code.

More recently it has been discovered that sugars attached to polypeptides in higher living things modify the mode of action of the polypeptide and that the functions of molecules at the cellular level depend both on polypeptide amino-acid sequences and upon the attached sugar.

A simplified presentation of the molecule studied by Sanger, which does not show the three dimensional aspects of the molecule, is shown overleaf.

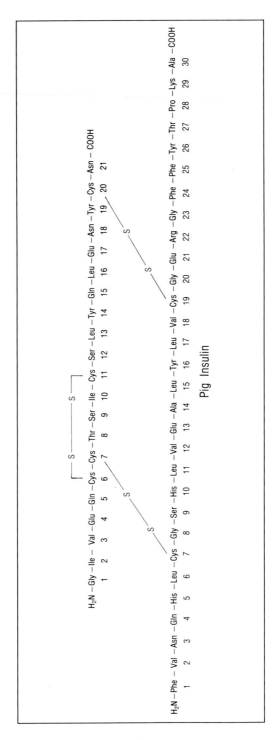

The analytical methods of Sanger, with improvements, have been used by many teams of scientists during the years since 1955 and polypeptides can now be analysed quite quickly. Amino-acid data banks for an increasing number of polypeptides are kept in several places. Among them are The Biochemical Research Council, Washington, D.C., and San Diego University, California.

It occurred to research scientists that some of the polypeptides such as cytochrome c (part of the electron transport system of aerobic respiration) and the globins (used for oxygen attachment and transport in animals) are widespread among many species. They have analysed the sequences of amino-acids in these polypeptides taken from a wide range of species and compared the similarities and differences of each of them. Many other widespread polypeptides are being analysed, and the collective evidence inferred by the amino-acid sequences in as many polypeptides as possible will provide an increasingly reliable method of relating the evolutionary trees of those species that are alive today.

At one time it was believed that the only means of inferring evolutionary pathways was by observing the fossilised imprints of ancient species in rock strata and relating them to other fossils by their similarities and differences. It now seems that there is retained, in some of the molecules of *living* species, considerable information about their evolutionary pathways, and this can be compared with the fossil evidence.

The distances of the evolutionary divergences of living species from each other can now also be worked out by comparing polypeptides common to all of them. The way this is calculated using simultaneous equations is discussed later in this chapter.

The Consequences of Substituting One Amino-acid for Another at Sites on a Polypeptide

It is convenient to give a defining number to each amino-acid in a polypeptide, starting at the $-NH_2$ end and continuing in sequence to the $-COOH$ end:-

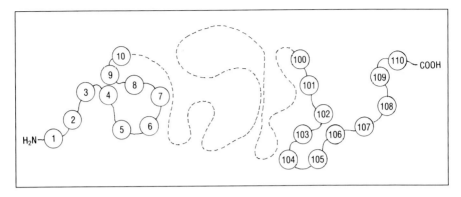

The insertion of any one amino-acid at its position on the polypeptide depends on a sequence of three bases in a gene that codes for the polypeptide. A sequence of triplicate codes in the gene therefore determines the sequence of amino-acids in the polypeptide, put there by transcription and translation. If mutations occur in the gene for any one amino-acid there may be the substitution of one amino-acid for another. Substitution does not always occur because the genetic code is degenerate. However each point mutation in the DNA will have a different consequence for the polypeptide site it controls:-

a) Some single base mutations produce codes that, although different from the original code, specify the insertion of the same amino-acid. For example:-

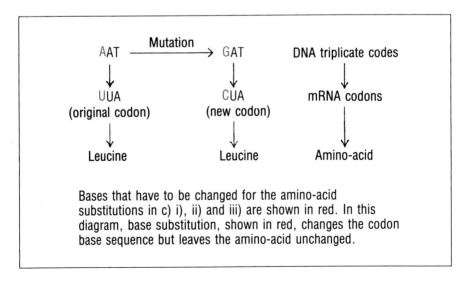

Bases that have to be changed for the amino-acid substitutions in c) i), ii) and iii) are shown in red. In this diagram, base substitution, shown in red, changes the codon base sequence but leaves the amino-acid unchanged.

It should be noted that the codon is carried in mRNA, but this ultimately depends on the triplicate sequence of bases in the DNA of a gene, from which transcription inserts each triplicate codon into the mRNA.

b) For any single base mutation which inserts a new amino-acid into a site, there can also be reverse mutations at some time later to reinsert the original amino-acid at that site. The chapter on mutations suggested that reverse mutations are, in general, much less likely than those which cause the forward mutation. However reverse mutations must be taken into account if accurate evolutionary trees are to be deduced, with correct time intervals calculated. At this distance in time from many mutations that have occurred during evolution, it will only be possible to interpret the molecular evidence that points to reverse mutations by using site sequences taken from many different, widespread molecules and correlating the collective data. Most sites will show the forward mutations and these can be used to demonstrate the exceptions in which successful, rare, reverse mutations have occurred during the long period of evolution on earth.

c) i. The substitution of some amino-acids for others requires only a single base mutation:-

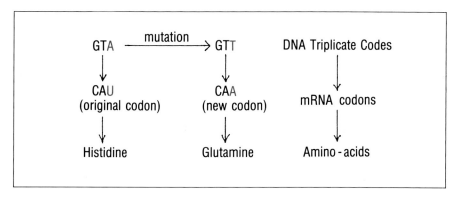

ii. The substitution of some amino-acids for others require two point mutations:-

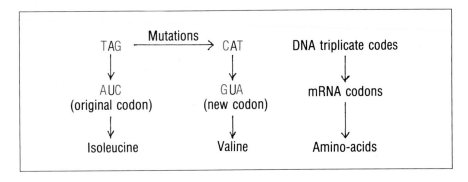

iii. The substitution of some amino-acids for others requires three point mutations:-

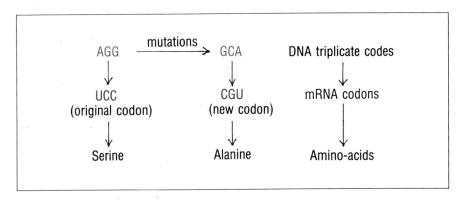

In natural populations, even those subjected to ultra-violet radiation from the sun, or other mutation-inducing agents, mutations of type i. are more likely than ii., which in turn are more likely than iii. Having an established amino-acid at any site therefore constrains the probability of any other amino-acid substituting for it, because of the decreasing probability of mutations as the number of base substitutions increases.

d). Any amino-acid that is distinctly different from one that previously occupied a site on a polypeptide will cause more of a structural change to the polypeptide than the substitution of an amino-acid with similar properties to its predecessor. Natural selection of the phenotypic effects caused by a point mutation is likely to favour the substitutions of amino-acids similar to the original one, in most cases. Occasionally a new amino-acid increases the fitness of the phenotype. Selection in favour of such a mutant may allow, subject to genetic drift as well as natural selection, the mutation to spread from the individual in which the mutation arose until, after a considerable period of selection, it becomes the established wild-type allele for the gene throughout the species.

e). There have been, and continue to be, a very large number of point mutations in each individual of each species. The comparision of cytochrome c found in widely differing species shows that, even after diverging from common ancestors many millions of years ago, there are many similarities between the cytochrome c molecules as they are found in a wide range of contemporary species. The inference is that there has been very powerful selection against the vast number of individuals, or cells produced in each species, in which damaging point mutations in the cytochrome c gene have taken place over the millenia.

f). Some regions of the tertiary structure of a polypeptide are critical for its function. This is especially true:-
    i. In its functional region. This may only be a small proportion of the whole polypeptide.
    ii. In the locality of those amino-acids, such as cysteine, that form sulphur bridges.
    iii. In the locality of glycine which, by having no side chain, greatly assists in the sharp folding of polypeptide molecules wherever glycine is found in them. Glycine therefore often has a critical effect on tertiary structure and therefore on the 'active site'.

The comparison of the sites in cytochrome c taken from different species emphasises the degree of local variability or invariability that can be tolerated by a polypeptide with regard to the constraints imposed by f)i, ii, and iii.

g). Amino-acid sites in polypeptides can be divided into those that are:-
  i. Invariable

    Invariable sites are absolutely critical for the active site of the polypeptide. Invariable sites either cause critical, local, chemical associations or they cause critical, tertiary folds without which the polypeptide is useless. For example, cysteine is always found at positions 14 and 17 on cytochrome c and the properties of cysteine's sulphur in this region of the polypeptide are critical for providing the correct association of the haem- ring of haemoglobin with this polypeptide:-

Species	\multicolumn{7}{c}{Sites in Cytochrome c}						
	13	14	15	16	17	18	19	20
Man	Lys	Cys	Ser	Gln	Cys	His	Thr	Val
Rhesus monkey	Lys	Cys	Ser	Gln	Cys	His	Thr	Val
Pig	Lys	Cys	Ala	Gln	Cys	His	Thr	Val
Dog	Lys	Cys	Ala	Gln	Cys	His	Thr	Val
Chicken	Lys	Cys	Ser	Gln	Cys	His	Thr	Val
Fruit fly	Arg	Cys	Ala	Gln	Cys	His	Thr	Val
Wheat	Arg	Cys	Ala	Gln	Cys	His	Thr	Val
Neurospora crassa	Arg	Cys	Ala	Gln	Cys	His	Gly	Glu
Baker's yeast	Arg	Cys	Glu	Leu	Cys	His	Thr	Val

Adjacent to one of the invariable cysteines, histidine is always found at site 18. Histidine in this position is critical for providing an electron used for a valency association which allows the iron atom of the haem- ring to accept and donate electrons for the oxidation and reduction functional properties of this molecule (most textbooks on biochemistry include an explanation of the role of cytochromes in electron transport during aerobic respiration).

There is an invariable sequence of amino-acids, in those species studied so far, that is always found between sites 70 and 80 (inclusive) in cytochrome c:-

etc - Asn- Pro - Lys - Lys - Tyr - Ile - Pro Gly - Thr - Lys - Met- etc
      70   71   72   73   74   75   76   77   78   79   80

An Invariable Sequence in Cytochrome c

This region of the folded polypeptide is also associated with the functional haem- ring. Glycine, in sites −1, 6, 29, 34, 41, 45, and 77, is critical for the tertiary folding of cytochrome c and is invariable, in those species studied so far, at these sites. Glycine at site 77 is part of the haem- associated sequence mentioned above.

It should be noted that the site numbers are those defined for a standard number of amino-acids in a polypeptide. For example the standard number of amino-acids in cytochrome c is 110, and this number is obtained by examining a large number of cytochrome c molecules taken from several species. In some species there may be some extension beyond site 1 and also beyond the highest standard site number. For example in cytochrome c some species have additional amino-acids to the left of 1 (these are given minus numbers) and some have additional amino-acids to the right of 110. Some species have additional amino-acids at both ends of the molecule:-

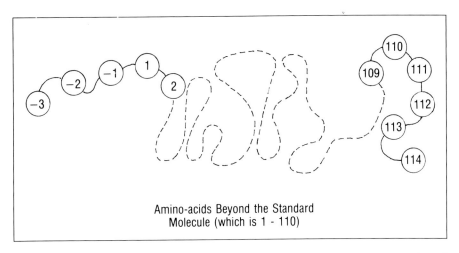

Amino-acids Beyond the Standard Molecule (which is 1 - 110)

Allowance must also be made in case there are any mutational additions or deletions of sites within the standard sites, and the site numbers have to be adjusted to take account of the addition or deletion of sites.

ii. Functionally Conservative

If an amino-acid has several of the properties of another amino-acid, it may be substituted for it in certain sites of a polypeptide, without damaging, and on rare occasions improving, the function of the polypeptide. Sites such as these require one of a small number of amino-acids to be in them and the substitution of one of these similar amino-acids for another has been at least of no selective disadvantage, if it has become established in the site for an entire species.

iii. Variable

Some sites are much less crucial than others for the correct functioning of a polypeptide. For example sites 89 and 92 of cytochrome c each have nine alternative amino-acids in those species studied so far. Of course each of the species has only one of the nine alternatives.

The Mathematical Probability of One Amino-Acid Substituting for Another to Give Rise to A New Predominant Amino-acid in Any One Site on a Polypeptide

i. Forward Mutations

The mathematical analysis, which is not yet perfected and requires a profound understanding of physics, chemistry and mathematics (all beyond the scope of this book) for predicting the probability of a new amino-acid being accepted at any site in an established polypeptide, has to take account of several variables. Among these are:-

a). The number of point mutations needed for the substitution.

b). The chemical structures of the amino-acids. The sulphur bridges in cysteine and the sharp tertiary folding at glycine make it very difficult to substitute these amino-acids successfully.

c). The physical and chemical properties of each amino-acid, such as solubility and electric charge.

d). The nature of the active site of the polypeptide. It is this aspect of the mathematical variables that will present the most difficulty. Every difference in overall tertiary structure, as well as every difference in local structure makes a difference to the function of a polypeptide. General mathematical rules concerning the substitution of one amino-acid for another, in the vast range of polypeptides found in living things, will be very difficult to find. However, mathematical predictions about the probabilities of acceptance, or functional improvements made by amino-acid substitutions in any one polypeptide may become possible and this could have some significance for genetic engineering.

B. Clarke has attempted to calculate the probability of one amino-acid successfully substituting for another in a polypeptide, with the results shown in the following scatter diagram:-

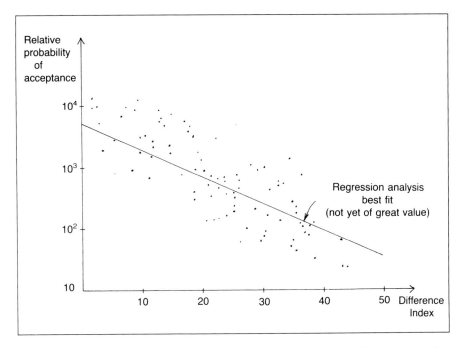

No allowance is made in the scatter diagram for the particular properties of each polypeptide structure related to its function. It is simply based on the degree of similarity or difference between the amino-acids.

ii. Reverse Mutations

The mathematical probability of a reverse mutation, in which the original amino-acid substitutes for the one that had previously taken its place in the site, is not quite as straightforward as it may, at first, appear. Once a mutation inserts a new amino-acid that withstands the pressures of natural selection, there may be further mutations at other sites that insert new amino-acids, so changing the properties of local and tertiary structure. This may occur in such a way as to alter the probability of the viable acceptance of the reverse mutation at the site under study.

iii. Quaternary Structure

The associations of several polypeptides to make a complex of large molecules with a localised functional region, introduces a further variable for the mathematical equations needed for predicting the probability of the *viable* acceptance of one amino-acid for another at any particular site on those polypeptides which associate to form the quaternary complex.

The Incorporation of a Tautomeric Point Mutation into the Somatic Cells of Sexual Diploids

In order that a tautomeric point mutation can have evolutionary significance, several criteria have to be fulfilled.

The first criterion is that the point mutation must occur in a cell line that leads to the formation of gonads or to at least a proportion of the gamete forming tissue.

A second criterion is that DNA replication must take place so that tautomeric shifts can occur.

In the following diagrams, a pair of bases is shown at the same position in a gene, on homologous chromosomes. The wild-type pair of bases persist in one of the homologous chromosomes in both cells arising from the first somatic division following a tautomeric shift.

In the example

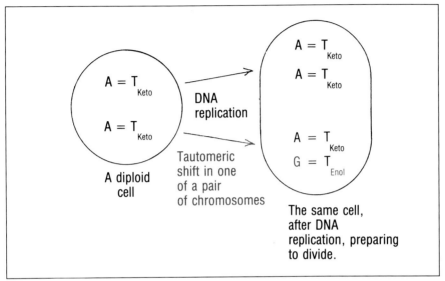

Into one of the two cells is passed the original wild-type allele in both homologous chromosomes. Into the second is passed the half-tautomeric mutation:-

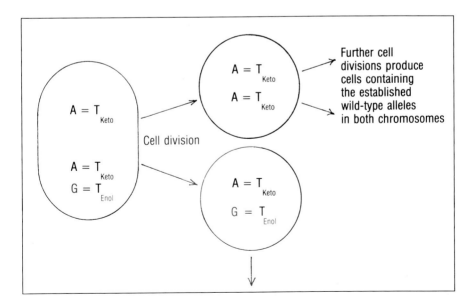

Further cell division has to occur before the full tautomeric mutation arises:-

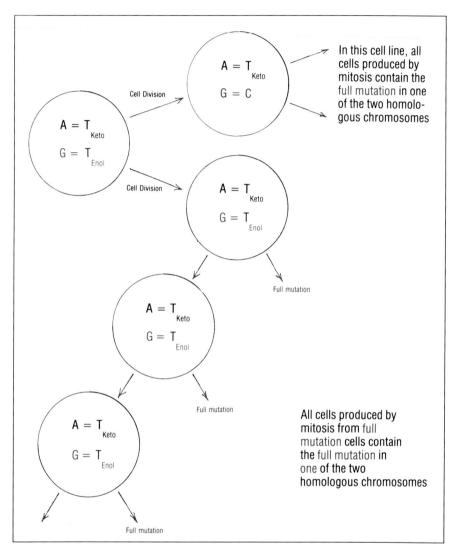

One out of the four cells arising from the cell in which tautomeric shift occurred will contain the full point mutation. One out of the four cells will contain the half-mutation:-

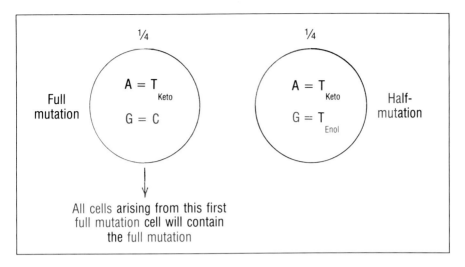

The cells arising from the half-mutation cell inherit the following base pairs during mitotic cell divisions:-

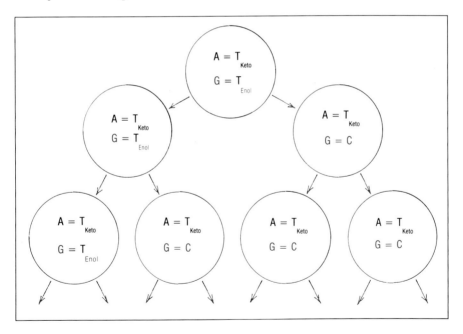

The following diagram shows that the proportion of cells containing the full mutation increases as the number of cell divisions increase

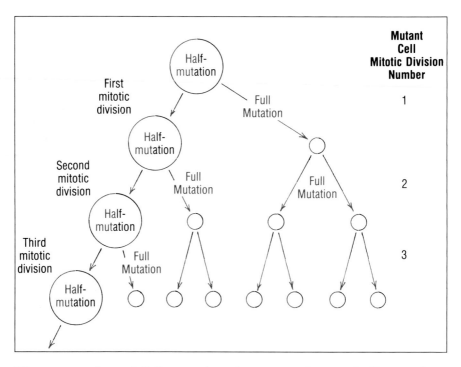

The proportion of full mutations by comparison to half-mutations in the cell arising from the one in which tautomeric shift occurred can be calculated from the formula

$$\text{Full mutation} : \text{Half mutation}$$
$$\Sigma (2^n - 1) : n + 1$$

Σ is the symbol for 'the sum of'.
n is the number of mitotic cell divisions

The assumption with this formula is that the rate of mitotic cell divisions is the same for every line of cells. In reality this is rarely the case.

The application of the formula for one mitotic division gives the ratio:-

$$2^1 - 1 : 1 + 1$$
$$\text{or} \quad 1 : 2$$

The ratio after two mitotic cell divisions is:-

$(2^2 - 1) + (2^1 - 1)$: $2 + 1$
or $\quad\quad\quad 3 + 1$: 3
or $\quad\quad\quad\quad 4$: 3

After three mitotic divisions the ratio is:-

$(2^3 - 1) + (2^2 - 1) + (2^1 - 1)$: $3 + 1$
or $\quad\quad\quad 7 + 3 + 1$: 4
or $\quad\quad\quad\quad 11$: 4

The application of the formula leads to the conclusion that the full mutation goes into nearly half of all the cells derived from the one in which the tautomeric shift originally occurred. This is because one quarter of the cells arising from the cell in which the point mutation arose contain the full mutation. Of the cells that develop from the quarter of the cells containing the half-mutation, a very high proportion will have the full mutation if several generations of cells are produced.

Tautomeric Shift in Asexual Species

A similar mathematical approach would, of course, apply to asexual diploid cells that arise by mitosis from a tautomeric mutant cell. However, in the case of unicellular, asexual diploids, the success of each independent cell would be subject to greater variations in genetic drift and natural selection than those in multicellular organisms. These two factors would cause unknown variables to alter the straightforward mathematics, which assumed a zero mortality of cells and a uniform rate of mitotic divisions in every cell line. The calculation of the probability of any one point mutation having an evolutionary effect is therefore a complex one.

Mutations Derived from Gamete Mother Cells

One of the types of point mutation that can be passed into gametes is the one that occurs during the mitotic development of an organism and which becomes incorporated in some or all of the gonad mother cells from which gametes are derived.

As has already been seen, a tautomeric point mutation will only be found in one of a pair of chromosomes in gamete mother cells.

Replication and crossing-over during the first meiotic cell division can place the tautomeric mutation on a chromosome different from the one in which it was found in the mother cell. The different fates of the replicated point mutation during meiosis will be determined by a number of factors:-

i. Whether or not all gamete mother cells contain the mutation.
ii. Whether or not one or both mutations go into an infertile product, such as a polar body.
iii. Whether or not the gamete is fertilised.
iv. Whether or not organisms that inherit the mutation in subsequent generations are viable in all aspects of their development from the zygote.
v. Whether or not genetic drift allows the spread of the mutation.
vi. Whether or not natural selection favours the mutation.
vii. Whether or not the mutation becomes the predominant allele in the local or general population.
viii. Whether or not the mutation becomes the predominant allele of the species.

Mutations Arising During Gamete Formation

Whereas a high proportion of gametes derived from gonad mother cells containing the full point mutation will themselves contain the full point mutation, mutations that arise during meiotic DNA replication itself may only become incorporated as a half-point mutation in one fertile gamete. The probability of such a small number of gametes being one of those that are fertilised is low, especially for male gametes which are produced in enormous numbers simultaneously.

For a half-point mutation that is passed into the zygote to be converted into the full mutation, the zygote must divide. The proportion of cells containing the half mutation and full mutation follow the rules of somatic cell division:-

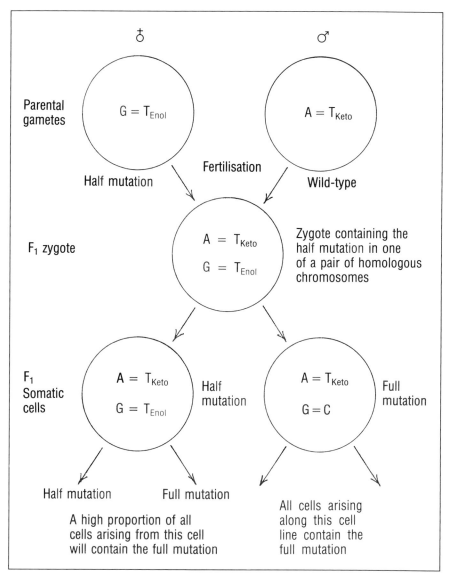

In fact, a high proportion of cells produced by mitotic cell divisions from a half-point mutation zygote will contain the full mutation. The probability of the gonad mother cells that develop in such an individual containing the full mutation is therefore high. Of course only one of two chromosomes in the gonad mother cells will contain the point mutation.

Genetic Drift

When a half-mutation occurs during gamete production the chance factor of genetic drift demands that individuals containing the full mutation in the second generation produce offspring that are homozygous for the full point mutation:-

	♀	♂
P	+ −	+ −
gametes	+ or − ×	+ or −
F$_1$ genotype	+ + : + − : − −	
F$_1$ genotype ratio	1 : 2 : 1	

Homozygous full mutants are then tested by natural selection. So long as there continue to be homozygous full mutants or heterozygotes produced in every subsequent generation, the mutation will persist. Many millenia may be needed to make a point mutation more successful than the established wild-type allele, thereby displacing it as the new wild-type allele for the species.

The Rate of Amino-acid Substitutions to Make New Genes

From the limited information available it is difficult to be precise about the rates at which amino-acid substitutions give rise to new genes. A new gene in this context is a mutant allele that becomes the predominant, wild-type allele for the entire species. Once a successful new allele has been made by amino-acid substitution it has to spread through a population, then throughout the species to become a new, invariable gene.

Some geneticists believe that the rate of creating new genes of this sort is as high as one new gene in each species every two years. Others believe that the rate is as low as one new gene for every 300 generations of a species. Environmental factors may have some influence on the rate of new gene formation by point mutations, and there is then genetic drift and natural selection acting on the mutants.

It should be noted that short generation times for unicellular species make one new gene per 300 generations faster than one new gene every two years. One new gene per 300 generations is only a slow process for those species with several years between each generation.

When more evidence becomes available from the amino-acid sequencing of widespread polypeptides it will be possible to take account of:-

i. The total number of genes in each species.
ii. The mutations and reverse mutations indicated by amino-acid sequences in widespread polypeptides.
iii. The dating of the mutations by comparison studies, and attempting to relate existing species to each other and to their fossil ancestors.

These factors will allow the mathematical calculation of a better and better approximation to the true rate of producing new genes by point mutations.

Unit Evolutionary Time

This is the average time for one viable new amino-acid to arise in a polypeptide per 100 amino-acids in the polypeptide. The study of some widely spread polypeptides has provided the information in the table.

Polypeptide	Unit Evolutionary Time ($\times 10^6$ years)
Fibrinopeptide	1·2
Haemoglobin	6·1
Cytochrome c	21·0
Histone H_4	600·0

The arising of a new viable amino-acid is usually by the substitution of bases in the material of inheritance. Sometimes it can occur by the addition of bases.

Evolutionary Trees Deduced from Differences in Amino-acid Sequences in Widespread Polypeptides

Comparisons can be made between cytochrome c taken from a wide range of species. The first attempt at constructing an evolutionary tree assumes that there are an equal number of successful single base mutations per unit of time, that there are no successful reverse mutations, and that the successful forward mutations lead to the establishment of a new gene which replaces the old one throughout the species. There are too many assumptions here and it will be necessary to collect comparative amino-acid sequences from a wide range of different, widespread polypeptides to verify or refute these assumptions. Composite evolutionary trees, improved by the information provided by each polypeptide study, will give a better perspective on the validity of the assumptions. The composite information should reveal reverse mutations, and give better approximations for the period of evolution in which the mutations took place.

The first attempt at constructing an evolutionary tree is made by comparing the amino-acids which occupy each site in a polypeptide that is widespread. In general, the larger the number of amino-acid differences at comparable sites, the further is the distance of a common ancestor, and the further apart is the relationship of living species from each other:-

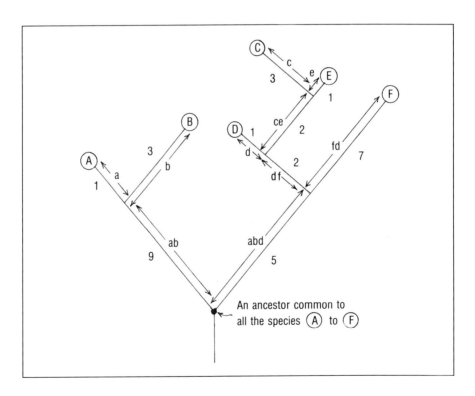

The capital letters indicate species. The small letters indicate distances from common ancestors that have to be worked out using simultaneous equations. The numbers represent the number of differences in the amino-acid sites of a polypeptide that is common to species Ⓐ to Ⓕ.

A study of the amino-acid sequences in a polypeptide common to species Ⓐ, Ⓑ, Ⓒ, Ⓓ, Ⓔ and Ⓕ, showed the following numbers of differences:

Species	Number of Amino-acid differences	Letters Used on The Evolutionary Tree
Ⓐ and Ⓑ =	4	a+b
Ⓐ and Ⓓ =	18	a+ab+abd+df+d
Ⓐ and Ⓒ =	22	a+ab+abd+df+ce+c
Ⓐ and Ⓔ =	20	a+ab+abd+df+ce+e
Ⓐ and Ⓕ =	22	a+ab+abd+fd
Ⓑ and Ⓓ =	20	b+ab+abd+df+d
Ⓑ and Ⓒ =	24	b+ab+abd+df+ce+c
Ⓑ and Ⓔ =	22	b+ab+abd+df+ce+e
Ⓑ and Ⓕ =	24	b+ab+abd+fd
Ⓒ and Ⓔ =	4	c+e
Ⓒ and Ⓓ =	6	c+ce+d
Ⓒ and Ⓕ =	14	c+ce+df+fd
Ⓓ and Ⓔ =	4	d+ce+e
Ⓓ and Ⓕ =	10	d+df+fd
Ⓔ and Ⓕ =	12	e+ce+df+fd

Simultaneous Equations Using the Information Supplied By The Amino-acid Sequence Study For Finding The Evolutionary Distances Between the Species

The following rather tiresome sequence of simultaneous equations is worth following because they show how the number of differences between amino-acids found in different species can be used to infer the evolutionary distance between species.

$$a+b=4 \quad ----- \quad (1)$$

and

$$a+ab+bd+fd=22 \quad --- \quad (2)$$

$$b+ab+bd+fd=24 \quad --- \quad (3)$$

Subtracting (3) from (2) $a-b=-2$

From (1)

$$\begin{array}{r} a-b=-2 \\ a+b=4 \\ \hline \end{array}$$

Adding

$$\begin{array}{l} 2a = 2 \\ a = 1 \end{array}$$

and

$$\begin{array}{l} b = 4-1 \\ = 3 \end{array}$$

From the amino-acid sequence study:-

$$c+e = 4 \quad --------- \quad (4)$$

$$c+ce+d = 6 \quad --------- \quad (5)$$

$$ce+d+e = 4 \quad --------- \quad (6)$$

Subtracting (6) from (5) :-

$$c-e = 2 \quad --------- \quad (7)$$

Adding (4) and (7) :-

$$\begin{array}{r} c+e = 4 \\ c-e = 2 \\ \hline \end{array}$$

Adding

$$\begin{array}{r} 2c = 6 \\ c = 3 \end{array}$$

Putting this value of c into (4) :-

$$\begin{array}{r} 3+e = 4 \\ e = 1 \end{array}$$

From the amino-acid sequence study:-

$$d+ce+c = 6 \quad \text{-------} \quad \text{⑧}$$

But we know that $c = 3$.
$$d + ce = 3 \quad \text{-------} \quad \text{⑨}$$

Also from the amino-acid sequence study:-

$$b+ab+abd+df+ce+e = 22 \quad \text{-------} \quad \text{⑩}$$
and $$a+ab+abd+df+d = 18 \quad \text{-------} \quad \text{⑪}$$

But we know that $a = 1 : b = 3 : e = 1$

From ⑩ $3+ab+abd+df+ce+1 = 22$
$$ab+abd+df+ce = 18 \quad \text{-------} \quad \text{⑫}$$

From ⑪ $\quad 1+ab+abd+df+d = 18$
$$ab+abd+df+d = 17 \quad \text{-------} \quad \text{⑬}$$

Subtracting ⑬ from ⑫ :- $\quad ce - d = 1$

$$\begin{aligned} ce-d &= 1 \\ ce+d &= 3 \quad \text{from ⑨} \end{aligned}$$

Adding $2ce = 4$
$\therefore \quad ce = 2$

and $\quad d = 1$

From the amino-acid sequence study:–

$$a+a+abd+fd = 22 \quad \text{-------} \quad \text{⑭}$$
and $$a+ab+abd+df+d = 18 \quad \text{-------} \quad \text{⑮}$$

We know that $a=1$ and $d=1$

Substituting for a in ⑭:-

$$1+ab+abd+fd = 22$$
$$\therefore \quad ab+abd+fd = 21 \quad \text{-------} \quad \text{⑯}$$

Substituting for a and d in ⑮:–

$$1+ab+abd+df+1 = 18$$
$$\therefore ab+abd+df = 16 \quad ⑰$$

Subtracting ⑰ from ⑯:–

$$fd-df = 5 \quad ⑱$$

We also know from the amino-acid sequence study:–

$$d+df+fd = 10 \quad ⑲$$

and we have already worked out that d=1

$$\therefore df+fd = 9 \quad ⑳$$

Solving the simultaneous equations for ⑱ and ⑳:-

$$\begin{aligned} fd-df &= 5 \\ fd+df &= 9 \\ \text{Adding } 2fd &= 14 \\ fd &= 7 \\ \text{and } df &= 2 \end{aligned}$$

From the amino-acid sequence study:-

$$a+ab+abd+fd = 22 \quad ㉑$$

We know that a=1 and fd=7

$$\therefore ab+abd = 14 \quad ㉒$$

It is not possible to work out ab or abd without continuing the study and extending the tree back to a more ancient common ancestor, and making further amino-acid sequence comparisons between species which diverged from the more ancient common ancestor.

In the example of an evolutionary tree on page 363 the figures for ab and abd have been derived by using simultaneous equations obtained from an extended study of the evolutionary tree.

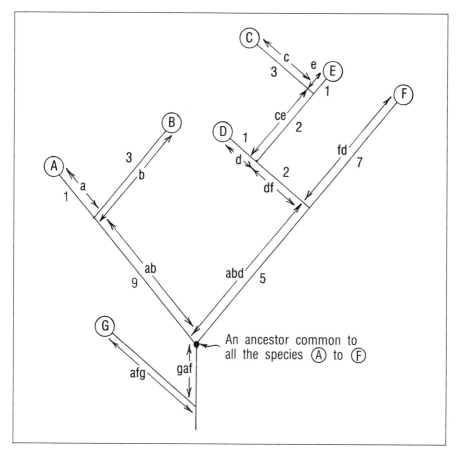

An extended evolutionary tree, for which the amino-acid differences between Ⓖ and Ⓐ, and between Ⓖ and Ⓕ are known, allows simultaneous equations to show that ab = 9 and abd = 5.

Using the assumptions made at the beginning of this chapter, an evolutionary tree has been made relating several species by counting the number of differences in the amino-acid sequences of cytochrome c. It is not thought to be suitable to include the tree here (most university books on genetics or biochemistry include it) as much refinement is needed before any reliability can be placed upon its accuracy. In a broad sense the degree of divergence between the phyla and between the species, as indicated by the amino-acid sequences, agrees with the divergences indicated by the fossil record.

Among other polypeptides studied so far are:-

Insulin A and B
α haemoglobin
β haemoglobin
Foetal haemoglobin
Ribonuclease
Fibrinopeptide A

Evolutionary Mutations Inferred by Comparing The Globins That Carry Oxygen in Different Species

A comparison of the oxygen carrying globins in different species has provided some inferred evidence about the evolutionary formation of the genes that code for the globins. For example in humans there are four different haemoglobins – α, β, γ, and δ. These polypeptides form tetramers (quaternary structures made of four associated haemoglobin molecules) in which two different pairs of haemoglobin molecules associate together, usually providing four haem- rings for the attachment of oxygen. Humans also inherit the gene for myoglobin, which is a molecule that carries oxygen from red blood cells to where the oxygen is used for aerobic respiration in cells.

A more detailed examination of human globins shows the associations of α, β, γ, and δ haemoglobin at various stages of the human life cycle. The globins in humans are then compared with the globins in other phyla, or other species, to infer the ancestral distance in time of mutations which gave rise to the different globin genes.

Human Haemoglobins

The four haemoglobins found in humans associate, in the ways shown on page 365, to form three different tetramers. All the tetramers contain two molecules of α haemoglobin, but each one has a different pair of haemoglobin molecules in the other two positions in the tetramer.

i. Haemoglobin A

This haemoglobin is made of two α polypeptide chains and two β polypeptide chains:-

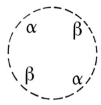

The α chain contains 141 amino-acids.
The β chain contains 146 amino-acids.

ii. Haemoglobin A_2

About 2½% of adult human haemoglobin is haemoglobin A_2. It differs from haemoglobin A by replacing β haemoglobin with γ haemoglobin:-

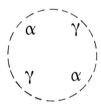

γ haemoglobin has the same number of amino-acids in it as β haemoglobin, but there are differences in 10 of the amino-acid sites.

iii. Haemoglobin in The Human Foetus

About 75% of the haemoglobin tetramers in the human foetus, and in a new born baby contain haemoglobin δ in place of β:-

The δ polypeptide differs from β in 39 positions. There have to be a number of control genes to switch off the δ gene and switch on the β gene after birth.

Human Myoglobin Compared to Myoglobin in Other Species

Myoglobin is made of a single polypeptide chain. Comparisons of the amino-acid sequence in human myoglobin with myoglobin taken from a wide range of other species, some of very ancient descent, indicates that myoglobin itself is of very ancient origin. It is thought that the gene that gave rise to all the oxygen-carrying globins in humans existed as a myoglobin gene in a primitive animal species more ancient than lampreys.

The Comparison of Human Globins With Globins From Other Species to Infer the Phyla in Which the Evolutionary Mutations of the Globin Genes Occurred

Myoglobin is common to a wide range of contemporary phyla. The gene for myoglobin, was, it is believed, the ancestral gene from which the genes for the different haemoglobins evolved. In addition to myoglobin the following species or phyla, from which man eventually evolved, contain the following haemoglobins:-

- Lampreys – only haemoglobin α
- Bony fish – haemoglobins α and γ
- Mammals – haemoglobins α, β and γ
- Primates – haemoglobins α, β, γ and δ

It is thought that the duplication of the original myoglobin gene was followed by different modifications to each of the duplicates. There then evolved the quaternary structure tetramer haemoglobins, followed by further duplications and further selection of genes through the selection of the phenotypes expressing the genes, to give rise to the several globins found in primates, including humans, today.

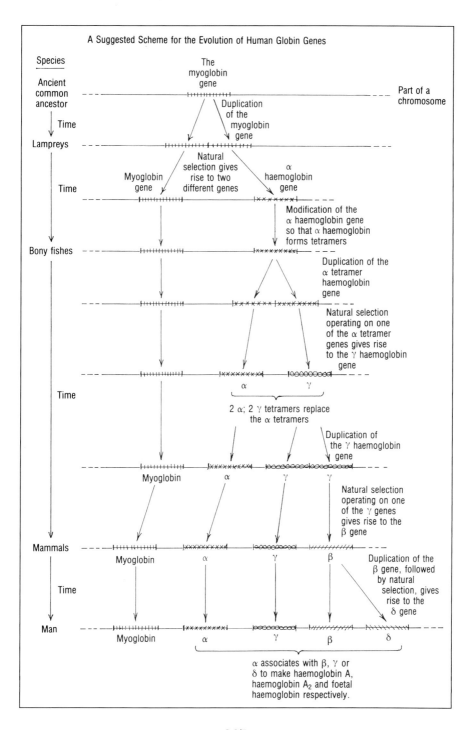

Conclusion

The comparative study of the globins shows how the evolution of new genes might arise after gene duplication, mutations, genetic drift and natural selection. Comparisons of genes that have arisen from a common origin can be made by examining some of the widespread polypeptides taken from contemporary species, and in some cases inferences can be made to relate contemporary species with the common ancestor in which the arising of new genes occurred. This type of study adds to the information obtained from comparisons of other widespread polypeptides taken from different species.

Chapter 4 made a brief reference to repeating genes. The globin example of this chapter shows how repeating genes can be of benefit for evolution. It would seem that mutations following gene duplication allow the original gene to continue fulfilling its function and also allow the creation of new viable genes. New viable genes create divergence between species and also increase the complexity of nature as a whole.

Chapter 11 Questions

1. What categories of change to polypeptides can result from point mutations?
2. Do all point mutations cause changes to the polypeptide governed by point mutation genes?
3. Which amino-acid substitutions in a polypeptide are likely to be viable and which inviable?
4. Which localities in a polypeptide are most likely to be critical for its function?
5. Has a satisfactory mathematical model been found for evaluating the probability of one amino-acid successfully substituting for another in a polypeptide?
6. Is sufficient evidence available to allow a good estimate to be made of the rate of new allele creation by point mutations?
7. How are evolutionary trees deduced from differences in amino-acid sequences found in functionally identical polypeptides widespread among many species?
8. What evolutionary information has been provided by comparative studies of the globins?

12

EVOLUTION

The list of references for this chapter is divided into two parts. Evolution is a complex study and has many different aspects. There is, therefore, a fairly long list of references within the main chapter 12 bibliography, and this constitutes the first list. If students would prefer to limit their reading in evolutionary studies, they could concentrate their reading on the five books marked with an asterisk in the main chapter 12 bibliography. The second list of references in evolution is presented as a supplementary reading list for chapter 12. This second list is divided into specialised evolution topics, the headings of which are among the most important to consider for finding specific examples of each aspect of evolution. The supplementary reading list for chapter 12 can be found at the end of the chapter 12 bibliography.

Ecological and laboratory population studies that have identified some of the processes of evolution have only been carried out for about a century. The astonishing variety of life on earth has given rise to an enormous number of genetic events that have the potential to both increase variety among individuals and to separate groups of individuals reproductively. The difficulty for ecological geneticists has been to identify those aspects of evolution that can be given a proper scientific study.

A number of population studies are given later in this chapter to demonstrate some of the principles by which evolution can occur. Inevitably, some of the discussion on evolution includes some modern discoveries about the properties of the material of inheritance.

The famous principle of 'environmental selection' as a mechanism for evolutionary change was, of course, put forward in 1858 by Charles Darwin and Alfred Wallace. Sadly, the obscurity of Gregor Mendel's work on 'inheritance particles' did not allow either Darwin or Wallace to consider inheritance at the cellular level. Now, more than a century later, considerable progress has been made in understanding the nature of the material of inheritance. We also know something about the passing of DNA from one generation to another. Discoveries in the science of genetics have shown that there are several factors other than 'environmental selection' that determine the biological inheritance of wild populations. These factors therefore control evolution and each of them needs careful consideration.

Mutations

Chapter 10 discussed the modification of DNA by mutations. Some mutations are very local (point mutations). Some other mutation categories involve quite large changes to a chromosome. Other mutations, such as a change in ploidy, can be very large indeed.

Sometimes chromosomes are rearranged by translocations, fusions or fissions. Even though all the original genes are retained, and no extra ones are added, rearrangements can lead to changes to the expression of some of the genes because of gene deregulation.

Additions, deletions (rarely) and inversions can all be viable mutations. Examples of each mutation type have been found in natural populations.

Recombination and Best Sets of Alleles

Even in the absence of mutations, the scrambling of existing alleles at meiosis in sexual species can lead to the natural selection of 'best sets' of alleles. This can lead to the arising of super-alleles such as those that control the light and dark colouration of peppered moths (Biston betularia) and the colouration of snails (Cepaea nemoralis).

The So-called Invariable Genes

The question arises 'should the invariable genes in a species be regarded collectively as a super-super-gene?' A second question then is 'does species viability depend upon the super-super-gene'? Certainly there seems to be a large proportion of the genome of each species that is invariable.

The so-called invariable genes may not be as invariable as was once thought. A further discussion on the species genome occurs later in this chapter.

Changes to Allele Frequencies in Populations

One pre-requisite for evolutionary change in a population is a change of allele frequency. If the evolutionary change is to persist with time, the allele frequency change must not be reversed later.

In order that evolution can occur, there must either be viable mutation (rare), or there must be a proportion of genes that have scramblable alternative (or several) alleles. Many of the scramblable alternative (or several) alleles will have been inherited from previous generations of the same species. Some, perhaps many, will have come from ancestral species. A small number of the alternative (or several) alleles are created within the time-span of the species itself.

The Creation of a New Invariable Gene

A few of the alleles that arise by mutation during the life-span of a species will become exclusively successful. Such alleles will replace all other former alleles for the gene. In cases like this, the new all-excluding alleles will join the large tribe of genes without alternative or several alleles. It will become part of the species' 'super-super-gene'.

The Evolution of Races and Species

It has already been seen in Chapter 10 that most mutations are damaging to the individual in which they arise, or to those individuals that inherit the mutation. Those mutations which are not damaging are often neutral in their immediate effects, even if they may improve the best set of alleles later by recombination. Very occasionally beneficial alleles arise through mutation.

Whether the cause of changes in allele frequency is mutation or the environmental selection of recombined alleles or super-alleles, there are often two main stages in the evolution of a new species:-

i. An existing species is divided into two or more races.
ii. There may be sufficient divergences of race, caused by mutations, genetic drift and natural selection, to lead to the emergence of a new species.

Genetic Drift

Genetic drift is the spread of an allele through a population. The allele may arise in the population either by mutation or by the arrival of a migrant with an allele that is new to the local population. The spread of the new allele through a population is a chance event. Genetic drift is discussed in more detail in relation to population size later in this chapter.

Classification

Nature is both rather untidy and vexatiously subtle. Many difficulties have had to be overcome by taxonomists when they attempt to classify the millions of species that inhabit our planet. For example, emerging cytologocial evidence has forced taxonomists away from the traditional animal/plant kingdom classification, and arguments continue about the exact number of kingdoms, and which species should be allocated to each kingdom.

Many early attempts at species classification have had to be changed. Systematic modern classification of living things depends on morphology, physiology, cytology, ethology (all aspects of behaviour), gamete compatibility and hybrid viability. Even so, the 'joiners' and the 'splitters' still have much to do to establish the true classification of many species, sibling species, semi-species, sub-species, super-species and races.

A classification system using a graded scale for each of the criteria mentioned in the previous paragraph would not necessarily have to replace the present system of Linnaean classification, which uses Latin names to distinguish between groups. However the rapid and enormous accumulation of data on computer tapes could be used for additional understanding of the relationships between groups of individuals in a way that would be more satisfactory than the words 'race' or 'sub-species'.

Ultimately all living things are descended from successful individuals which were themselves descended from a common ancestor, or from a small number of common ancestors. The relationship of all living things stretches back more than three-and-a-half thousand million years. The original efforts by taxonomists to classify the relationships by morphology can now be much improved by the combined studies of cell biologists and observers of population genetics.

Population Genetics and Evolution

The remainder of this chapter attempts to give a logical and simple introduction to the rather difficult and elusive subjects of population genetics and evolution. To begin there is a discussion about races and species. There follows a brief summary of the processes of speciation. Some examples of species that cause difficulties in their classification are then described. A discussion about establishing a new mutation by genetic drift is next. This is followed by a discussion about the breeding mechanism and its significance for the establishment of a new mutation. Population size, genetic drift and natural selection are then described in their inter-relationships. Finally, some examples are given of field and laboratory studies of evolution at work.

Races

Races can and do exchange genes from time to time. The isolation of races geographically does not necessarily create a permanent isolating barrier for gene exchange. Migration can throw individuals from each race together again after a period of isolation from each other, so allowing the exchange of genes between the races once more.

The races of man have emerged from geographical isolation followed by genetic drift and environmental selection. Modern migrations of large numbers of individuals from different races between the continents and islands have resulted in a considerable number of 'mixed' marriages and gene exchange between human races. The children of mixed marriages usually show less obvious racial characters than their parents.

For many human characters there are little or no differences between the races. The mean values of a distribution curve produced for each race for very many characters show almost no differences at all. However, the spread of the distribution curve for the same characters within each race if often quite large, certainly very much larger than the differences between race mean values. Nonetheless, there are some human populations which used to live in isolation, that do show extraordinary differences from other races. The blood group proportions in each race show considerable differences, each blood group having a different resistance to endemic diseases in different parts of the world. Many other examples of obvious racial distinctions are known. Among these are the diminutive size of Congo (Zaire) Basin pygmies, the size and shape of buttocks for fat storage in a hostile region (steatopygia) among the indigenous tribes of southern Africa, and skin pigmentation for protection against ultra-violet radiation and skin cancers caused by the sun. These local characters do not prevent cross-breeding between the races.

For centuries man has been selectively breeding races within those species that are of use to him. The varieties of plants and the breeds of animals can nearly always give rise to hybrids when crossed with other varieties or breeds of their species. Many of the hybrids can give rise to viable offspring.

It is not known how many groups of individuals, now classified as species, could and would interbreed with other closely related groups that are now classified as different species. The rejunction of groups, at present isolated from each other, is a necessary pre-requisite for hybrid production. Such rejunction may never happen in the wild. If rejunction of isolated racial groups never happens, the emergence of two species from two genetically isolated races is likely.

Species

The species concept needs to be considered under two main headings:-

   i. The composition of the species genome – the 'species gene' concept.
   ii. Speciation in the absence of gene exchange.

i. The Composition of The Species Genome – The 'Species Gene' Concept

Many of the genetic properties and potential genetic properties of the non-allelic genes of a species' genome are unknown at present. For example, is the so-called invariable nature of these genes really invariable? Or would it be more accurate to say that, in higher species, the exons which encode the functional region(s) of a gene's polypeptide are invariable?

The base sequences in those parts of exons that do not encode the functional regions of polypeptides have been shown to be quite variable in the genes of several polypeptides studied up to now. However, the variability of the DNA that encodes for the polypeptides under study (examples were given in Chapter 11) tends to vary between different species rather than within one species for the functional region codes.

Even if the exons in non-allelic genes of one species only show very limited variation for the functional codes, the non-functional codes are known to show much more variation in the exons. Even more so can the introns of so-called non-allelic genes vary, so long as the introns are never used for encoding amino-acid sequences, without disturbing the inheritance property of the gene.

Are introns never used for inheritance? For example, are the base sequences in the introns of genes that overlap other genes used in a different way by the overlapped gene, by becoming part of an exon in the overlapped gene?

There is then the question of what can happen to the base sequences in repeating genes? In many cases, mutations in one of several repeating genes will be neutral in their effects. This is because the remaining repeating genes can be used to produce sufficient polypeptides, even at times of cellular stress. Occasionally, a mutant allele arising from repeating genes may have a damaging effect. However, mutations in repeating genes also give the possibility of creating advantageous new alleles that add variability and complexity to a species. Such an advantageous gene could also start genetic processes that eventually lead to speciation. Seemingly invariable repeating genes could, therefore, conceal a considerable variety of base sequences in those parts of each repeating gene that do not encode for the functional part of the polypeptide product.

All living things evolved from very primitive forms of single cell organisms. Among the descendants of those primitive species, there must have been additions to the genome of some individuals of some species in some generations. Additions have been made by adding DNA fragments taken from other chromosomes, by adding repeating genes or by adding spacer regions which seem to code for no polypeptide products at all. If additions allow viability both to the individuals in which they arise and to subsequent generations, since life began there has existed a mechanism for increasing the length of genomes. In general this has led to the increased complexity of the living things which inherit the elongated genomes.

If, therefore, mutations occur to alter base sequences in a non-functional part of a gene, or if repeating genes can make good the damaging effects of a mutation in one of them, the apparent homozygosity of genes may be illusory.

Even if there is homozygosity in a gene, some genes change their polypeptide encoding properties by being mobile. When a mobile gene is found in one part of a genome it encodes for one particular polypeptide. When a mobile gene moves to another part of the genome it can encode for another polypeptide, or be switched off.

Mobile genes can be very beneficial when in one place on the genome, and very damaging (for example cancerous) when in another. Although there may be the full complement of so-called invariable genes still present, the genes adjacent to the mobile genes in their new locus are different. Adjacent genes are often controlling genes, so the control of translocated genes will be different in different loci.

Cells can amplify a proportion of their genes under genetic and other cellular controls. The production of a large number of repeating genes by gene amplification can alter the expression of the amplified gene, without any need for alternative or several alleles. A seemingly non-allelic part of the genome once again gives rise to differential gene expression. Cancerous genes can arise as a result of gene amplification. Once again differential gene expression can come from seemingly homozygous genes.

With so many difficulties associated with the definition of a truly non-allelic region of a genome, is it right to try to define the large, seemingly invariable region of a species' genome as a 'super-super-gene' – the 'species gene'? Furthermore, it must be questioned whether or not the set of invariable genes in each species is the factor which separates one species from another? It would follow that allelic genes are those genes which confer a degree of variability on the expression of the species gene.

At the time of writing this book, no-one has analysed all the invariable genes of two closely related eukaryotic species. Only by doing so for several pairs of closely related species will it be possible to find out if the invariable genes do, in fact, determine the genetic isolating mechanism of speciation. Perhaps, like much of nature's perverseness, in some cases of speciation they do, and in others they don't.

ii. Speciation in the Absence of Gene Exchange

The definition of species by reference to differences in external morphological features is always a temptation. However, in modern times, it has become apparent that the morphological differences between groups of individuals do not always supply the best ways to distinguish between species. It is more correct to define species by establishing that there are two groups of individuals which never exchange genes in the wild, even if they were to meet after a period of isolation from each other.

Of course, nature's vexatious anomalies have allowed some exceptional viable hybrid production in the wild. For example, in plants, there are some very surprising viable cross-pollinations between individuals that are classified as different species. The plant Valeriana is largely a European mainland species. However on the edge of its range in England, it has produced an octoploid. Octoploidy has allowed chromosome compatibility with a different local plant species in England. Viable hybrids are produced from this cross between two 'species'. The accepted definition of species is clearly transgressed in this case.

Another reference to anomalous hybrid production is given in chapter 16, in which, it seems, that viable hybrids can be produced by matings between monkeys of different 'genera'. Again the conventional views on the definition of species are flouted.

Reproductive Barriers that Cause Speciation

Whatever the problems are for defining races and species, there are some distinct reproductive barriers that give rise to speciation. These can be divided into primary reproductive barriers and secondary reproductive barriers.

i. Primary Reproductive Barriers

Primary reproductive barriers are those that *prevent fertilisation*. Some of the more common pre-fertilisation barriers are:-

a) Geographical
If there is geographical isolation, without rejunction, during the speciation process, an effective reproductive barrier is created. Rejunction later, after divergence, may not reverse the genetic divergence caused by the geographical barrier.

b) Seasonal
The divergence of two groups of individuals genetically can occur without much physical separation geographically. Differences in photoperiodic responses by two groups that arise from common ancestors are not unusual. Different seasons of sexual maturity can lead to genetic isolation and speciation. Several flower species have the potential to cross-fertilise each other in the wild, but never do so because of seasonal isolation.

c) Ethological
Ethology includes all those behavioural aspects that are important for inducing mating by animals. Courtship rituals, mating calls and feeding habits can all operate, either as inducers of mating or as opposers of mating. The production of specific chemicals can have similar inducing effects. Even though fertilisation, development and the fertility of offspring are all possible (laboratory cross-fertilisations can demonstrate this), ethological barriers to hybrid production can be effective causes of speciation.

d) Mechanical
Some plant species which can be cross-pollinated artificially are isolated genetically in natural conditions because they require different cross-pollinating animal vectors to carry the pollen. As with other barriers to fertilisation, mechanical barriers can cause speciation.

e) Barriers Caused by the Failure of Male and Female Gametes to Form a Zygote
Mating between individuals with the instincts or vectors to pass gametes between them may not result in the fertilisation of eggs because the gametes are incompatible.

ii. Secondary Reproductive Barriers

Secondary reproductive barriers are those that occur *after fertilisation*.

a) Inviability of The Hybrid
There may be the failure of every hybrid to pass successfully through every stage of the life cycle to maturity. There are, of course, an enormous number of reasons why hybrid viability cannot ever occur.

b) Infertile Hybrids
Infertile hybrids fail to pass on their genes. For example, horses and donkeys can cross-fertilise to produce mules. Hybrid mules are sterile, so horses and donkeys must be regarded as different species.

Co-adaptive Peaks of Fitness – A Method of Defining Species

Taxonomists have been squabbling for quite a long time over the definitions of species, sub-species, semi-species and sibling species. To these can also be added super-species and races. Dobzhansky, T. has made a good attempt to define the species concept. A slightly modified version of his definition is as follows:-

In a broad sense, each true species has a peak of co-adaptive fitness which allows each species to be successful among the other species that occupy its habitat and range. The co-adaptive peak of fitness of each species gives some of the individuals in it a viable co-adaptiveness with their physical, chemical and biological environment. Between the co-adaptive peaks of each species will be genetic gaps. The gaps represent those unsuitable gene combinations, in whatever ways they are expressed, that separate each species from its closest relatives. Barriers to hybrid production are to be found in the gaps. If such gaps exist or arise between co-adaptive peaks in wild populations, from whatever causes, then the gaps represent barriers to cross-breeding and give rise to separate species.

The discussion about the species gene has not, so far, mentioned mobile genetic elements. In some cases, it seems, genes can be taken from the cellular genome of one species, attached to a mobile genetic fragment such as a virus, and transported to cells of other species. The evolutionary implications of inter-species DNA transfers of this sort are discussed in Chapter 13. The conventional views of species being genetically isolated must be revised to take account of the transfers of DNA base sequences from one species to another. Dobzhansky's definition of species no longer seems tenable, without qualification.

The Origins of Genetic Isolation

Genetic isolation can arise either when populations are geographically separated from each other (ALLOPATRIC isolation), or when they inhabit the same place (SYMPATRIC isolation). PARAPATRIC (contiguous but not overlapping) genetic isolation can emerge from allopatric or from sympatric isolation.

i. Allopatric Isolation

Geographical isolation may lead to considerable differences in genetic drift and natural selection. Genetic divergence in isolation may be sufficient to create one or other of the barriers to reproduction, if ever the geographical separation were to discontinue.

Isolation may not at first create one of the breeding barriers. However, if rejunction occurs and hybrids are less viable than the individuals of formerly isolated groups, selection against the hybrids occurs, wherever they are produced. This can lead to more complete genetic isolation of the two groups. A complete barrier to reproduction may arise later.

Allopatric isolation, such as occurred in isolated, dry cra forest refuges for some tropical African and South American bird and butterfly species, can lead to a parapatric distribution pattern of species if rejunction occurs after a period of isolation. Competition between closely related species occupying similar habitats and ecological niches is avoided by parapatric species.

ii. Sympatric Isolation

Sympatry presents a number of problems for its accurate definition. For example many parasites occupy specific parts of their hosts. Many insects feed on specific plants. At a very local level, these parasite species could be considered to be allopatric.

There are nonetheless true sympatric genetic isolating mechanisms. Changes of photoperiod, changes of behaviour and cytological changes can all create barriers to reproduction. In particular, chromosome rearrangements can set up immediate and permanent sympatric genetic and therefore reproductive isolation.

Not all the cellular changes that cause genetic isolation need be as dramatic as chromosome rearrangements. There are sometimes very small genetic changes that set up sympatric barriers to breeding. For example, in maize, barriers to cross-fertilisation have been recorded after only a few generations of selective breeding, without any obvious cytological or morphological changes.

Sympatric genetic isolation is often followed by the redistribution of ranges of closely related species, once again resulting in parapatry and the reduction of competition.

Allopatry, Sympatry and Parapatry

i. Allopatry

Attempts to define allopatry are often defied by biological populations. For example, allopatric populations are conventionally thought of as those that are separated by a physical barrier to breeding. The separation of groups of individuals by wide expanses of ocean is an obvious example of allopatry. The chance arrival of Darwin's finches at the Galapagos Islands cut these birds off from their South American mainland contemporaries. The distances across water between the islands then isolated some birds from others. The subsequent evolution of the finches occurred in the unique environments of the several islands in what may have been total genetic isolation from the South American mainland.

What about parasites and their hosts? For example, man is host to many species of lice. Three of these parasite louse species occupy different parts of man's body. Phthiris pubis are found only in the axillae. Pediculus humanus is found only on more exposed skin surfaces and on clothing. Pediculus capitis is found only on head hairs. Are these three species allopatric? At the very local level, taking the louse eye view, it can be argued that the three louse species are largely allopatric. Many insect species only breed on specific host plants, in some cases rarely leaving their hosts. Such insects are allopatric, or largely so, at the local level at which they breed.

The first stage in the arising of host specificity is that genetic variation in a species allows host specific races to develop sympatrically. Bush (1975) studied tephritid fruit flies and was able to show that mate selection and host plant selection are both genetically controlled in this species. Only a few genes need to be different to cause differences of mate and host plant selection. It is thought that many host-parasite specificities have arisen quite recently by co-evolution.

In a diverse habitat there can develop considerable polymorphism in a sympatrically distributed species. This allows some differences in the ecological niches which can be filled by the species in the locality. The allele (or super-allele) frequencies in different ecological niches will be different, so establishing different sub-populations in each ecological niche. In some cases, the divergence of each morph from the others could lead to speciation. Allopatry can therefore develop from sympatry because of a heterogeneous habitat occupied by several morphs of a species.

ii. Sympatry

Sympatric species are those that share the same habitat, with overlapping ranges. However, care must be taken to establish which species are truly sympatric, taking account of three dimensional space and seasonal migrations. For example, sea-birds sweeping across the surfaces of oceans have an entirely different habitat from species that have the same geographical range, but live on the bottom of the ocean below them. It should also be noted that some closely related species which occupy the same habitat, do so at different times of the year, thereby reducing the interaction between them to a very low or non-existent level.

The simultaneous occupation of grasslands by intermingling species belonging to the great African herds, in places like the Serengeti of Tanzania or beside the Chobe River in Botswana, are examples of sympatric species.

iii. Parapatry

Some species have habitats and ranges that meet, but do not overlap. Wingless grasshoppers (Vandiemenella viatica) in contemporary Australia exhibit parapatry. It is believed that chromosome rearrangements occured at several different times among these morabine grasshoppers. Grasshoppers of different karyotypes, but closely related, could only survive if one or both of the karyotypes shifted their ranges so as to reduce competition in previously shared habitats. These species of different karyotypes never share the same range, even though the arising of a mutation must have enforced a shared habitat for a while. The occupation of contemporary ranges by these karyotypically different grasshoppers may be very different from the ranges of their common ancestors. Contemporary ranges give no clues about the places where chromosome rearrangements arose. In 240 species of these morabine grasshoppers about 100 different chromosome rearrangements have been found.

Another example of parapatric species' distribution is that of forest birds in South America. The mechanism for creating parapatry in this case was not chromosome rearrangement. Pipra aurelia, Pipra filicauda and Pipra fasciicauda are found in clearly defined, but not ecologically different, forest regions that do not overlap. With the exception of one small region of eastern Peru, where P. filicauda and P. fasciicauda do occasionally produce hybrids, these three species do not interbreed. There are no obvious ecological barriers to cross-breeding in the boundary zones. The likely cause of breeding barriers in this case was the retreat of the South American forest during the Pleistocene period. Isolated groups of birds, formerly of one species, diverged sufficiently in isolation to create barriers to breeding when a wetter climate later allowed trees to spread as uninterrupted forest. Competition between these closely related species forced them to occupy ranges that are parapatric.

The Broad Processes of Sympatric, Parapatric and Allopatric Speciation

There are quite a large number of allopatric, sympatric and parapatric sequences which can lead to speciation.

In addition to the processes for giving rise to allopatric, parapatric and sympatric speciation, there is a further genetic process which can lead to speciation. This is the creation of clines.

Clines are found in many populations. The term 'cline' describes a directional change of genotype and phenotype across widely distributed contiguous populations.

The genetic differences between adjacent individuals in a cline are usually (but not always) small. However, the differences become greater as the distance apart of individuals becomes greater. Even when there is considerable gene flow between adjacent individuals along the cline, there can, in some cases, be quite sharp (stepped) clines across a narrow geographical band. An example of a cline in North American frogs (Rana pipiens) is given later, to show that taxonomy can be made rather difficult by the existence of clines. Clines don't show the sharp divisions of parapatric speciation.

Of course, a large number of variations on the three types of speciation can be envisaged, depending on their order of occurrence and on the number of species that arise from an ancestral species. It must also be remembered that a high proportion of species become extinct.

For the purposes of making an illustration fairly simple, the diagram that follows shows an ancestral species that persists among its descendant species. Descendant species can arise sympatrically, or allopatrically and may become parapatric in their ranges later. Parapatric species can also result from rejunction after isolation.

It is difficult to estimate the relative importance of allopatric, sympatric and parapatric barriers to reproduction as the means of evolving new species during the long history of this planet. Not enough extinctions are shown in the diagram, the purpose of which is to summarise the processes of speciation.

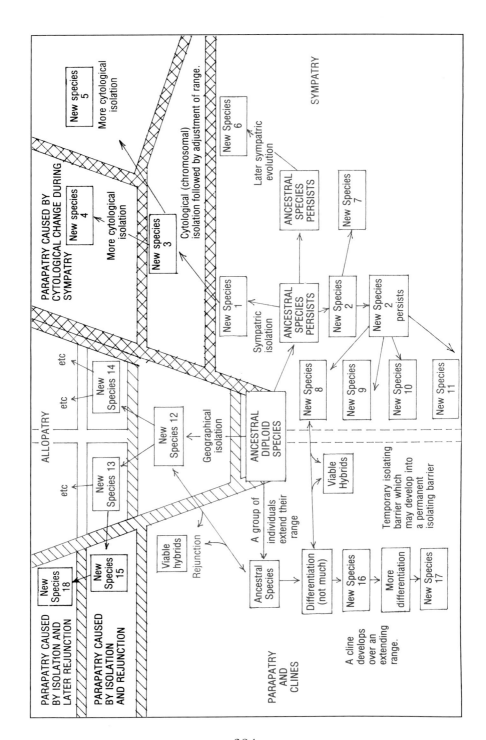

The number of species arising from an ancestral diploid species will, of course, vary for each of the three categories of speciation. There may be no new species arising in some of the categories. Over a long time period, most lines will become extinct.

Once the processes of reproductive isolation have started, the rate at which it actually occurs for the production of a new species will vary. The emergence of hybrid zones from shallow, steep or stepped clines, or from the readjustment of parapatric ranges, or from rejunction (if this occurs) will also take just that period of time it takes. The exact instant in history when each new species becomes or became reproductively isolated from its closest contemporaries, or specifically different from its ancestors, would be interesting to know. The unlikely discovery of such a precise instant in time is disappointing, but not very important. The study of a small isolated population may, at some time in the future, allow the accurate fixing of the moment in history when a new species separates from an ancestral species.

When a new species comes into being, or when an established species is modified, it has to adjust its range in the ecological situation of its co-evolution with other species in its habitat. Each new species and the ancestral species from which it emerged will either be successful, with varying degrees of genetic stability or diversity, or they will become extinct. Most of them will become extinct.

Classification Using Modern Data

The difficulties faced by taxonomists in ascribing an accurate classification to every individual living organism are enormous. There has to be continual revision of traditional classification methods to take account of more recently recognised criteria for distinguishing between groups of living things. For example, the study of animal behaviour has reduced the apparent number of bird species from about nineteen thousand at the end of the nineteenth century to an internationally agreed number in the late twentieth century of about eight thousand species.

As well as recognising races and species, it has also been necessary to recognise sibling species, sub-species and super-species.

Sibling Species

If there are pre- or post-mating barriers to cross-breeding between groups of individuals that are morphologically indistinguishable, the groups of individuals are divided into sibling species. Sibling species do not produce hybrids in the wild.

The medical profession has been caused considerable problems in distinguishing between sibling species that are vectors of disease. For example, it took some time to identify nine sibling species of European mosquitoes, Anopheles maculipennis, and to further identify which of these transmit malaria to humans. In Africa, five out of six mosquito sibling species can transmit malaria to humans and some of these can transmit elephantiasis as well. Anopheles gambiae is now known to be the most efficient vector for spreading both these diseases in humans. There is usually a distinct difference between the capacity of male and female mosquitoes to act as vectors of malaria, the female being the more efficient carrier of disease.

In tropical Africa, river blindness is spread by blackflies, Simulum damnosum. Cytological studies have shown that there are at least twenty-four sibling species of blackflies in tropical Africa. It is not yet known how many of these, nor of which sex, are vectors of human river blindness. Sandflies, protozoa and helminths are further examples of human disease vectors with sibling species.

There are, of course, many sibling species that are not vectors of disease.

The orders Orthoptera, Crustaceae, Nematoda and Acarina have many examples of sibling species.

Sibling species can be separated by detailed examination of:-

    i. behaviour.
    ii. seasonal differences.
    iii. mating partner selection.
    iv. cytological differences, often karyotype differences. Karyotype differences are differences in the chromosomes themselves. Staining reveals differences in karyotype.
    v. amino-acid sequencing as discussed in Chapter 11.

Sub-species

There is still considerable uncertainty about the classification of some groups of living things. 'Sub-species' has come to be used as the term for describing groups of individuals that have many of the characteristics of individuals belonging to other groups, or to one other group, with which they occasionally exchange genes. The rarity of cross-fertilisation or of producing viable, fertile hybrids together with clear morphological, behavioural or chemical differences between groups sometimes makes the term 'races' less than adequate while the term 'different species' is too definitive to describe them. The study of Rana pipiens, a frog that inhabits North and Central America (page 399) illustrates the difficulties faced by taxonomists in classifying groups of living things that show local differences, but which change gradually from one locality to another.

In the future it may be possible to improve the present classification system by taking account of:-

 i. The degree of viable genetic exchange between groups of individuals that occurs in wild populations.
 ii. The degree of potential hybrid production between groups of individuals, as shown by successful cross-fertilisation experiments in laboratories, even if they never cross-fertilise in the wild.
 iii. The degree of morphological differences between groups.
 iv. The degree of differences between polypeptides that are common to different groups of individuals.
 v. The degree of difference in behaviour such as mating calls, feeding preferences and photoperiodic responses.
 vi. Cytological differences such as differences in karyotype.

Sub-species are often connected by hybrid, transitional populations. European mice are divided into the light bellied sub-species, Mus musculus, which lives in Denmark, Scandinavia and Eastern Europe. Dark bellied mice of the sub-species Mus domesticus live in Western Europe. A narrow hybrid zone runs from the west coast of Denmark on the North Sea to the shores of the Adriatic, near Trieste.

Oak trees in California provide examples of plant sub-species. Three oaks, Quercus douglasii, Q. dumosa and Q. durbinella usually do not cross-pollinate. However, in extremely disturbed environmental conditions, such as when land is cleared of vegetation, these oaks show the characteristics of sub-species by cross-pollinating to produce viable hybrids. Fossils show that these sub-species have persisted for at least ten million years.

Super-species

When several very similar species arise from a common ancestor, they are often found as parapatric or allopatric species. Sometimes they have hybrid zones.

Crows belong to a world-wide super-species. Of these, two semi-species which live in Europe are hooded crows, Corvus cornix, and carrion crows, Corvus corone. A hybrid zone runs from eastern Denmark to Nice. Another hybrid zone runs from Edinburgh through the central lowlands of Scotland to Glasgow. Carrion crows inhabit the temperate regions of western Europe. Southern Britain and Spain are included in their range. Hooded crows live in Eastern Europe and Scandinavia. These two European crows have contiguous or overlapping ranges with other crow sub-species that extend the range of the crow species into Asia and Africa.

It is possible that super-species diverge into distinct species, no longer having hybrid zones where they meet.

Explosive Evolution

Explosive evolution results in the production of a very large number of related species from a common ancestor. Usually, individuals must spread quite widely from the range of their common ancestor to occupy a variety of ecological niches. Groups of individuals must then be isolated in several localities. Each locality must have quite a large number of potentially suitable habitats so that there is also a variety of ecological niches to be filled by each isolated group.

Cichlid fish in Africa provide an excellent example of explosive evolution. Before the geological events that formed the rift valleys of eastern and central Africa, these fish were distributed in some of the great African rivers and their tributaries. As a consequence of rift valley formation, cichlids could swim into the rift valley lakes and into other lakes formed during rifting but not in the rift valleys themselves. Many of the lakes were later cut off from the drainage systems of the great African rivers and from many other drainage systems. Isolation in their lakes has led to the evolution of about 520 species of cichlid fish in Africa, 500 or so of which are endemic in their own water beds.

The diagram on page 389 shows that cichlids constitute a high proportion of African freshwater fish. The information in the diagram is arranged as follows:-

    i. Figures in brackets show the proportion of endemic species in each water course.
    ii. Figures in circles show the number of non-cichlid fish *families* in each water course.

Explosive Evolution – Cichlid Fish in African Rivers and Lakes

(R)
NILE

C : 10 spp. (20%)
NC : ⑯ : 105 spp. (20%)

(R)
NIGER

C : 10 spp. (20%)
NC : ㉕ : 124 spp. (4%)

(L)
TURKANA

C : 7 spp. (40%)
NC : ⑭ : 32 spp. (16%)

(R)
CONGO
(ZAIRE)

C : 40 spp. (65%)
NC : ㉓ : 650 spp. (80%)

(L)
VICTORIA

C : 170 spp. (99%)
NC : ⑪ : 38 spp. (42%)

(L)
TANGANYIKA

C : 126 spp. (100%)
NC : ⑬ : 67 spp. (70%)

(L)
MALAWI

C : 200 spp (99%)
NC: ⑧ : 42 spp. (62%)

C = cichlids
NC = non-cichlids

Cichlids were probably among the earliest invaders of the large non-rift and rift associated valley lakes, when the lakes were at an early stage of being filled. Their invasion water courses from other drainage systems were later slammed shut by geological movements, so allowing the extraordinary preponderance of endemic cichlids in the great African lakes.

Explosive evolutionary divergence to produce species flocks is not uncommon.

Having discussed some of the broader aspects of evolution, it is now necessary to examine some of the more detailed aspects of establishing a new mutation or new set of recombined genes in a population by genetic drift. The breeding mechanism – sexual or non-sexual – and the cellular requirements for establishing a new mutation are both of importance.

The Breeding Mechanism

Some geneticists believe that asexual reproduction, or inbreeding always reduces the probability of the long term success of a species. They would argue that the lack of variation in asexual species reduces their capacity to co-adapt to changing environmental pressures. Certainly, in asexual species, there is likely to be a reduced adaptability by comparison to sexual species, because in asexual species there is the absence of recombination of allelic genes at meiosis. However, if enormous numbers of individuals are produced in each generation of an asexual species, mutations alone can sometimes create sufficient viable variation to allow the success of the species, at least for a time. Each different mutant line offers the species a degree of adaptability. In addition, the absence of meiosis reduces the problems of cell viability when chromosome rearrangements occur without the loss of genes. Asexuality may not, therefore, be quite as disadvantageous for the temporary success of the species as is sometimes supposed.

Many species, previously considered as asexual, are now known to have more than one breeding mechanism. Such species have quite a high capacity for creating variation and therefore have quite high adaptability. We already accept that sexual outbreeding, with crossings-over at meiosis, is an excellent mechanism for creating variation and through variation gives considerable adaptability to a species.

Inbreeding Mechanisms

i. Asexual Species

It is generally believed that those species that are truly asexual, without any alternative form of reproduction nor the capacity to generate any alternative form of reproduction, are doomed to extinction quite quickly. A low level of adaptability at a time of species crisis caused by environmental changes would encourage extinction.

ii. Asexual Species With Alternative Reproductive Mechanisms

Many species that were, at first, thought to be asexual are now known to have the ability to donate and receive DNA from other individuals of their species and sometimes from individuals of other species. For example, some bacteria species have a high level of adaptability by having alternative or several methods of reproduction. Many species of fungi and plants that have an asexual reproductive method can also reproduce sexually. Such species, especially bacteria, seem to be able to persist in their rather primitive forms for a very long time indeed.

iii. Parthenogenic Species

Aphids are well-known species that give birth to live young. The young emerge from the spontaneous mitotic divisions of totipotent (having the genetic, growth and development properties of the cell from which an entire new individual arises) cells in the bodies of females. This process is called parthenogenesis. Aphids have no sexual method of reproduction.

By producing an enormous number of offspring, there may be sufficient variation, caused by viable mutations, to give a degree of adaptability to aphid species. However, their key to long term success may be a reversion in the sexual mode of reproduction that was probably found in their ancestral species.

iv. Hermaphroditic Species

Hermaphrodites produce both male and female gametes in the

same individual. Inevitably there is some reduction in variability caused by very strict inbreeding in hermaphroditic plants and animals. In some cases, hermaphrodites have outbreeding alternatives for creating variability, thereby increasing adaptability. Hermaphrodites with outbreeding alternatives have a better chance of avoiding extinction than true hermaphrodites.

The Cellular Requirements for Establishing a New Mutation

1. Sexual Species

A few examples of viable mutations were given in Chapter 10. It is evident that, in an outbreeding sexual species, when a mutation occurs, it will only be viable in a first generation hybrid if:-

   i. The chromosomes donated by the mutant parent are compatible with the chromosomes donated by the other parent. Successful mitotic cell division, after the formation of the zygote in which a mutation exists, depends on chromosome compatibility.
  ii. The hybrid can produce gametes that are compatible with the gametes produced by non-hybrids, with which the first generation hybrid may have to cross-fertilise if the mutation is to be passed into subsequent generations.
 iii. The hybrid contains chromosomes for which there are no undesirable consequences of meiotic crossings-over.
  iv. The hybrid contains chromosomes for which there are no problems of independent segregation at meiosis.

It should be noted that hermaphroditic species are more likely than species with true males and females to accommodate the compatibility of chromosomes after mutations, and it is for this reason that viable multiple mutations, such as the arising of a new set of chromosomes, are found in self-fertilising species such as some flowering plants.

2. Asexual Species

Viable mutations in cells that give rise to the next generation arising from asexual species are inevitably passed into the cells of subsequent generations. Mutant individuals that arise by way of mitotic cell division from the parent mutant therefore contain chromosomes that are compatible at mitosis, having been tested in the mutant parent.

The Establishment of New Mutations in Populations of Sexual Species

The discussion about the establishment of new mutations is necessarily one which tries to identify general rules that could be applied to every sexual species of plant and animal. Of course there will be different forces of selection operating on the phenotypic expression of a new mutation every time a new mutation comes into existence. There will also be different chance factors that will either encourage or discourage each new mutation in its establishment in a population.

The argument on the following pages concentrates on the establishment of new mutations in populations, but the same argument could be applied to the introduction of an allele, which is already established in a population of a species, by a migrant into a local population where the allele had previously been absent.

i. The Simultaneous Effects of Natural Selection and Genetic Drift on the Evolution of Sexual Species

Once the chromosomal requirements for the viability of a mutation have been met, first generation hybrids will face two main problems which have to be overcome if the mutation is to be successfully transmitted into subsequent generations.

The first problem is the familiar one of selective pressures imposed by the environment. Each new hybrid has to be able to survive to maturity. Natural selection imposes rigorous demands on the hybrid, and its success depends to a considerable extent on the phenotypic expression of the mutation, at every stage of its life cycle.

The second problem is that of finding a mate, or of cross-fertilising a gamete from the opposite sex, and successfully breeding from it. This is a chance event, so the transmission of a successful mutation into a local population only occurs in the earliest generations if chance allows. The viable mutation will be lost if cross-fertilisation by chance does not occur. The spread of a viable new mutation in a population is called 'genetic drift'.

It is not possible to make a general statement that specifies the relative effects of natural selection by comparison to genetic drift. Each new mutation in every species is subjected to different forces of selection and there will also be different local factors affecting cross-fertilisation from every new hybrid. Two very important factors are:-

a) The size of the population in which the mutation occurs.
b) The degree of isolation of the local population from other individuals of the same species.

ii. The Number of Individuals in the Local Population in which a Mutation Occurs and the Relative Importance of the Founder Member of a New Local Mutation

In some species, individuals restrict their matings to a limited number of individuals in a local group. Other species choose their mates or are cross-fertilised from a much larger group of individuals. Mobility, or lack of it, is not necessarily the factor that determines the potential number of individuals with which cross-fertilisation will take place. Animal herds can be mobile over great ranges, but matings may be restricted to a small number of individuals within each herd. There may also be the isolation of small groups of individuals from the main body of a species, either on the periphery of its habitat or for some other reason within the habitat range of the species. If cross-breeding is restricted to a small number of individuals, a viable new mutation is likely to have a much greater effect per generation in changing the average genotype, and therefore the average phenotype, of the small, isolated group, than would occur if the mutation occurred in a large population. This general rule does not always apply because small, isolated groups are often more likely than large populations to die out at times of falling numbers, so that a proportion of viable mutations are lost for reasons other than the viability of the new mutation. At times of stable or increasing numbers, viable mutations in small populations can lead to a rapid change in the proportions of new phenotypes, provided that the early generation problems of genetic drift are overcome.

The success, or otherwise, of a mutation in a local population will depend on its selective advantage, neutrality or selective disadvantage. A mutation's success also depends on the stability or degree of change imposed by the environment in maintaining or altering the forces of natural selection on all the genes of individuals in the local group, and also on genetic drift.

In a very small population, a high proportion of matings or cross-fertilisations favour the spread of a viable mutation through the local group. In these circumstances genetic drift can be of greater importance than natural selection in establishing the mutant allele, even though the same mutation in a large population would be extinguished by the forces of natural selection. In some cases successful mutations in small populations spread quite quickly and become the homozygous allele for the mutant gene, so creating a change with considerable evolutionary significance.

The next stage of the evolutionary process is for meiotic recombination to 'scramble' locally established alleles. Natural selection then operates so that a 'best set' of alleles becomes established in each locality. Each local 'best set' may eventually become homozygous for some of the new, local mutant alleles, and each local 'best set' will be different from the others because each locality has different mutations and different forces of natural selection operating. This process can give rise to new races and eventually to new species. The diagram overleaf gives a summary of the processes of speciation:-

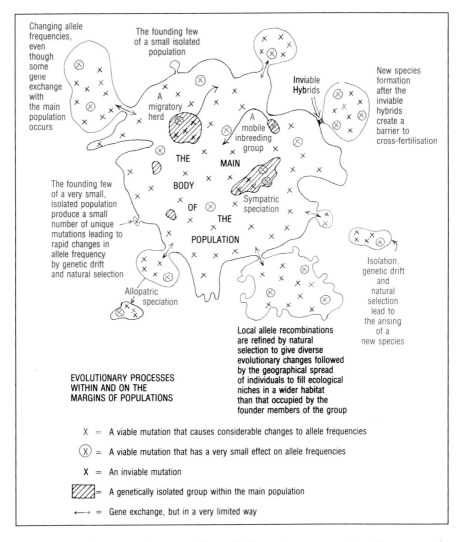

EVOLUTIONARY PROCESSES WITHIN AND ON THE MARGINS OF POPULATIONS

There are three main sets of conditions that control either genetic stability or genetic change in populations:-

i. Those in which all individuals of a species experience the full range of the habitat of the species. Each population or group will tend to resemble each other in such cases and, if there is widespread interbreeding between populations or groups, and if the environment in the habitat of the species stays the same, there will tend to be an equilibrum of allele ratios for all genes in the species. Such a situation is probably rare. Usually there are slow consequences of genetic drift resulting from mutations creating new viable alleles. In addition, natural selection nearly always operates in a continuously changing environment. For large, interbreeding populations, whose habitats stay largely the same year after year, there will be very slow changes in allele frequencies, and evolutionary change is therefore likely to be very slow within them.

ii. Those in which populations are restricted to distinctly defined localities or to small groups (which may be migratory), so leading to distinct local races in which there will be different local changes in the allele frequencies of a proportion of the genes. Genetic drift and natural selection are the causes of allele frequency changes. Allele frequencies can change as much as 25% per generation in very small, local groups whereas the rate of change of allele frequencies in large populations is very small per generation.

iii. Those which allow individuals of a species to adapt to a gradual, but directional change of environment. Changes of latitude can lead to gradual climatic changes over long distances. Such changes lead to different ecological and environmental pressures of selection. A gradual change of allele frequencies leads to changes in gamete compatibility, changes in phenotype and changes in behaviour. Adjacent groups can, and do exchange genes from time to time, with varying degrees of hybrid success. Adjacent groups are different races, all originating from one species. However, if individuals from groups that are more widely separated never exchange genes, then the widely separated groups must be regarded as different species. The gradual changes in groups over a wide geographical range in this way is called a 'cline'.

ECOLOGICAL GENETICS

Ecological genetics is the study of biological inheritance in wild populations. A selection of studies is given on the following pages. Having examined the selected studies, it will become evident that there are, as yet, only very simple criteria that have been used for supporting the theories of genetic drift and natural selection. Indeed, the separation between the ideal of following genetic drift and natural selection to the exact point when speciation occurs and contemporary populations studies is very wide indeed. Only the example of Agrostis tenuis (a grass) in the Welsh mine tip study, among those on the following pages, gives any direct evidence of laboratory tested evolutionary change within a known time span. Even in the study of Agrostis tenuis, there is a degree of artificiality because it was as a direct result of man's activities that a new grass species may, and only may, be forming by evolutionary processes. Nonetheless, the Welsh mine tip study and the other studies given on the following pages are quite informative about the processes of evolution.

SOME ECOLOGICAL STUDIES OF THE GENETICS OF NATURAL POPULATIONS

The ecological studies, sometimes supported by experiments in laboratories, outlined in the pages which follow, have been divided into five broad categories:-

   1. The first category includes some studies of large populations that give indirect evidence which suggests that several existing races, sub-species and species originated from common ancestors which were genetically quite similar to them.
   2. The second category includes a study of hybrid inviability, causing the genetic isolation of one group from the other because selection against hybrids prevents the hybrids from passing on their genes into viable offspring.
   3. The third type of population study includes some observations of phenotypic changes, and therefore allele frequency changes, in small populations over a period of time.
   4. Mimicry and camouflage.
   5. An example of evolutionary change during a known period of time.

1. Studies of Large Allopatric Populations That Imply The Evolution of Races, Sub-Species or Species from a Common Ancestor

Studies of large populations provide very few clear indications of the genetic and selective processes which formed races, sub-species or species. However they do sometimes give indirect evidence that races, sub-species and species are related to each other and are descended from a common ancestor.

i. **An Example of a Cline in the Large, Widely Distributed Population of Rana pipiens, a Frog found in North and Central America**

An extensive study, carried out by J. Moore and lasting for more than thirty years, of twelve species of North and Central American Frogs, of the Genus Rana, suggests that they all evolved from a common ancestral species which was not greatly different genetically from the twelve species found today. The study concentrated on Rana pipiens, which is probably divided into four species. The study was carried out by observations of wild populations and also by conducting genetic crosses in controlled laboratory environments. Some of the important factors studied were:

a) **The Compatibility between Male and Female Gametes**

The further apart the latitude from which males and females were taken, the less compatible were their gametes, as indicated by laboratory cross-fertilisation experiments. The gametes taken from individuals at opposite ends of the range of the species failed to cross-fertilise.

b) **The Degree of Phenotypic Abnormality in Laboratory Hybrids Resulting from Crosses between Frogs taken from Different Latitudes**

The further apart the latitude of the origins of frogs that were mated in the laboratory, the higher was the degree of phenotypic abnormality. In the cases of those matings between frogs that allowed viable hybrids:-

More Northerly females × more Southerly males gives enlarged heads in the F_1 generation

and

More Southerly females × more Northerly males gives reduced heads in the F_1 generation

Although frogs of widely separated latitudes can sometimes produce hybrid offspring, there are often serious problems of different rates of development in different body parts leading to the inviability of the hybrids.

c) Crosses between Frogs Taken from Different Regions of Similar Latitude

There is a wide East to West distribution of frogs belonging to the genus Rana across North and Central America. It is of significance that cross-fertilisation of eggs and sperm, taken from frogs of different species that live at similar latitude, can produce hybrids in the laboratory. They never do this in the wild because of factors other than the compatibility of the gametes. This, of course, indicates a recent common ancestor for these 'same latitude' frogs that have evolved into different species.

d) Temperature as a Factor that is either Important or Unimportant in the Speciation of Rana

Northern embryos of Rana pipiens can develop quite fast in cool temperatures (about 10°C), whereas Southern embryos develop very slowly at cool temperatures. At the other end of the temperature range, Northern embryos cannot withstand temperatures of 28° C or above, whereas Southern embryos develop very rapidly at temperatures from 28° C to 35° C. A temperature induced cline, depending on latitude, which alters many ecological factors that subject Rana pipiens to differing pressures of natural selection, has been established along the Eastern side of North America and through Central America. Ten races of Rana Pipiens have been established with varying degrees of hybrid viability. The gradual changes in the frogs along the cline from North to South eventually leads to sufficient differences between individuals which are widely separated by latitude, to enable their classification as different species.

By contrast, the East to West distribution of the Rana genus has given rise to the evolution of different species along lines of latitude where temperature does not seem to be such a significant factor. Although there is still chromosome compatibility and the normal development of hybrids, as shown by laboratory cross-fertilisation of different species taken from similar latitudes, such frog species of the Rana genus never exchange genes in the wild. The causes of speciation at similar latitudes are not at all clear.

It is of interest that mountains, which lower temperature with altitude, reduce the latitude effects of the North/South cline. In particular, Puerto Rican mountain frogs, in laboratory experiments, can cross-fertilise and produce viable hybrids more readily with frogs from near the USA/Canada border than they can with frogs taken from low-lying country of similar latitude on the mainland of Central America.

The following diagram shows the races of the North/South cline and also shows the approximate East/West lines on which live frogs of genetic compatibility:-

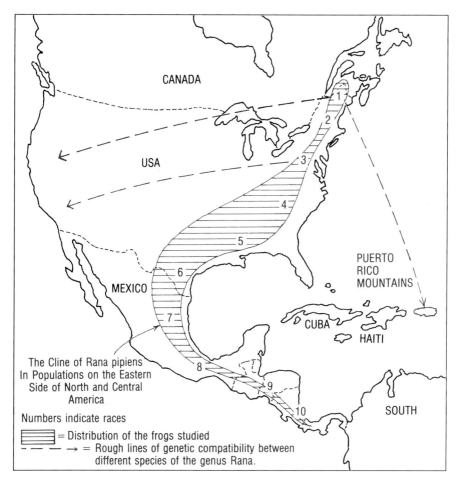

The Cline of Rana pipiens In Populations on the Eastern Side of North and Central America

Numbers indicate races

▬▬ = Distribution of the frogs studied
– – – → = Rough lines of genetic compatibility between different species of the genus Rana.

e) Rana Mating Calls

The mating and other calls of Rana pipiens in different geographical locations are different. Mating calls, and other aspects of behaviour, are important factors in isolating groups of individuals of similar morphology. If an individual mates only in response to the pattern of calls similar to its own, there will be the genetic isolation of groups of individuals from each other, the isolating factor being calls. By this criterion there are four 'species' of Rana pipiens distributed through much of the Americas, North of the Panama Canal.

J. Moore's study of the very large Rana population shows that there are many difficulties in establishing clear definitions of race, sub-species and species of the genus.

ii. An Example of a Number of Sub-species that are Geographically Distributed so as to Form a 'Ring Species' – Larid Gulls (Herring Gulls and Lesser Black-backed Gulls)

Once again the definitive classification of groups of these birds into distinct species is difficult. There are also difficulties that prevent the testing of hypotheses which attempt to guess the evolutionary processes that led to the formation of groups of individuals, some of which eventually evolved into different species.

An example of a 'Ring Species' to illustrate the results, without any details of the processes of genetic drift and natural selection, is that of larid gulls. These birds include herring gulls (Larus argentatus) and lesser black-backed gulls (Larus fuscus). They are widely distributed in the Northern hemisphere in a ring around the foodless ice-cap of the Arctic Ocean.

The range of Larus argentatus and of Larus fuscus, as shown in the following diagram, shows that for parts of the distribution of this Ring Species there are divergences that give Northerly and Southerly ribbons:-

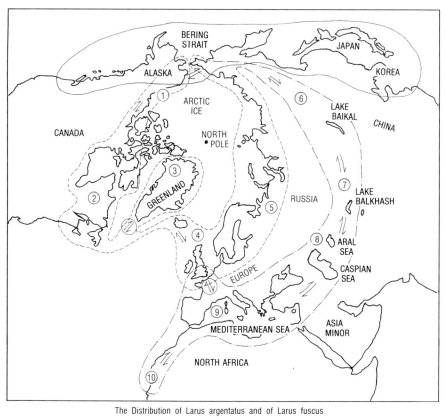

The Distribution of Larus argentatus and of Larus fuscus

```
[---] = The range of Larus argentatus      [   ] = Possible geographical origins of Larid gulls
[---] = The range of Larus fuscus          Ringed numbers are 'sub-species'
  ⇌  = Gene exchange between races
  ⇄  = Very limited gene exchange
```

The ten so-called sub-species that form all parts of this Ring Species may have been formed by genetic drift and natural selection in ice-free refuges, isolated from each other during the last ice-age, when fingers of ice spread Southwards. When the ice retreated, birds in adjacent geographical regions could once again exchange genes and there was the establishment of hybrid zones.

The spread of Larus fuscus from the North Pacific by way of the large inland seas of Asia took them to the Mediterranean Sea and beyond after the ice retreated. Some of these gulls probably reached the same regions of Western Europe as individuals of Larus argentatus which had spread to Western Europe from North America and by way of Arctic Russia. There are known cases of fertile hybrids being produced in Holland resulting from matings between small numbers of Larus argentatus and Larus fuscus. Normally in Western Europe each of these two 'species' of gulls have their own nesting colonies and there are very few cases of cross-fertilisation.

It has to be stated that the route by which (East-about or West-about) the gulls of the Northern ring spread to Scandinavia, Western Europe and Arctic Russia, is by no means certain. In addition the Greenland group of birds has been classified as Larus glaucoides, but this species probably evolved from the same common ancestor as Larus argentatus and Larus fuscus.

Studies of the large populations of larid gulls show once again how much uncertainty exists in classifying races, sub-species and species. Very few conclusions can be made for certain about the spread of these birds from their ancestral range, nor about the processes of genetic drift and natural selection which have caused the grouping of these gulls into what seem to be two species and ten sub-species.

The difficulties encountered during studies of large populations have led many ecological geneticists to study small or local populations. A few examples of such studies that yielded valuable information about the evolutionary processes of population genetics are given on the next few pages. Readers are once more urged to make use of the references given for this chapter to obtain a wider understanding of evolutionary genetics.

2. An Example of Rigorous Selection Against a Hybrid

An example of natural selection at work is that of the meadow brown butterfly, Maniola jurtina. In this case there is multifactorial variation, caused by many interacting genes, leading to the loss of fitness where two types of the butterfly interbreed, no matter what the environment where interbreeding occurs.

The genes which govern the number of spots on the underside of the hind-wings of the meadow brown butterflies are closely linked to genes that govern their rate of growth and development. The spots themselves may be of little value in determining fitness.

Different types, identifiable by their spot patterns, are distributed in the South-West Peninsula of England as shown in the following diagram:-

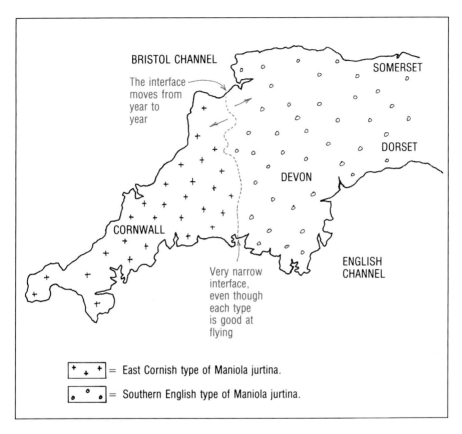

The interface where interbreeding occurs does not seem to depend on environmental factors for its position. It therefore has to be concluded that there are genetic reasons why cross-bred butterflies are selected against in the narrow interface region. Further to this there seems to be a 'reverse cline' (clines are gradual changes in one direction) at the interface. The characteristics of the two types become more pronounced close to the interface, indicating that individuals resulting from cross-breeding are less fit for survival and are therefore eliminated:-

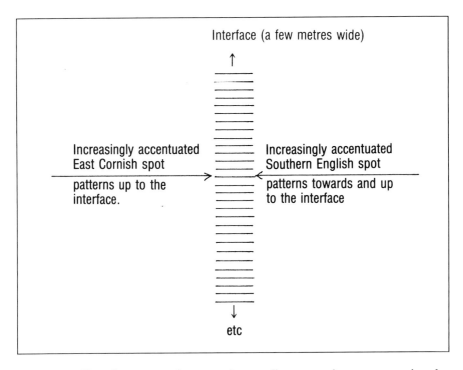

Interbreeding between the two butterfly types is common in the interface area, but no blends of the two types are found there.

This is an example of rigorous natural selection against the hybrid.

3. Some Examples of Small, Wild Populations that Support The Theories of Evolution

If small populations become isolated from other individuals of the same species and are then subjected to changes in the forces of genetic drift and natural selection, changes in the morphology of the isolated populations can occur quite quickly.

A general presentation of how this can occur is as follows:-

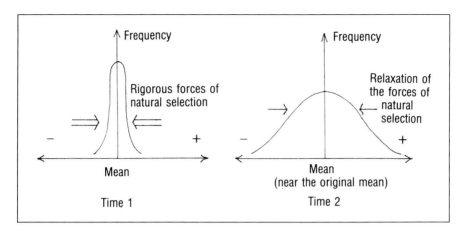

A change in the environmental factors causes a relaxation in the forces of natural selection between Time 1 and Time 2. At Time 1 the powerful forces of selection on a small, isolated population result in a small number of invariable, or almost invariable individuals.

A relaxation by the forces of selection at Time 2 allows a larger total number of individuals and new recombinants created by crossings-over allow a wider range of morphological features to be viable in a more favourable environment.

There may then be another period of rigorous environmental selection, but the forces of selection are different from those that existed at Time 1.

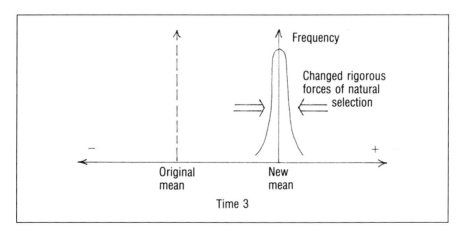

In addition to the directional changes illustrated in the two previous diagrams, selection can also act to stabilise the phenotypes, and therefore in many cases, the genotypes, of a population:-

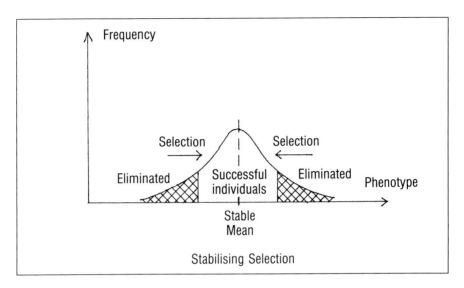

This type of selection operates on large populations that mate or cross-fertilise at random, for which the environment does not change much from year to year.

In some cases selection can be disruptive, thereby introducing the possibility of evolutionary divergence:-

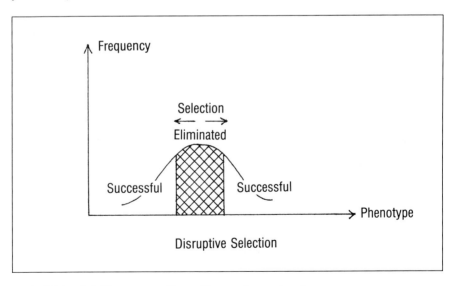

Hybrid inviability may allow disruptive selection.

(a) A specific example of these principles was observed by H.D. Ford and his son E.B. Ford, both of Oxford University, between 1912 and 1926, on a small isolated population of butterflies – the marsh fritillary, Euphydryas aurinia – in Cumberland, following up a study that had started in 1881 and which ended in 1935. For eight years up to 1920 the isolated colony of butterflies was few in number and almost invariable. Between 1920 and 1924 there was a reduction of selection against new recombinants and there were more butterflies in the colony with a wider range of phenotypes.

In 1925 there was an unidentified environmental change which led to rigorous selection. The pressures of selection were different from those of the years leading up to 1920 and there were once again few butterflies present. Those that were found after 1925 were morphologically similar to each other, but distinctly different from those found during the years before 1920.

(b) Another example is that of the meadow brown, Maniola jurtina. The biologists W.H. Dowdeswell and E.B. Ford studied this species on Tean, one of the smaller Scilly Isles. In the first period of their observations they recorded the number of spots on the wings of these butterflies from 1946 to 1951. In 1951, after a long occupation of the island, the cattle were removed.

The effect of removing the cattle was to change the female meadow brown from having bimodal maxima for spots of 0 and 2, to a stable population having a unimodal maximum of 2 spots. The removal of cattle changes so many ecological factors that it was too difficult to define the causes of the changes in the butterflies.

(c) Another example of natural selection is that of the snail Cepaea nemoralis. In the case of this species there are clearly discontinuous forms of the snails. Some of each form are found together in the same locality, but in different proportions depending on the locality. The name applied to the different forms of the snails is 'polymorphism' and the discontinuous forms of the snails are determined by 'super-genes'. Because the genes in super-genes are close together in a continuous line there is little crossing-over within a super-gene and, because the usual rules of dominance and recessiveness apply to super-alleles there are no intermediate forms of the snails.

Shell colour in Cepaea nemoralis, or rather apparent shell colour – the colour when the shell has an inhabitant – is determined by a super-gene. The colours are:-

(i) Dark Brown.
(ii) Pinkish.
(iii) Yellowish (green with an inhabitant).

In addition to their basic colours, the shells have a number of blackish bands drawn onto the background colour. Modifying genes determine the number of bands and these modifying genes are found on the DNA close to the super-gene for shell colour:-

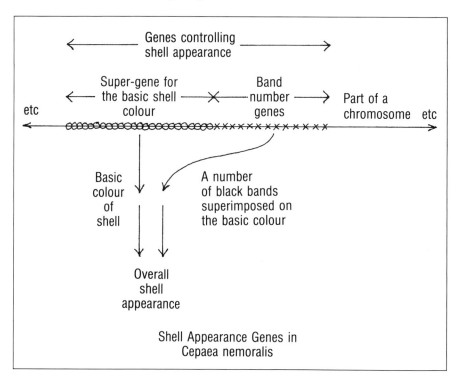

Shell Appearance Genes in Cepaea nemoralis

In most cases there is no crossing-over within super-genes of an individual snail at meiosis. However there is sufficient crossing-over within the super-genes of the species to allow some variation upon which natural selection operates, and this may give rise to new super-genes and therefore new genetically controlled polymorphic forms. The colours, modified by bands, give camouflage against predation by thrushes when on appropriate backgrounds:-

Yellow (green) predominates on grass.

Brown predominates on woodland leaf litter.

Banded shells allow camouflage upon a wide variety of backgrounds. (In some localities the physiological effects of calcareous soil is as selective as camouflage for determining the predominating phenotype in Cepaea nemoralis). Some exampes of genetic crosses in Cepaea nemoralis are given in Chapter 5.

(d) The heterostyle of primroses, Primula vulagaris, is composed of the male and female projections in a flower – the stigma and style is the female projection; the anthers on the filaments are the male projections.

In primroses two super-alleles exist in the population and they control the relative positions of the male and female projections:-

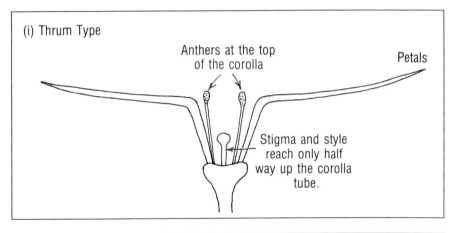

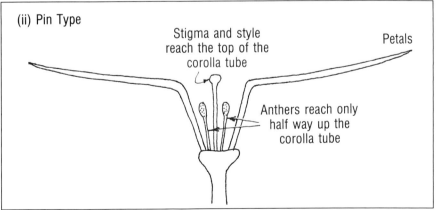

Usually these two types are not self-fertile so they are usually outbreeding, depending on insects for cross-pollination. Modern insecticides have been effective in removing the cross-pollinating vectors and these two primrose types are in decline in some areas.

Rare crossings-over in the heterostyle super-gene have led to recombinants in which there are long projections for both the male and the female reproductive organs. This homostylic type is self-fertile and insecticides are therefore encouraging the suppression of new gene combinations that could have resulted from out-breeding.

(e) The classic study of the peppered moth, Biston betularia (which has two forms only – dark and light), by H.B.D. Kettlewell, showed that the proportion of each form found in a locality depends on the general colour of the background on which they rest.

Super-genes control the dimorphism, and the dark super-allele dominates over the light super-allele. Nineteenth century soot from industrial and domestic chimneys darkened the countryside in many populous areas of Britain. The soot revealed the resting, light moths and concealed the dark ones, and predation by birds caused an increase in the proportion of the dark form in these areas.

In unpopulated areas upwind of the smoking chimneys, the light colour of most resting places allowed the proportion of the light form to remain high by comparison to the dark form. As backgrounds became cleaned of soot in industrial and populated areas in the years after the Clean Air Act, the proportion of the dark form began to fall.

It is of interest that when the dark British form of the moth is crossed with a closely related North American moth, the genes controlling the expression of the dark super-allele are segregated in a way that breaks down the dominance of the British dark super-allele. Wing colour and pattern is variable from such a cross.

4. Mimicry and Camouflage

a) Mimicry

Mimicry involves an organism (the mimic) which simulates the signal properties (usually visual) of a second species (the model), perceived as signals of interest by a third species (the operator) such that the mimic gains in its probability of survival because the operator identifies it as an individual of the model species (*based on the definition of R.I. Vane-Wright, with permission.*).

There are quite a large number of ways in which mimetic species operate. Each mimetic system is defined by its components. There seem to be five predominant ways in which the components combine:-

    i. All the components are of different species.
    ii. The predator and mimic are conspecific.
    iii. The model and mimic belong to the same species (for example females mimic males).
    iv. The mimic and operator are conspecific.
    v. All three components belong to a single species.

Models, mimics and operators can interact in about eight different ways. Five mimicry systems, each having the possibility of acting in eight different ways give forty theoretical mimicry combinations. About twenty have been observed in nature. Batesian and Müllerian mimicry seem to be prevalent.

Batesian Mimicry

H.W. Bates spent several years of the mid-nineteenth century in the Amazonian forests of South America. He collected an enormous number of previously unknown species and sent them back to England. Among the papers he wrote for the Linnaean Society of London was a paper on mimicry in insects.

Batesian mimicry occurs when a palatable species mimics an unpalatable species (the model). The genetic and evolutionary sequence leading to Batesian Mimicry are:-

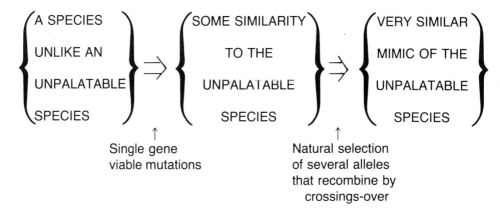

The model gives a protective advantage to the mimic provided:-

  i. the model is unpalatable or poisonous to predators.
  ii. the ratio: model/mimic is high enough.

If these two conditions are fulfilled, predators believe, and by experience continue to believe, that the entire population of individuals is unpalatable.

Batesian mimics are often polymorphic. For example, the mocker swallowtails of Uganda (Papilio dardanus) have five morphs. Each morph mimics danaid and acraeid unpalatable species. Many tropical butterflies exhibit mimicry in females more often than in males. Since non-mimetic males mate with several females, the loss of an individual male by predation is less damaging to the species as a whole, than the loss of each female.

Müllerian Mimicry

A number of unrelated, distasteful or poisonous species resemble each other so that predators believe, correctly, from tasting one individual from one species of the mimetic ring, that all individuals from similar species are distasteful. Müllerian mimics are only rarely polymorphic.

In South America the isolation of groups of butterflies in forest refuges, during drier climates than now, caused unique differentiation in each forest refuge. Each refuge would have had a dominant poisonous model that tended to 'capture' the outward appearance of other poisonous butterfly species by Müllerian mimicry in the forest refuges. In today's continuous forests of the Amazon Basin, butterfly descendants from groups once isolated in different refuges, meet in hybrid zones and produce a large number of racial hybrids. However, the distribution of mimic rings of butterflies probably had their origins in isolated forest refuges, and the overlapping of existing ranges of mimic rings came about by the spreading of forests when wetter climates returned. Local mimic rings of Heliconius spp., about forty species altogether, display Müllerian mimicry in the tropical Americas.

Müllerian mimicry tends to reduce the variety of aposematic (warning) patterns. However, each model species and its mimics are likely to inhabit a different forest stratum, from forest floor to near the crowns of trees.

Transvestism and Mimicry

The greatest barrier to cross-breeding between some African butterfly species such as Papilio dardanus and P. phorcas seems to be mate selection rather than cytogenetic incompatibility. Both species have some females that resemble males. In the local absence of effective Batesian mimicry, because suitable model species are absent, the colonisation of a new habitat could give some advantages to females which look similar to males. Males of these species either act aggressively towards other males or act cooperatively with them. Flying objects with the appearance of males induce males to fly towards them. If the flying object turns out to be a female in male-mimicking clothing, male aggression turns to more amorous pursuits. Transvestism in male mimicking females may give them a better chance of luring a potential mate.

A Plant that Mimics one Phase of an Animal's Life Cycle

South American passion flower vines (Passifloraceae) can produce swellings on their leaves that mimic the eggs of heliconid butterflies. The females of this butterfly group only lay eggs on the leaves of passion flower vines where eggs are absent. Egg mimicking swellings persuade female butterflies to look for another plant upon which to lay her eggs. The reduction in caterpillar grazing gives some survival advantage to individual passion flower vines.

A *Agaura sp.* A Katydid from the Costa Rica rain forest camouflaged as a leaf with brown senescent patches.
Oxford Scientific Films, with permission.

B *Biston betularia* the peppered moth. A camouflaged light form, England.
Oxford Scientific Films, with permission.

C *Hymenopus coronatus,* a mantis from Borneo, camouflaged as part of a flower.
Oxford Scientific Films, with permission.

D Model *Bematistes (Acraeinae spp.)* butterflies (left) and their *Pseudacraea eurytus* mimics (right).
Oxford Scientific Films, with permission.

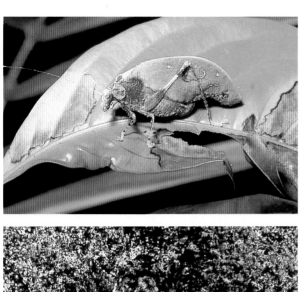

A

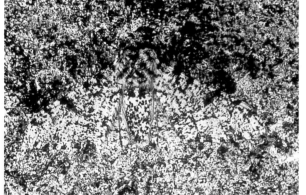

B

D

C

Mimicry and Population Size

Mimicry depends on being conspicuously similar to another species. Does mimicry allow a larger number of individuals of the species to survive and pass on their genes? The answer to this question is uncertain. Some mimic species are very numerous; others are very rare. Of course, factors other than mimicry also control the total population size in mimetic species. Each stage of the life cycle has its perils and the supply of food is always a factor that has to be considered.

The fundamental question is whether or not the total number in a mimetic ring is greater or smaller than it would be for a number of species occupying the same ecological niche without mimicry? To some extent at least, the answer depends on the relative time spans of each stage of the life cycle. For example, the mimetic forms of butterflies occur in the adults. In many mimetic tropical butterfly species, adult females (males are usually not mimetic) live for some months and have important functions in maintaining the eggs they have laid. Clearly there is an advantage to the species by having longevity in the females, and mimicry will allow a larger number of egg-maintaining females to survive, so allowing a larger brood of her offspring. However, this cannot apply strongly to those mimetic species that are very rare in the wild. In rare mimetic species, factors other than mimicry must play a part in population control. Whether or not mimicry in female adults of rare species gives larger numbers of butterflies than without mimicry, is not known.

Mimicry – A Summary

Although the genetic mechanisms which bring about the 'capture' of mimetic species by a dominant model species are quite well understood, it is not yet established whether or not mimicry increases the total number of individuals in a species. Mimicry probably increases the temporary viability of a species. It is uncertain that it increases the evolutionary time span of those species which are mimics and models.

(b) Camouflage

It is of selective advantage to prey if they are well camouflaged from their predators. Natural selection in favour of the camouflaged individuals is not unlikely and many species are very well hidden against their habitat backgrounds – leaf and stick insects, the caterpillars of several Australian moths and trout are well camouflaged species.

Predators also have a selective advantage if they merge with their background when hunting for prey – jaguars, leopards and tigers are examples of camouflaged predators.

Whether, or not, camouflage encourages a larger number of individuals in a species is not certain. Camouflage may only apply to one phase of the life cycle, and factors other than camouflage will often control the total population number.

Similarly, it is not known if camouflage is the critical factor for the time span of a species in its co-evolution with other species.

5. An Example of Evolutionary Change During a Known Period of Time – Sympatic Evolution observed in the Study of Agrostis tenuis (a grass) Growing on Spoil-heaps of Heavy Metal Mines in North Wales – Copper Tolerance of Agrostic tenuis Plants and of Their Seeds

Although the extraction of the mine materials was a man-made artefact, the ecological events that occured after the abandonment of the Drws-y-Coed copper mine have not been much interfered with by Man. The environmental factor that caused the changes in the plants – copper in the soil – is known.

Observations and controlled experiments have shown that there are changes in the copper tolerance of both the vegetative form of the plants and of the plants that emerge from the seeds, depending on where the plants or seeds are found in relation to the spoil heap and in relation to the prevailing wind:-

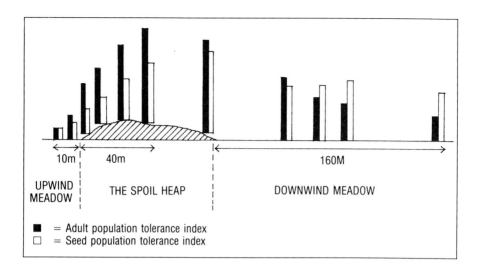

(a) Upwind Meadow Plants

These were largely unaffected by the presence of the spoil-heaps.

(b) Spoil-Heap Plants

Cross-pollination from the upwind meadow plants causes the seeds produced by the spoil-heap plants to be less copper-tolerant than the plants that give rise to them. The plants that grow out of the cross-pollinated seeds are less tolerant to copper than is needed for their survival. The plants growing on the spoil-heaps therefore tend to be those that result from the pollination of copper tolerant plants by pollen from other copper tolerant plants.

(c) Downwind Meadow Plants

Cross-pollination from spoil-heap plants increases the proportion of copper-tolerant seeds in the meadow down-wind from the spoil-heap. However the plants that grow out of these seeds compete less well than the normal plants on normal soil, so the normal plants predominate.

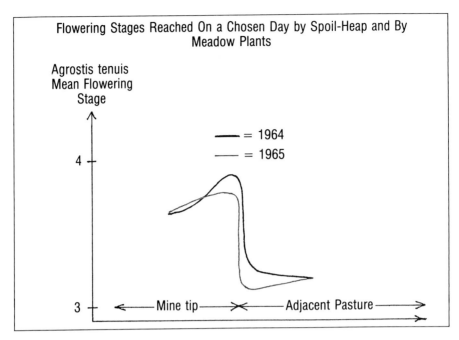

These results are corroborated by the flowering stages reached in a greenhouse, when all plants were subjected to ceteris paribus.

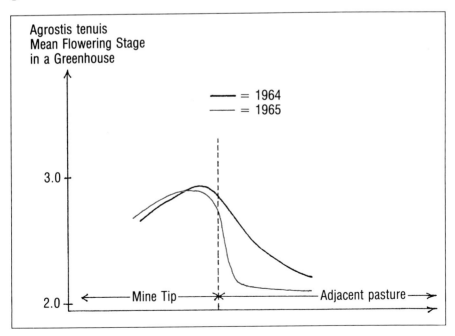

Similar studies have been carried out at other heavy metal tips in North Wales. Some references to these are given in the list of references that is associated with this chapter.

Some Further Examples of Evolution

a) An Example of the Early Stages of Sympatric Speciation in The Moth Hyponomeuta padella

Hyponomeuta padella is a moth species whose caterpillars feed on apple trees and hawthorns. Adults that develop from eggs laid on both apple trees and hawthorns look similar, with a slight increase in the proportion of darker forewings on hawthorn derived moths.

Matings between individuals tend to be between adults which were raised on the same food plant. Egg laying is nearly always on the plant species from which the females emerged as caterpillars. The laying of eggs on the 'wrong' plant tends to make individuals that emerge from the eggs sterile.

Although not yet completely isolated genetically, there has been the apparent divergence of one species into two sub-species, the divergence being caused by different feeding preferences.

b) An Example of an Animal Species in the Later Stages of Sympatric Evolution – The Bug Pysella mali

Pysella mali is a bug that feeds on apple trees and hawthorns. It has been observed that these differential feeding groups never cross-fertilise each other, even though their differences in morphology at each stage of their life-cycles are only slight.

Again, the problem of defining what is truly sympatric and what is truly allopatric presents itself. By the time there is a complete difference in host/parasite relationships, allopatry occurs at the local level.

c) An Example of Rejunction After Isolation

Breeding groups originating from a common ancestor which have been geographically and genetically isolated for a considerable time sometimes rejoin and overlap their ranges later. The consequences of rejunction and renewed sympatry depend upon a number of factors. Among these are:-

i. The Willingness and Capacity to Cross-breed

If this is high, there may be the spread of alleles among hybrids to allow the re-emergence of a single breeding population, with fewer and fewer differences between all individuals. Eventually the population is sufficiently uniform so that hybrids disappear.

ii. The Degree of Differentiation Between the Two Groups

If the degree of differentiation between rejoining groups is low, there will be problems of competition, perhaps leading to a parapatric distribution if no hybrids are produced.

An example of rejunction has recently occurred in Madagascar. In isolation on this large Indian Ocean island there evolved the grebe Tachybaptus rufolavatus, a co-adapted descendant of the widespread little grebe, T. ruficollis. The little grebe has recently re-invaded Madagascar and some hybrids from crosses with T. rufolavatus have been observed. It will be of interest in the future to record the genetic and habitat changes as these three groups of birds evolve in their close associations in Madagascar.

Conclusion

The science of population genetics is quite young and it must be stated that there is yet to be found a clear instance of speciation occurring during a period of study.

The Agrostis tenuis study shows some distinct differences between plants that originated from a less variable population and the greenhouse results show that the differences are heritable. Grasses growing on the heavy metal tips of North Wales also show that geographical isolation is not always a pre-requisite for starting the processes of speciation. The Drws-y-Coed spoil heap has only been in existence for about 150 years, so the selection of copper tolerance is known to have occurred quite recently in terms of geological time.

It is hoped that readers have been given enough information about the material of inheritance, and some of the ways it causes variations in populations, so that they can allow at least the probability that

DNA (nearly always) is the template that allows variations among the vehicles that carry it from generation to generation, and upon which genetic drift and environmental selection operate.

In due course the genetic studies of populations will be improved by the analyses of amino-acid sequences in polypeptides that are common to many races, sub-species and species. The eventual analyses of base sequences found in genes, alleles, super-genes, super-alleles and the complex inter-actions of sets of these inheritance units will contribute further information about the relationships between individuals, races and species.

It is hoped that population genetics studies over a long period, by several generations of ecological geneticists, will eventually allow the observation of speciation in the wild. In the meantime we also have to hope that discoveries in areas of science other than biology will not lead to mistakes of judgment that destroy, or catastrophically alter the marvellous variety of viable DNA molecules which determine the nature of those species that inhabit our small world.

Chapter 12 Questions

1. Distinguish between 'races' and 'species'.
2. Is 'sub-species' a useful classification?
3. What barriers are there to reproduction among individuals arising from a common, recent ancestor?
4. What is the difference between ALLOPATRIC and SYMPATRIC speciation? What processes lead to PARAPATRY?
5. What are the cellular requirements for establishing a new mutation in:-
   a) sexual species?
   b) asexual species?
6. How is a mutation established in a population? Emphasize the difference between 'genetic drift' and 'natural selection'.
7. What is the significance of population size in establishing a viable mutation?
8. Why are the studies of large allopatric populations unsatisfactory in providing evolutionary information?
9. Are morphological differences necessarily the best tests for distinguishing between species?
10. What is a cline?
11. How do ring species arise?
12. What is the significance of the hybrid in evolution? Give an example of selection against a hybrid.
13. What studies of small, wild populations support the theory of evolution?
14. Do camouflage and mimicry support the theory of evolution?
15. Give an example of evolutionary change to a plant species during a known time span.

13

MOBILE GENETIC ELEMENTS – THEIR INFLUENCES ON EVOLUTION AND THEIR POTENTIAL USEFULNESS TO MAN

An enormous number of mobile genetic fragments exist in our contemporary world. Brief references were made to some categories of mobile genes in Chapters 1, 4, 6, 8, 10 and 12. It is the intention to use this chapter for expanding upon the theme of mobile genetic elements, and to imply that they have the ability to transform cells either temporarily or permanently.

Some mobile genetic fragments are made of DNA; others are made of RNA. The genetic fragments are often associated with protective coats made of polypeptides, but some fragments have no protection at all.

A proportion of these mobile genetic particles may have quite ancient origins. Others arose at later times during evolution. A small percentage, among which HIV could be an example, may have arisen as mobile genetic elements very recently indeed. There is quite a high probability that some prokaryotic and some eukaryotic host cells in each generation have been permanently modified after being invaded by either mobile RNA or by mobile DNA. Permanently modified host cells that give rise to subsequent generations of individuals of the species pass the genes of mobile genetic fragments on from generation to generation, from evolved species to evolving species and from era to era. Some of the modifying RNA and DNA from mobile genetic elements may be lost in individuals that fail to reproduce some of their own species and in species that become extinct. However, even in individuals that fail to reproduce their own offspring, genes can be taken from them before they die by mobile genetic elements and transferred both to other individuals of their own species and occasionally to individuals of different species. Chance incorporation of these mobile fragments into the permanent genetic composition of germ cells in individuals of the same species or in different species hosts may allow them to become passed on once again as part of the permanent genetic material of individuals.

Such events have many implications for evolution.

Categories of Mobile Genetic Elements

Among the categories of mobile genetic elements are viruses, viroids, virinos, prions, plasmids, transposons, shuffled genes, spliced regions of mRNA, transferred parts of chromosomes, accidentally excluded RNA or DNA fragments from genomes, and whole chromosomes (sometimes single stranded, if they are made of DNA).

Some of these mobile RNA and DNA fragments are known to cause disease. Others seem to do little damage to the cells they invade or in which they arise. Yet others drag pieces of DNA or RNA out of their normal positions on genomes and take them for insertion to new positions on the genome. While it is certainly true that disease itself has played an important role in evolution, what may have been of more significance for evolution was the restless movement of mobile genetic elements within and between cells, taking with them various arrays of genetic material. Such variation in genetic material from cell to cell and from individual to individual is the right stuff for genetic drift and natural selection.

Mobile genetic elements can also alter gene controls in host cells without becoming part of the permanent genetic make-up of the cell. Such alterations can sometimes be to the advantage of the host cell. Usually, they are not.

The co-evolution of mobile genetic elements with their host cells has sometimes been as a result of the mobile fragments becoming incorporated into the host cells' genomes. Sometimes it has not. Some examples of how viruses and their host cells are inter-related demonstrate, later in this chapter, the ways that mobile genetic elements can exert their influences either by incorporation into the host cell genome or by remaining as free-floating packets.

Plasmids were discussed in Chapters 4, 6, 7 and 8. Some examples of transfers between chromosomes were given in Chapter 10. An example of splicing in the immune system of mice will be given in some detail in Chapter 14.

This chapter now considers some other categories of inheritance fragments that can direct evolutionary changes.

Mobile Intracellular DNA Genetic Elements – Transposons

There are short DNA base sequences in the genomes of both prokaryotic and eukaryotic cells that can wriggle into and out of their own cell's genome. These mobile fragments are called TRANSPOSONS. They can become incorporated at several loci on the genome. It is not clear whether the sites are determined by the equivalents of 'sticky ends', under enzymic control, or if the incorporation sites are more randomly available throughout the genome.

Two further questions which remain unanswered are those which ask whether or not transposons carry their control genes with them, and whether or not they modify the encoding properties of genes adjacent to where transposons become inserted into the genome?

Some transposons transcribe DNA copies of themselves. These copies can become integrated into the genome elsewhere. The additive effects of extra, integrated transposons is not yet clear.

It would seem, therefore, that the evolutionary significance of transposons is likely to be in their capacity to mutate in such a way that they can give rise to completely mobile genetic elements that can travel from cell to cell. Certain base sequences found in transposons are very similar to complementary sequences found in some retroviruses (page 433). These sequences occur in transposons, close to where they become integrated into the genome. In their free-floating state, the genetic codes in transposons could have given rise to the transcribed RNA sequences that evolved into retroviruses.

Alternatively, some retroviruses could have encoded the production of some DNA fragments that later developed the capacity to be mobile in a single cell as transposons. An explanation of the mode of action of retroviruses, which have RNA as their inheritance material, is given later.

The evolutionary direction of movement to transposons from DNA produced from codes in retroviruses is no more proven yet than the transition of RNA encoded out of the genes of transposons to become retroviruses.

Mobile Inter-cellular Genetic Elements

Four broad categories of mobile genetic elements that have the potential for causing evolutionary genetic change by inter-cellular movements, sometimes by way of disease, have been identified.

i) Prions (infectious polypeptides)

Some scientists have been forced to resort to speculative theories to explain some diseases. For example, some diseases have eluded the search for the particles that cause them. Some theorists have come up with the rather improbable notion that there exist infectious polypeptides. Such molecules could act in two ways (at least):-

    a) As controls for genes in a host cell's genome. The genes they control encode the production of the infectious polypeptides themselves.

    b) As templates for reverse translation into mRNA. The mRNA could then encode for the production of more infectious polypeptides, using the components of the host cell.

The co-evolution of infectious polypeptides, if they exist, with host cells would only seem to be of benefit in sustaining their own existence. But nature is subtle, and further inter-relationships between molecules such as these and their host cells need examination.

The evolutionary roles of such theoretical molecules, if they do indeed exist, with the characters described above, is not yet known. If they are more substantial than figments of frustrated scientists' imaginations, they may well act in an evolutionary selective capacity, simply as a disease.

Infectious polypeptides are candidates for scrapie, a disease of sheep.

ii) Virinos (DNA or RNA)

Virinos are probably extremely small DNA or RNA paired base sequences. There can only be a very small number of genes in these tiny genetic fragments. The function of virinos may be to switch on genes in the genome of their host cells. The switched on genes then encode:-

    a) The production of the virinos' DNA or RNA by very local DNA or RNA replication in the virinos themselves.

    b) The production of the polypeptides needed for the flimsy protective coat around the virinos' material of inheritance.

DNA virinos do not need to become incorporated into the genome of their host cells for their transcription and translation. Neither the origins of virinos, nor the ways in which they alter the direction of evolution are known.

Virinos may also act in an evolutionary selective capacity as a disease.

It is possible that scrapie, the disease mentioned in (i) above, is caused by a virino and not by an infectious polypeptide.

iii) Viroids (RNA)

Very small mobile RNA particles have been found without any protective polypeptide coats around them. When they invade host cells it may be that their RNA is too small to encode for the production of polypeptides. So far they have only been found in plants. Sometimes they harm their host cells; sometimes they do not.

If they cause damage to their host cells, there are two possible ways in which the damage occurs:-

a) Viroids bind to specific sites on the host cells genomes, so preventing desirable gene expression (deregulation).

b) Viroids consume nucleic acid components to such an extent that the transcription of host cell genes cannot be properly completed.

Again, there seems to be no incorporation of viroid-encoded nucleic acids into the genome of host cells.

Both the origins of viroid particles and their evolutionary effects in plants are uncertain.

iv) Viruses (DNA or RNA)

Viral particles have been dodging in and out of genomes in host cells or passing from cell to cell in prokaryotic and eukaryotic species for periods of time that may have been very long for some of the particles and much shorter, even recent, for others.

Some of these particles are very poorly protected by a flimsy outer coat of polypeptides, and some have no coat at all. Such particles disintegrate very quickly when they are outside the cells that sustain them. Other particles have better protection from their coats and can persist for considerable periods of time outside their host cells.

The inheritance material in some viruses is DNA. In others it is RNA.

Whether or not the particles have ancient origins, the persistence of viruses through the ages will have resulted from the following of one of a number of pathways:-

a) Ancient, very simple life forms, similar to viruses, co-evolved with more complex cells to become obligate parasites of the more complex cells.

b) Ancient cells became obligate parasites of other cells, in the same ways as mitochondria may have done, and then needed fewer and fewer genes of their own because their host cells provided many of the materials needed for the sustenance and multiplication of the parasitic invaders. Unnecessary genes could be discarded by mutational fragmentation, leaving behind only those small genetic fragments that continued to depend on host cell products.

c) Fragments of DNA and RNA in prokaryotic and eukaryotic cells acquired an independence of mobility that allowed them to hop into and out of cells, and sometimes into and out of the host cell's genomes as well.

There are several possible cellular components from which viruses could have evolved.

The origins of RNA viruses could have been mRNA, tRNA, rRNA, or snRNA. It would have been necessary to join together more than one RNA fragment in order to make an RNA particle long enough to encode a few genes.

In the case of mRNA, splicing occurs in eukaryotic species. Splicing discards parts of mRNA before translation (an example is given in Chapter 14). Some of these discarded fragments could, perhaps, be stitched together by chance over a period of time, evolving viral properties in a small proportion of cases.

The origins of DNA viruses could have been those regions of genomes with the ability to wriggle out of the genome to become mobile and independent DNA genetic elements. It is known that there exist in modern cells, plasmid-like fragments of DNA that do wriggle in and out of the genome. It is possible that mutations can occur in these transposons, changing them into viruses.

Whatever the origins are of DNA and RNA viruses, these genetic fragments often cause damage to their host cells. In some cases the damage is so great in very many cells that the host organism dies. As with other parasites, viruses can afford to kill a proportion of the species that provides them with host cells. But to wipe out the entire population of a host species would also be suicide for viruses. So viruses spread their plagues in a way that does not usually wipe out their host species. Many viruses have co-evolved with their hosts without causing much damage to them.

Before considering any further evolutionary aspects of viruses, it is necessary to summarise the categories of viruses that have been found in the contemporary biological world. All viruses depend on their host cells for the consituents of their materials of inheritance, for the constituents of their polypeptides, and for the enzymes that allow viral multiplication. The different categories of viruses arrange for the provision of their constituents and for their multiplication in different ways.

For those viruses studied up to now, there seem to be six ways in which the viruses operate. Two viral types contain DNA. Four types contain RNA.

DNA Viruses

i) Double Stranded DNA Viruses

After entering a host cell, viral DNA can either float free in the watery medium of the cell or it can become incorporated into the host cell's genome. Transcription can occur from both free and incorporated viral DNA. mRNA is produced by conventional transcription. Translation then depends on the host cells' mechanisms for these processes.

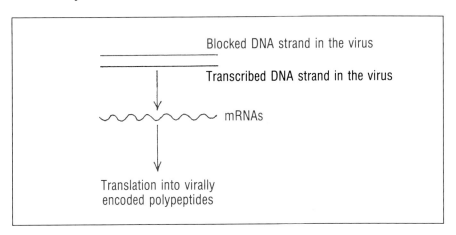

ii) Single Stranded DNA Viruses

After entering its host cell, single stranded viral DNA is used as a template to encode a second complementary single strand of DNA. It is from this complementary strand that transcription occurs:-

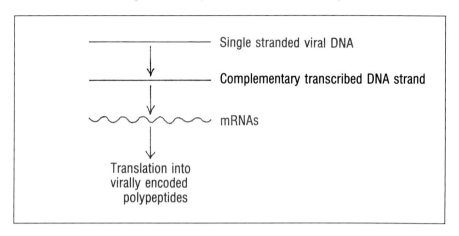

RNA Viruses

i) Double Stranded RNA Viruses

Except for using RNA as the material of inheritance, the transcription of double stranded RNA into mRNAs is similar to that from double stranded viral DNA. Viral double stranded RNA does not become incorporated into the host cells' genomes.

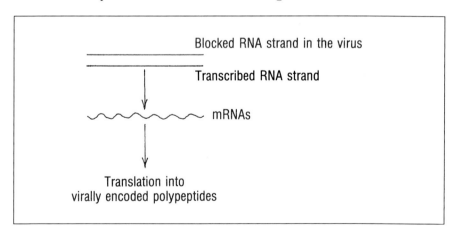

ii) Single Stranded RNA Viruses – Type 1

The transcription of Type 1 RNA viruses follows a similar pattern to the transcription of single stranded DNA viruses:-

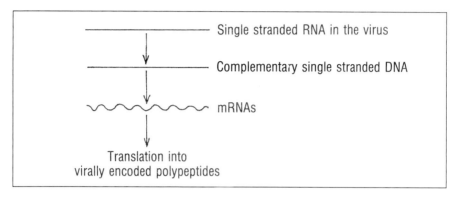

```
──────────────── Single stranded RNA in the virus
       ↓
──────────────── Complementary single stranded DNA
       ↓
   ∿∿∿∿∿∿   mRNAs
       ↓
   Translation into
virally encoded polypeptides
```

(iii) Single Stranded RNA Viruses – Type 2

Messenger-RNAs, encoded by the single viral strand, are transcribed directly from the virus particle;-

```
──────────────── Single stranded RNA in the virus
       ↓
   ∿∿∿∿∿∿   mRNAs
       ↓
   Translation into
virally encoded polypeptides
```

(iv) Single Stranded RNA Viruses – Type 3

This type of virus has been given the collective name RETRO-VIRUSES. The name is derived from the fact that, in the presence of an enzyme called REVERSE TRANSCRIPTIDASE, the single strand of the virus acts as a template for encoding the synthesis of a complementary strand of DNA. The single DNA strand has the pairing DNA bases brought to it and the second DNA strand appears. This double stranded DNA can now be inserted into the host cells' genome. The 'genes' in retroviral RNA can only be transcribed into mRNA when the double stranded DNA has been incorporated into the host cells' genomes. Retroviruses are known to cause nasty cellular behaviour – cancer and AIDS are both known to be caused by retroviruses.

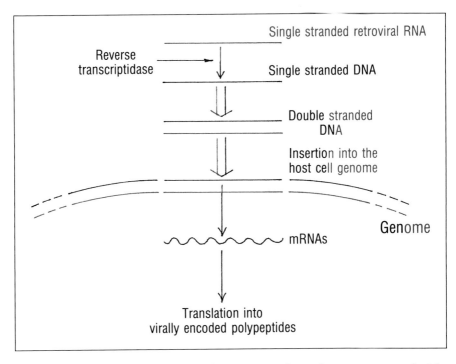

The insertion of retrovirally encoded DNA into the genome probably has regulatory or deregulatory effects on genome genes, as well as encoding for the production of polypeptides of its own.

The Parts of Host Eukaryotic Cells Where Viral Genes are Transcribed

DNA viruses are usually transcribed in the nucleus of eukaryotic cells. So is DNA derived from retroviruses.

RNA viruses are usually transcribed in the cytoplasm.

The translation of viral genes into polypeptides takes place in the cytoplasm of eukaryotic cells, in the same way that other mRNAs, derived from other genes, are translated there.

The Interaction of Viral Genes and Host Cell Genes

Viruses carry a few of the genes needed for their coat polypeptides and for their own multiplication. So far as multiplication is concerned, retroviruses carry an RNA base sequence that encodes the enzyme reverse transcriptidase.

However, most of the genes needed for making the constituents of new viral particles are found in the genome of the host cell.

The Assembly of Viral Particles

In a similar way to the self-assembly of ribosomes, as described in Chapter 3, the juxtaposition of the viral material of inheritance with their polypeptides in a watery medium allows them to assemble by the rules of lowest free energy. In the cytoplasm of cells, the lowest free energy of viral polypeptides and the viral material of inheritance is achieved when the polypeptides encapsulate the viral DNA or RNA in the desired way.

Of course, when viruses move out of their preferred cytoplasmic medium, their lowest free energy in transit may well be in a disintegrated form. Hence the rapid breakdown of many viral particles in the drier transit media which they must cross to find new hosts.

Base Sequence Comparisons of Viruses, Viroids and Host Cell Genomes – Some Evolutionary Implications

When comparing base sequences in viruses with those in the genomes of host cells, it is of little importance whether the sequences are found in viral DNA or viral RNA. This is because RNA sequences are complementary to DNA sequences and similarities in complementary sequences, in the case of RNA viruses, could still imply an evolutionary connection.

Comparisons have been made between base sequences in host cell genomes (chromosomes and plasmids) and sequences found in viruses. Some base sequences in those naked RNA particles, called viroids, that infect plant cells, have been found to be complementary to those found in exon/intron boundaries of eukaryotic DNA. In some viroids, there are nucleotide sequences which are very similar to those found in both transposons (from genomes) and in some retroviruses.

It is known that some bacterial plasmids have a gene that encodes a polypeptide that causes the formation of a passage between cells. The passage allows the movement of plasmids from one bacterium to another. A gene such as this could change by mutation so that the polypeptide it encodes encapsulates DNA or RNA particles. New viral particles could, perhaps, arise in this way from a gene of this sort.

Retroviruses and transposons have base sequences in some of them that are very similar to analogues in the DNA of host cells. These DNA analogues are parts of genes that encode a polypeptide that encourages the insertion of DNA derived from retroviruses or derived from transposons into the genome.

These first clues to the origins of mobile genetic elements are, of course, only the hazy beginnings of a long stretch of scientific detective work. The inter-connections between the mobile genetic flotsam and the cellular seas will become increasingly clear as more and more base sequence analogues are found.

The unanswered questions at the time of writing this book are:-

> 'Did the joining together of mobile genetic elements give rise to more complex genetic components which led to the formation of cells?'
>
> or:-
>
> 'Did viruses arise out of the genetic constituents of cells?'

Nature seems to perpetuate some aspects of very ancient life in many modern cells. There is therefore nothing too remarkable about suggesting that very ancient genetic elements did combine to cause the formation of cells. The cells retained some of these genes from ancient forms of life. Within cells there have always been contemporaneous mobile genetic elements. The mobile elements derive partly from those ancient life forms which never became fully integrated into cellular genomes. No doubt some of the modern mobile genetic elements arose *de novo*, either from previously integrated genomic DNA sequences in cells, or *de novo per se*.

Further Evolutionary Considerations

Both from the point of view of evolution and in the study of the transmission of disease, the mobility of the material of inheritance in various categories of genetic fragments is of immense interest. The movement of the AIDS virus, HIV, into human populations seems to be of very recent origin, probably some time in the 1950s. However, contact between this virus and humans may tell us little about the full history of the HIV virus.

It is becoming clear that the wobbling of viral DNA, for example, or retrovirally derived DNA into and out of a host cell's genome can grab a section of adjacent DNA in one cell and deposit it in another cell. This can happen between the cells of one individual. Alternatively, the transfer can take place between individuals of the same species. Less expected, but perhaps not uncommon, is the fact that mobile genetic elements can carry genes that are taken from the genome of individuals of one species and inserted into the genome of individuals of other species.

The ability of mobile genetic elements to transfer genes from one species to another makes the definition of species even more difficult than ever. Even Dobzhansky's definition in Chapter 12 is no longer tenable for all species. A flexible approach to speciation and species, to take account of mobile genetic elements, is therefore advisable.

Mobile genetic elements therefore have the ability to create genetic variation within a species. They can also transfer genes from one species to another to increase the genetic variation of the other species. As usual, genetic drift and natural selection operate upon any new genes or new gene combinations.

Long Term Consequences of Invasion by Mobile Genetic Elements

Apart from the parasite/host relationships between mobile genetic elements and their host cells, which often involves disease, a long term consequence of invasion by these elements can only occur when certain conditions are fulfilled.

DNA, however it is produced, must be incorporated into the host cell genome. In addition, the DNA must be incorporated into host cells in a way that it is not rejected at subsequent cell divisions. Further to this again, the newly incorporated DNA genes must be passed on to subsequent generations of the species into which it is placed.

For unicellular organisms, such as bacteria, these conditions can be fulfilled with less difficulty than with multicellular organisms. In the case of multicellular organisms, it is necessary for the new gene(s) to be included in the (generally) haploid genome of the germ cells. Similar rules apply here to those described in Chapter 11 for the inclusion of point mutations into germ cell genomes. New genes must either be put into embryonic cells that will develop into gamete mother cells; or new genes must be inserted into the genomes of germ cells directly.

Further unanswered questions which arise are:-

a) 'What proportion of individuals in each species carries a part of their genome that had origins as mobile genes?'

b) If the answer to a) is 'all of them', a subsidiary question is 'do all individuals of the species carry all those genes of mobile origins that are spread throughout the entire population as part of the so-called invariable species gene?'

c) 'What proportion of transposons in a species have viral origins?'

d) 'How much variation is there, and has there been, in the proportional change of integrated mobile elements in the life span of each species?'

e) 'To what extent has speciation depended upon the insertion or removal of mobile fragments into or from the genome of host species?'

f) 'To what extent has extinction been caused by these mobile genetic elements, beyond the obvious one of a devastating plague?'

A Summary of The Effects of Mobile Genes on Evolution

The insertion of mobile genes, with or without their attached genome genes, into a previously unencumbered genome, will not necessarily lead to the success of those cells or of those individuals which carry the insertions as part of their permanent genome. Mobile genes must have wriggled into an absolutely enormous number of genomes during the long biological history of this planet. Mobile genes not only alter gene expression directly but can also alter the effectiveness of adjacent promoters and repressors. They could also change the base sequences in exons wherever they cut into the host cells' genomes. Mostly, therefore, as with other ways of changing base sequences in genomes, mobile genetic elements which do so are likely to be damaging. A few genes found in a variety of species have quite close similarities to a few genes found in mobile genetic elements. The evidence for interchange between mobile genetic elements and genomes is quite strong. We do not know the extent to which this has occurred during the co-evolution of these two types of inheritance particles.

Of course, a large number of mobile genetic elements have failed to survive into modern times. A large number of primitive cell lines have been lost by extinction during evolution, as have many more complex species. Our answers to the questions posed in this chapter will inevitably only come from the survivors of evolution in our contemporaneous world. There is a very low probability that chance preserves still exist today, in which very ancient and very primitive cells have been locked away unaltered. There is just a faint hope, though, that such ancient forms of life, together with their parasitic invaders, may be found somewhere to allow the detailed molecular analysis of their structures. Let us hope that anyone who finds such treasure reports the fact to a specialist who can make best analytical use of such cellular contents.

Uses to Which Viruses can be Put

There are several ways in which viruses can be manipulated to be of benefit.

i) Bacteriophage Antibiotics

The genetic inheritance of some phage particles can be used for fighting some human bacterial diseases. For example, some phages cause the lysis (breakdown and destruction) of bacteria. Some medical applications of this type of therapy have already been successful. There are also agricultural and veterinary applications that have proved effective.

In particular, viral lysis and destruction of drug resistant bacteria are of particular value. Viral lysis of host cells is usually specific to one type of bacterium, so damage to human or farm animal cells is unlikely. A distinct advantage of such therapy is that the doses given to patients need only be very small. This is because the viruses multiply in large numbers in each bacterial cell, and lysis releases these viruses for the infection of further bacterial cells.

As always, there will very likely arise a phage resistant, drug resistant strain of damaging bacteria. Genetically engineered strains of the phages will be needed to destroy them.

Dysentery, bronchitis and pharyngitis have been successfully treated with bacteriophages in their antibiotic capacities.

ii) The Transformation of White Blood Cells

Viruses can be used to transform white blood cells. In particular, the disease-specific B-cells can be transformed by viruses so that they produce specific antibodies. These antibodies associate with invading disease and bind to those white blood cells that associate with the antibody/disease complex. Destruction of the disease follows quickly.

Epstein-Barr virus has been used to transform white B-cells for specific antibody production.

iii) For Making Genetically Engineered Vaccines

If genes that encode the antigens of disease-bearing viruses, bacteria or protozoa, are inserted into viruses that are otherwise harmless, such viruses can be used as vaccines. Vaccines stimulate the production of long-lasting, specific white blood cells for fighting specific diseases. Vaccination against many diseases will be possible in this way.

iv) Phages For Transferring Desirable Genes into Bacterial Genomes – Another Example of Viral Genetic Engineering

Instead of following the procedures outlined in Chapter 8 for inserting genes into plasmids, with all the difficulties of making 'sticky ends' and purification of genetically engineered products, viruses can be genetically engineered so that they contain desirable genes. Such viruses can enter bacterial cells and the genetic machinery of the bacteria, in concert with the virus, allows the incorporation of the desirable gene in the virus into the bacterial genome. Transformed bacteria then make desirable products encoded by the gene(s) received from such genetically engineered viruses.

v) Retroviral Genetic Engineering in Mammals Especially for Correcting Human Genetic Defects

It will gradually become possible, within the code of the Hippocratic Oath, to insert desirable human genes, together with appropriate promoters and suppressors, into the human genome. Retroviruses are a possible vector for introducing such gene combinations into humans or other mammals.

There are two categories of cells in the very early development of a mammal into which retroviruses could be introduced for the purpose of inserting desirable genes. The first is the germ cells; the second is early somatic cells after fertilisation. One very desirable aim of genetic engineering of this sort is to pass the desirable genes into the gonad mother cells from which future generations of the mammals arise. This aim can only be assured if the desirable gene is in the individual somatic cell from which all the gonad tissue arises. Such a cell already exists after a small number of mitotic divisions from the zygote. Four cell embryos are very good targets for genetic engineering.

In more developed mammals, some types of cell continue to divide in very large numbers both during growth and after maturity. Such cells can also be the targets for genetic engineering. In particular, white blood cells are being made continuously in human bodies, arising out of cell divisions in the stem cells of bone marrow. If the stem cells can be transformed by retroviruses, white blood cells that arise out of the stem cells will be able to make essential polypeptides that may be deficient in a patient with an inborn error of metabolism.

What is not yet clear is whether or not these retrovirally introduced genes will, in fact, persist in the desired way in large enough numbers to sustain the beneficial effects for long periods of time.

Of course, the genetic engineering of retroviruses that transform stem cells must have no undesirable side effects.

The genetic engineering of cells such as stem cells greatly benefits any individual who suffers from a genetic defect. However, for the viability of the human species as a whole, and for the purpose of reducing future pressures on the health services, it is more desirable to correct genetic defects in gametes or in very early embryos. Only by this type of genetic engineering will the proportion of the human race which inherits genetic disorders be reduced. Genetic engineering using retroviruses is, unfortunately, likely, in the short term, to increase the proportion of the human population that pass on inborn errors of metabolism.

vi) The Insertion of Both Drug Resistance and Other Desirable Genes Simultaneously.

If it necessary to use drugs to kill undesirable cells in a human, and it is also necessary to genetically engineer desirable genes into cellular genomes, these can both be attempted simultaneously using transforming viruses at some time in the future when such techniques have been perfected.

vii) Cell Targeting of Virus Polypeptides That Encapsulate Drugs

Some viruses carry polypeptides in their coats that only attach to the outer plasma membranes of the cells of particular tissues. Such polypeptides can be produced in large quantities by genetically engineered bacteria. These polypeptides can then be used as part or as all of an encapsulating cover around a drug molecule. The drug would, by this means, be delivered only to target tissues.

Other cells in the body would be largely unaffected by the drug.

For example, polypeptide antibodies can be made to encapsulate anti-cancer drugs. Genetic engineering can encode the production of antibodies that only attach to cancer cells, leaving other cells largely unharmed.

Conclusion

Man's ability to examine base sequences in DNA and in several different sorts of RNA has opened up an enormous number of routes for further discoveries in genetics.

Evolution has to be looked at again to take account of mobile genetic elements. In particular, there is probably much less genetic isolation of one species from another than had been supposed before the relationships between different categories of mobile genetic elements and cells become fully appreciated. No doubt, there is much more to discover.

Mobile genetic elements cannot always be direct agents for evolution. Some categories of these small genetic fragments cause great damage by disease, but the genetic materials of some of these mobile disease elements do not necessarily enter their host cells' genomes. Such genetic materials disintegrate when host cells or host organisms die, and seem to have little influence on long term genomic inheritance. However, they do cause natural selection among individuals in which they cause disease.

Finally, the benefits that may become possible from the genetic engineering of viruses, in particular, are very great indeed.

Chapter 13 Questions

1. What categories of mobile genetic elements have been identified so far?
2. Which of these mobile genetic elements are intra-cellular, and what are the characteristics of such intra-cellular mobile genetic elements?
3. Which of these mobile genetic elements are inter-cellular?
4. What are the characteristics of infectious polypeptides (prions)?
5. What are the characteristics of both DNA and RNA virinos?
6. What are the characteristics of viroids?
7. How many different categories of viruses are there?
8. How do each of these virus categories have their genes transcribed and translated?
9. Have viruses evolved from cellular components, or have cellular genomes depended on the amalgamation of viral particles?
10. How do mobile genetic elements make obsolete the notion that species never exchange genes with other species in the wild?
11. What evolutionary significance do mobile genetic elements have?
12. To what uses can viruses be put that are of benefit?

14

THE MODIFICATION OF GENE EXPRESSION BY DIFFERENTIATION, HORMONES AND AGEING

The changes that occur to the expression of genes in cells that arise by mitotic cell division during the life cycle of an individual have four main causes:-

1. Differentiation.
2. Hormones.
3. Ageing.
4. Somatic mutations (some of which accelerate the processes of ageing).

Mutations have already been discussed in Chapter 10.

This chapter discusses some aspects of the first three causes of alterations to gene expression.

Before any details of the modification to gene expression can be considered, it is necessary to understand the relationship between the zygote (the fertilised egg, from which all cells in an individual arise) and the somatic cells that are produced from it by mitosis.

1. The Modification of Gene Expression by Differentiation

Totipotency

A totipotent cell is one which has all the genetic information in it as that which is found in the zygote from which it ultimately arose. The differentiated form of each cell depends on the repression of some genes and the activation of others.

If cells taken from certain parts of animals and plants are isolated from each other, their nuclei can be extracted as for the amphibian described on page 449 and page 453. The extracted nuclei can be introduced into eggs that have had their nuclei removed. In these circumstances, the combination of repressed or activated genes is altered to return to that set of genes needed for starting the earliest mitotic cell divisions.

This allows the production of a new individual which has identical genetic inheritance to the individual from which the differentiated cell was taken. In another experiment carrot root cells were isolated from each other to remove the differentiating controls of the root cells on each other. Each isolated cell that is provided with suitable nutrients develops into a new carrot plant, complete with shoot, root and the capacity to flower and produce fertile seeds.

This ability of some cells to return to the 'capacity' of the zygote for producing a new and genetically identical individual is called 'totipotency'.

Not all cells can do this, and in those that cannot, it seems that alterations to the genetic composition of their chromatin have proceeded too far to allow the nucleus to revert back to the full capacity of the genes in the zygote from which they arose.

Cell Differentiation During Growth and Development

Development is the specialisation of cells during the growth of a living thing. Growth and development both depend considerably upon mitotic cell divisions in multiple-cell organisms.

CELL DIFFERENTIATION is the production of specialised cells by mitotic cell divisions. Cells having the same genetic information in their chromosomes therefore have to express that information in different ways in order to become specialised in form and function.

The present state of knowledge about the causes and controls of cell differentiation is not at all good. This chapter includes some simple generalities which apply to differentiation. It also states some simple experimental data on the subject.

It should be remembered that there is an inter-action between genes and cytoplasm. The inter-action operates in both directions, so that the chemicals in the cytoplasm control the activity of genes to some extent, as well as genes directing the activities of cytoplasmic chemicals.

The Controls for Differentiation

The norm is that mitotically dividing, somatic cells in a developing organism receive identical sets of chromosomes (exceptions are known where some insect species lose chromosomes from their cells during embryonic cleavage; gene amplification can occur to produce specialised cells in amphibia). In general, therefore, there has to be gene control in individual cells if cell specialisation is to occur.

Gene expression depends on:-

(i) Transcription of the DNA gene into mRNA.

(ii) The rate of mRNA production compared to the rate at which it is broken down or modified.

(iii) Translation of the mRNA base sequence into a polypeptide, which to some extent depends on the number of ribosomes available.

(iv) The influence of the cytoplasm on chromosomes.

(i) Control During Transcription

(a) Puffs in Drosophila melanogaster Salivary Glands

By examining the chromosomes found in the salivary glands of D. melanogaster, short bands along the lengths of the chromosomes have been observed to be much wider than the rest of them. The wider parts are called 'PUFFS'.

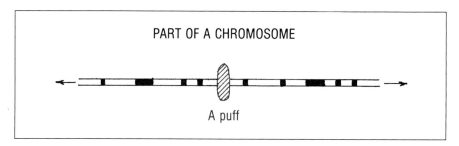

The number of puffs can be increased by raising the temperature from 25° to 37°C:-

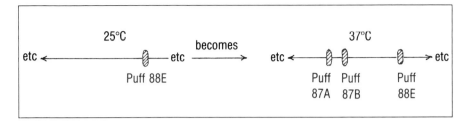

Whenever the number of puffs increases there is an increase in RNA synthesis associated with the genes at the puff sites, much more so than for other sites, and there is a considerable increase in polypeptides with a 70,000 relative molecular mass.

It is thought that puffing causes 'blocked' genes to become free to transcribe their mRNAs.

(b) The Analysis of mRNA Molecules in Differentiated Cells

In red blood cells the mRNA that codes for globin is found in higher concentrations than in other body cells. The number of mRNA molecules in a cell for any one polypeptide is important. The rate of mRNA synthesis, the life-span of mRNA and its rate of breakdown all influence cell differentiation.

(c) Inducers, Repressors, Final Product Concentrations and Hormones

These four types of molecules alter the rate of mRNA synthesis.

(ii) Control Between Transcription and Translation

(a) Breakdown of mRNA in the Nucleus

A considerable proportion of the mRNA produced by transcription is broken down within an hour without leaving the nucleus, and is therefore never translated into polypeptides.

(b) Modification to mRNA

Many mRNA molecules that move into the cytoplasm for translation require molecules to be added at both the 3' end and the 5' end to prolong the life of the mRNAs.

$$\text{7 methyl guanine} + \frac{\text{mRNA}}{\text{(unstable)}} + \text{200 adenine nucleotides} \longrightarrow m^7G \frac{\text{mRNA}}{\text{(stable)}} AA\ldots AA$$

The mRNAs coding for histones do *not* require the stabilising additions.

(iii) Control at Translation of the mRNA Codons into Polypeptides

(a) Actinomycin D inhibits the transcription of the codes in genes into RNA. Experiments have been carried out in fertilised sea urchin eggs to show that the rate of polypeptide synthesis following fertilisation is independent of the ability of the fertilised eggs to make RNA:-

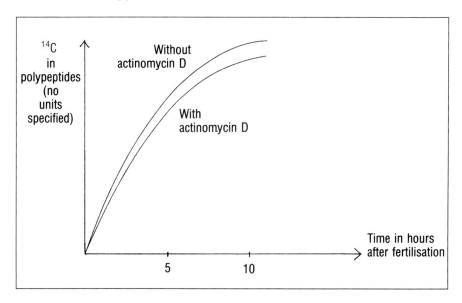

There must be factors other than mRNA concentration or other RNA concentrations controlling the rate of polypeptide synthesis during translation in this case.

(b) If mRNA from a specialised cell of one species is introduced into the cell of another species, the recipient cell in almost all cases translates the mRNA into a polypeptide. For example mRNA molecules coding for globin, extracted from rabbit red blood cells, were introduced into wheat germ. The wheat germ made globin.

Similar experiments have shown that there is not much control of what is made at the translational stage. Given mRNA from any source and ribosomes and tRNAs from the recipient cell, it seems that translation will occur. Thus there is little qualitative control at the translational stage, but there may be quantitative control.

Summary of Gene Control

A. Qualitative Control of Gene Expression

Which genes are expressed is mostly controlled at the transcription stage. Inducers and repressors operate for qualitative controls at transcription (see B (i) also).

B. Quantitative Control of Gene Expression

(i) Inducers and repressors, which may be controlled by 'end product' concentration or by hormones, control the quantity of mRNAs produced at transcription.

(ii) By DNA elongation making available many identical genes, as in puffs, more mRNA molecules can be made.

(iii) By producing 7-methyl-guanine and by attaching m^7G and many adenine molecules to stabilise mRNAs, the concentrations of mRNAs can be increased.

(iv) At translation, in order to control the rate of polypeptide synthesis, there may be product (polypeptide) feedback or hormonal controls to inhibit or increase the rate of polypeptide synthesis.

The Influence of the Cytoplasm on the Nucleus

If the nuclei of specialised cells are transplanted to replace the nuclei of unfertilised eggs (destroyed by ultra violet radiation) the genes of the specialised cells' nuclei are directed to behave in the same way as the genes operating in the original egg nuclei. This shows that substances in the cytoplasm are used to switch genes on and off, depending on the specialised function of the cell. For differentiation to occur this has to happen, as all the somatic cells in an individual contain the same complement of genes.

Transplants of this sort can develop into normal animals. In amphibia like Xenopus laevis, intestine epithelium nuclei or foot web nuclei can be transplanted into enucleated eggs and the new egg nucleus is then controlled by the cytoplasm to divide and develop into normal adults.

Gene Controlling Substances

A. DNA Non-histone Proteins that Control the Transcription Processes for Making Globin mRNA

In an experiment two types of differentiated mouse cells were taken:-

(a) foetal mouse liver cells (globin genes very active).
(b) mouse brain cells (globin genes inactive).

Chromosomes in each type of cell are closely associated with several substances including large quantities of polypeptides. These polypeptides are divided into two fractions:-

(i) Histones.
(ii) Non-histones.

These fractions can be separated. The following scheme demonstrates that the non-histone polypeptide fraction taken from foetal liver cells can switch on the globin genes of purified brain cell DNA, and that the non-histone fraction taken from brain cells, switches off the purified foetal liver cell's ability to make globin mRNA:-

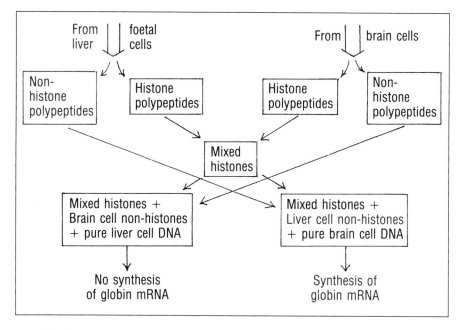

Purification techniques are not yet good enough to be sure of the precise molecule in the non-histone fraction that controls the globin gene transcription.

B. RNA Polymerases in Non-histone Polypeptides

Work is being carried out to try to identify *which* polypeptides in the non-histone fraction control *which* genes on the DNA. Some rather general groups have been identified:-

Group I controls the synthesis of rRNAs.

Group II controls the synthesis of mRNAs.

Group III controls the synthesis of some tRNAs and for some rRNAs.

This grouping is not of much value in this generalised form and more detailed analysis of the polypeptides in each group is needed in order to ascribe function to each of them.

C. Zinc-finger Polypeptides

Discoveries in the mid-1980s about the structures of certain DNA-associating polypeptides showed that they have finger-like protrusions in them. The protrusions depend upon the presence of a heavy metal. Zinc has been identified in several DNA associating fingers which project from the so-called 'zinc-finger' polypeptides.

A specific example of a zinc-finger polypeptide that has a function in determining some male characteristics in humans is given in Chapter 16. Zinc-finger polypeptides often have a gene control function.

Upstream and Downstream Genes

The products encoded by certain genes act as controls for other genes. Inducers and repressors are produced by transcription and translation from 'upstream' genes which control the switching on and switching off of 'downstream' genes. It is likely that there are several levels of the streams which govern gene control, and that there are very complex inter-action of genes and their products. It is this complex inter-action that leads to cell differentiation in growth, development, circadian rhythms, ageing, senescence and also to differentiation in response to environmental factors. There is much to be discovered about the streams that control differentiation.

The Use of Amphibian Eggs to Investigate Differentiation

The discussion so far has been about differentiated cells produced by mitotic cell division. However, fertilised eggs are the starting point for mitotic cell division. The study of developing unfertilised eggs, and of eggs after fertilisation has shown some of the processes of differentiation.

(i) Gene Amplification in Oöcytes – Extra rRNA Genes

During prophase of the first meiotic cell division in amphibia ovaries, each egg increases considerably in size. At the same time the chromosomes appear elongated with lateral loops called lamp-brushes. These lateral loops can reproduce large number of certain rRNA genes, especially for 28S and 18S rRNAs. The replicated rRNA genes migrate to a large number of nucleoli that appear at this stage of egg development – about 1,000 of them.

The very large number of extra rRNA genes are used to make the very large number of extra ribosomes needed for polypeptide synthesis, if the egg is fertilised, to complete two meiotic divisions and subsequently to be able to divide mitotically.

(ii) The Modification of Differentiated Cells' Genes When the Nuclei of The Differentiated Cells are Transplanted into Enucleated Unfertilised Eggs

If the nucleus of an unfertilised *Xenopus laevis* egg is destroyed and the nucleus of a differentiated cell such as the intestine of a tadpole is transplanted into the cytoplasm of the egg, the egg can develop by mitotic division, using the 2n chromosomes of the differentiated cell, into normal adult toads, in a small proportion of cases. The inference is that the cytoplasm of the egg has altered the 'switched on' genes of the specialised, differentiated cell so that the processes of *embryonic development* take place using a different combination of 'switched on' genes.

(iii) The Unequal Distribution of Substances in Eggs after Fertilisation

(a) When the egg of the tunicate ascidian Styella is fertilised, four regions of different colours appears in it soon afterwards. Each coloured region regularly gives rise to cells that later develop into specific body parts determined by the coloured regions.

(b) In amphibia the pattern of development depends on the unequal distribution of egg cytoplasm, such as the yolk concentration and the grey crescent contents. A typical sequence of events is as shown overleaf:–

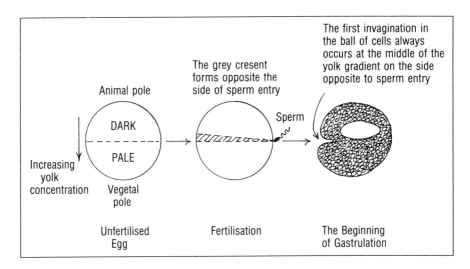

The distribution of cytoplasmic contents therefore seems to depend upon unequal distribution of yolk before fertilisation, and upon the side of sperm entry.

(iv) Specialised Regions of Amphibian and Dipteran Egg Cytoplasm That Determine the Development of Gonads – Germ Plasm and Pole Plasm

In amphibia (germ plasm) and Diptera (pole plasm) eggs, close to the vegetal pole, there is a special kind of cytoplasm containing clusters of granules, the size of ribosomes. This 'specialised' cytoplasm becomes associated with one or two of the first sixteen cells to be produced by mitotic cell division. The cell containing germ plasm (or pole plasm) divides at a different rate from the others of the cluster and eventually the single cell develops into the gonads, from which the gametes for the next generation are produced.

Destruction of the pole plasm regions of Diptera eggs, in eggs that are later fertilised, allows the development of animals which look normal, but they are sterile.

Genes That Control Development in Animal Embryos

The control of genes by induction and repression for the differentiation of cells is an area of genetics that has yielded few clear answers to the question – 'how is a standard set of genes in each cell expressed in different ways?' In the 1980s a team of American scientists began the process of identifying some of the genes that determine the spatial orientation and cell specialisation in very immature animal embryos, that least complex phase of animal development.

Genes containing common base sequences have been identified which determine some of the fundamental spatial organisation of cells in embryos. Several species, including Annelid worms (the ancestors of insects), Drosophila melanogaster, frogs, mice and man, are known to contain the common base sequence. Some of the genes involved in the control of embryonic development must have arisen more than 500 million years ago. It is also evident that these genes have remained intact to a considerable degree, surviving all the variable forces of natural selection in several species during the millenia when they were diverging from each other in other ways.

The name given to the common base sequence is the HOMEO-BOX. Each species usually has several homeobox genes. Each homeobox gene is quite large by comparison to most other genes, and the common homeobox base sequence is only a small proportion of the total length of each homeobox gene. By way of transcription and translation, the homeobox base sequence controls the synthesis of an amino-acid sequence in the active site of a polypeptide. The polypeptide controls some aspects of the spatial organisation of embryo cells. The sequence of amino-acids in the active site of the polypeptide has been named the HOMEO-DOMAIN.

The homeo-domain amino-acid sequence has been deduced from the sequence of triplicate codes in the homeobox. Several of the sites in the homeo-domain are occupied by amino-acids with basic properties – lysine and arginine are common. This allows the homeo-domain to associate readily with acidic DNA at specific positions on the genome where the homeo-domain polypeptide acts as an inducer or as a repressor for genes that control at least some aspects of the spatial orientation of embryonic cells. It should be made clear that polypeptides containing the homeo-domain are not the only molecules that control spatial orientation and differentiation. Some of the other molecules involved in these processes are discussed later.

The Homeobox

In Drosophila melanogaster most of the homeobox genes have been identified in clusters in those parts of the genome that are known to control the spatial and differentiating processes of dividing, embryo cells. These genes seem to specify which of the mitotically dividing nuclei and cells (pages 462/463) become part of the anterior insect, and which become part of the posterior insect. The cluster of genes that determines the anterior development are called the Antennapedia genes, and those that govern the posterior development are the Bithorax gene cluster.

Similar base sequences have been found in several animal species, including man, mice and frogs, and these homeobox genes are known to control similar embryonic development processes such as those observed in Drosophila melanogaster.

In those homeobox sequences studied so far, there is base homology in 60% to 80% (depending on the species) of the 180 bases in the homeobox. However, because of the degeneracy of the genetic code, the degree of homology in the homeo-domains in the polypeptides taken from several species is higher than that of the homeoboxes. For example 59 out of 60 amino-acids in the homeo-domains corresponding to the homeoboxes in the MM3 gene in Xenopus are the same as those in the Antennapedia homeobox of Drosophila melanogaster. The homeobox base sequence has also been compared with some genes in yeast which govern some aspects of sexual development. A considerable similarity has been found.

The following diagram shows the homeo-domain sequence of amino-acids that is produced by the transcription and translation of the Antennapedia homeobox gene in Drosophila melanogaster:

```
                1
        Arg – Lys – Arg – Gly – Arg – Gln – Thr – Thr – Thr
                                                            \
                                                          Arg 10
                                                            |
                                                           Tyr
         20                                                 /
        Phe – Glu – Lys – Glu – Leu – Glu – Leu – Thr – Gln
        /
      His
       |
      Phe                                     30
        \
         Asn – Arg – Tyr – Leu – Thr – Arg – Arg – Arg – Arg
                                                            \
                                                            Ile
                                                             |
                                                            Glu
                    40                                       /
         Glu – Thr – Leu – Cys – Leu – Ala – His – Ala – Ile
         /
       Arg
        |
       Glu                             50
         \
          Ile – Lys – Ile – Trp – Phe – Glu – Asn – Arg – Arg
                                                             \
                                                             Met
                                                              |
                    60                                       Lys
          Asn – Glu – Lys – Lys – Trp                        /
```

The Homeo-domain Corresponding to the
Antennapedia Homeobox in Drosphila

This sequence has been compared with some other homeo-domains, including a mutant Drosophila homeo-domain and wild-type homeo-domains in mice and frogs. In mouse homeo-domain MM 10, in Xenopus homeo-domain MM 3 and in three homeo-domains found in Drosophila melanogaster (Antennapedia, fushi tarazu and ultrabithorax) there are 22 variable amino-acid sites and 38 invariable sites.

It has been shown that homeobox genes lie in the same part of the genome as genes which:-

    a) govern the division of the embryo into segments and

b) govern the differentiation of segments.

These two processes follow the fundamental spatial organisation of cells that arise from the fertilised egg. A diagramatic summary of these processes is given on page 463.
The method by which the translation of several homeoboxes, distributed widely on the genome, controls the repression and induction of those genes which govern the three processes of embryonic development is not yet known.
Although few genetic facts emerge from the following pages, the description of a few aspects of embryonic development in more detail may help to put the action of homeobox genes into perspective.

The Control of Development in Drosophila melanogaster Embryos

There are three broad aspects of embryonic development that have been studied in Drosophila melanogaster.

i. Spatial orientation, governed by the chemicals produced under the genetic control of the female parent – Mother Effect.
ii. Segmentation, under the control of segmentation genes.
iii. Differentiation, under the control of homeotic genes.

i. Mother Effect

In Drosophila melanogaster, the development of mature eggs in the ovaries of the mother is in close association with nurse cells and epithelial cells:-

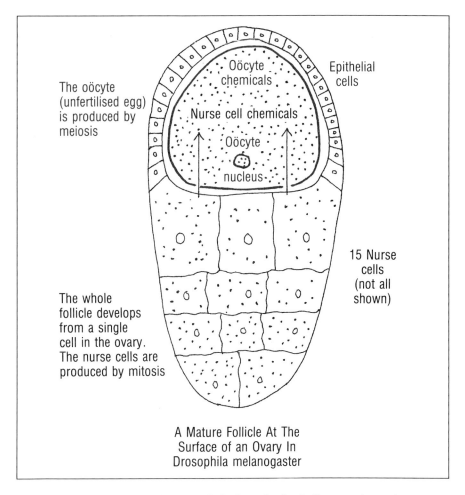

A Mature Follicle At The Surface of an Ovary In Drosophila melanogaster

Some of the mutations in an adult female fruit fly pass into the nurse cells and thereby into the oöcyte nucleus. In addition, the membrane between the nurse cells and oöcyte breaks down to allow some of the cytoplasmic contents of the nurse cells to become distributed in the oöcyte cytoplasm. The way in which the nurse cell chemicals are distributed in the oöcyte cytoplasm has an influence on the subsequent movements and specialisations of the nuclei which are produced by a form of mitotic cell division in the fertilised egg.

Several nuclei are produced in the fertilised egg before the synthesis of cytoplasmic membranes which separate the nuclei from each other. Several nuclei therefore float in the unevenly distributed chemicals of the egg cytoplasm during the earliest stages of embryonic development in Drosophila melanogaster. Unlike some species, such as Xenopus, which produce cells with an outer plasma membrane as soon as the fertilised nucleus completes its first mitotic cell division, the earliest mitotic divisions in Drosophila melanogaster produce new nuclei, but no membrane bound cells:-

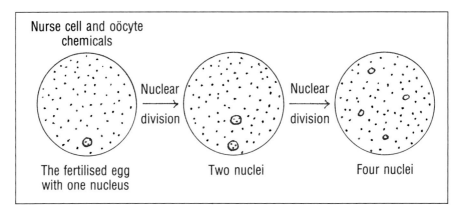

Nuclear divisions continue until there are 256 nuclei scattered throughout the egg cytoplasm. The nuclei then begin to migrate to the periphery of the egg, where they line up, still without their own cytoplasmic membranes:-

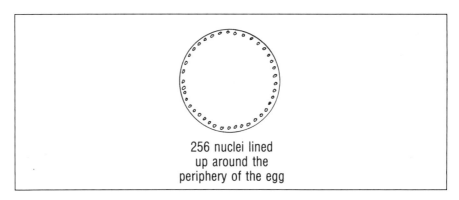

The exact mode of action of the molecules transferred from the nurse cells in altering spatial organisation and differentiation is not known. It is possible that maternal mutations can alter the developmental patterns of embryos by altering the chemicals which are passed into the oöcyte from the nurse cells, and by altering the genes which are put into the oöcyte nucleus. Mother effect can therefore fundamentally alter the polarity of the oöcyte. This in turn fundamentally alters the spatial organisation of the embryo.

ii. Segmentation Genes

When the nuclei which line up around the periphery of the egg generate membranes to cut off their local cytoplasm, they form the blastoderm. A small number of cells at one locality on the blastoderm eventually develop into the gonads. This cluster of cells are called the 'pole cells':-

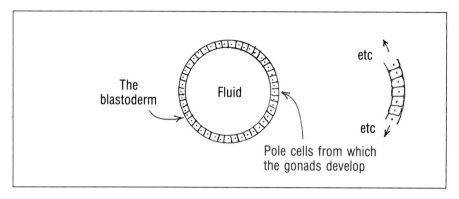

Streaming of the blastoderm cells then occurs together with mitotic cell division. It seems that the mother effect chemicals determine the spatial arrangement of the segments either by the spatial movement of the nuclei to the blastoderm, or by the direction of streaming of cells into specific segments. However, the segmentation genes only seem to determine how many segments there should be, without having any influence on the spatial arrangement of the segments. The normal number of segments in early embryonic development in Drosophila melanogaster is fourteen. A well-known segmentation mutation has been called 'fushi tarazu', which in Japanese means 'not enough segments'. This mutation results in only seven segments instead of fourteen.

iii. Homeotic Genes

The homeotic genes specify the differentiation of each segment. These genes therefore specify the parts of the adult animal such as the eyes, legs and antennae, and where they will grow. In some mutations the development of the wrong body part occurs from some of the segments. For example, in some mutants legs develop from the head segment instead of antennae. Such mutations alter the spatial expression of the homeotic genes.

Summary of i, ii, and iii

The positions taken on the blastoderm by Drosophila oöcyte nuclei seem to depend to some extent upon the chance distribution and composition of chemicals from the nurse cells. However, there must not be much variation caused by the distribution of these chemicals, otherwise there would not be a more or less standard development pattern in embryos of the species.

The positions to which cells stream depend on their positions of origin in the blastoderm. Segmentation genes then specify how many segments there should be at the end of gastrulation (the streaming process). During metamorphosis, homeotic genes specify the internal and external organs that grow in each segment.

At the present state of knowledge, it has to be said that none of the maternal effect genes have been found yet, nor is it known how the spatial and differentiating effects are coordinated by the homeobox master-genes.

The following diagram illustrates some differences between wild-type and mutant development in Drosophila melanogaster:-

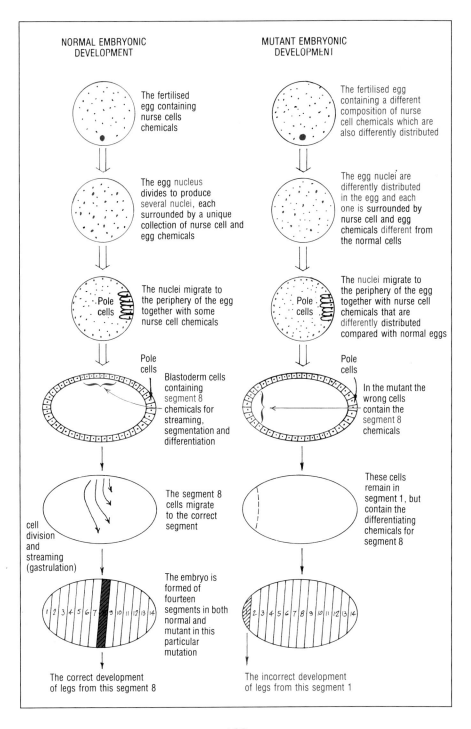

Variations in the Patterns of Early Embryonic Development

Although the genetic reasons have not yet been discovered, the patterns of early embryonic development seem to be divided into those that are:-

a) Highly Regulated

Some species have very precise patterns of development, each cell in the adult resulting from a clearly defined pattern of cell division starting with the fertilised egg. For example, differentiation in the segmented roundworm Caenorhabditis elegans is invariable, except in mutants, and the number of cells in the adult is always the same.

b) To Some Extent Variable

Many species have a degree of flexibility for the ultimate, differentiated fate of each of the cells contained in the early embryo. The spatial migration of each cell, together with the chemical composition of its cytoplasm, determine the ultimate fate of each cell. The switching on and off of the homeotic genes, under the control of the homeobox genes, determines the differentiation of cells in the segments of animal bodies. There seems to be a degree of flexibility in distributing chemicals, and in determining the positions of nuclei in the fertilised egg in Drosophila melanogaster. However, once the positions of cells, with their cytoplasmic contents, have been taken on the blastoderm, there are then fairly tightly controlled cell lineages, which develop first into segments and then into phenotypic expression.

Differentiation in the Production of Mammalian Antibodies Caused by Gene Mobility

Specialised, disease-combating cells produced in the lymph system of mammals have complex genetic systems for producing a large number of different antibodies. A surprisingly small number of genes control the production of about ten million different antibodies. Such an arrangement allows the economic use of DNA and avoids the need to provide a massively long genome containing ten million genes for antibody production alone.

In mice, antibody molecules are made from the transcription and translation of sets of genes, each gene set consisting of a small number of individual genes.

Antibody molecules are made of polypeptide chains, each polypeptide depending on different sets of genes for directing its synthesis. There are two types of light chain polypeptide called the κ (kappa) and λ (lambda) chains. There is also a heavy chain type of polypeptide, the H chain. Each antibody is an assembly of four polypeptides. There are always two H polypeptides present, each of which is either associated with two identical κ polypeptides or with two identical λ polypeptides.

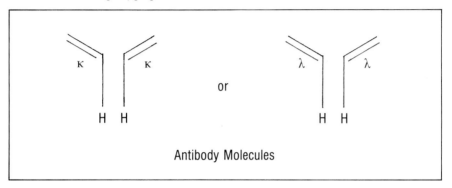

Antibody Molecules

The genetic system of mice can produce a large number of κ, λ and H polypeptides.

In mice the κ and λ light chains each have three sets of genes for the control of their synthesis. The H chain of this species is genetically controlled by four sets of genes. Light chain, antibody polypeptides are controlled by V, J and C gene sets. Heavy chains are dependent on V, J, C and D gene sets.

The following table shows where the gene sets are situated on the genome and how many genes there are in each gene set.

Antibody Polypeptide	Chromosome Pair	Number of Genes in Each Gene Set			
		V	C	J	D
κ	6	several	1	4	Nil
λ	16	2	2	4	Nil
H	12	several	5	4	10

The Translocation of Genes So That Light Chain Antibody Polypeptides Can be Correctly Synthesised

One of the λ light chain molecules from a mouse was studied in some detail. A comparison was made between the order of amino-acids in the polypeptide and the sequence of triplicate codes contained in the bases of the region of DNA where the gene sets for the λ light chain molecule are found. It was evident that only 97 out of the 110 triplicate codes in one of the V_λ genes were used to insert amino-acids into the polypeptide. The thirteen amino-acids at the carboxyl- end of the molecule were discovered to correspond with a DNA base sequence close to the C genes, in the J_λ region.

During the embryonic development of mice, the V_λ genes are quite widely separated from the C_λ genes. The J_λ genes lie close to the C_λ genes at all stages of development.

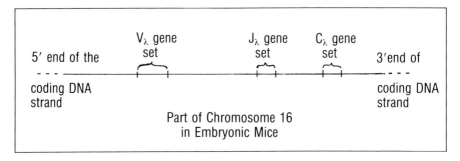

When an analysis of chromosome 16 was made in mature mice it was discovered that the V_λ gene for the polypeptide under study had been translocated. The result of translocation was that the 97 triplicate codes on the V_λ gene became continuous with 13 triplicate codes on the J_λ gene. The following diagram, which is at a greater magnification than the last one, shows the relative positions of the V_λ, J_λ and C_λ genes after the translocation of the V_λ gene.

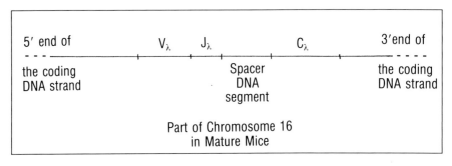

The deduction made from this genetic study was that, during the development of mice, a large part of one of the V_λ genes was translocated and joined to a segment of one of the J_λ genes.

This discovery about the translocation of antibody genes in mice demonstrated that embryonic mice are limited in their ability to make antibodies. It also led to further investigations to show the ways in which the transcription and translation of mobile antibody genes can give rise to an enormous number of different antibody molecules, using only a small number of genes.

The Synthesis of a Light Chain κ Antibody In Mice

The antibody genetic system responds to the invasion of a mouse's body by an antigen. Invading pathogens, such as disease bearing bacteria and viruses, usually carry specific antigens on their surface and the antigen acts as a trigger for switching on the appropriate, mouse antibody genes. The specific antibody produced by lymphocyte cells allows the lymphocytes to destroy the invading pathogen.

The relatively small number of genes that control the synthesis of antibodies can be shuffled by translocations. In this way a very large number of permutations can be made for arranging the order of adjacent genes. In the case of the synthesis of one of the light κ antibodies the sequence of events is as follows.

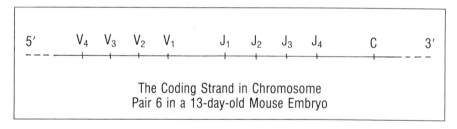

The Coding Strand in Chromosome Pair 6 in a 13-day-old Mouse Embryo

The shuffling of the genetic system during the later development of the mouse can place genes V_2, J_3 and C (for example) in a continuous line. This is partly done by removing the DNA between V_2 and J_3. It is not clear if this is done in the production of new lymphocytes by mitosis, or if the translocations are carried out in existing cells in response to invasion by a specific antigen. In any case, mRNA, corresponding to the sequence of bases between the 5' end of V_2 and the 3' end of C, is then transcribed.

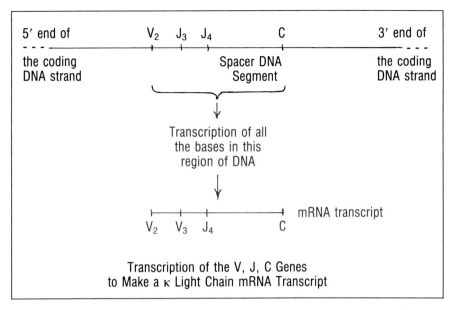

Transcription of the V, J, C Genes to Make a κ Light Chain mRNA Transcript

The sequence of bases in the mRNA transcript, corresponding to the sequence of bases between the 5′ end of the V_1 gene and the 3′ end of the C gene, is then spliced.

'Splicing' removes the bases from the J_4 and spacer regions of the mRNA transcript.

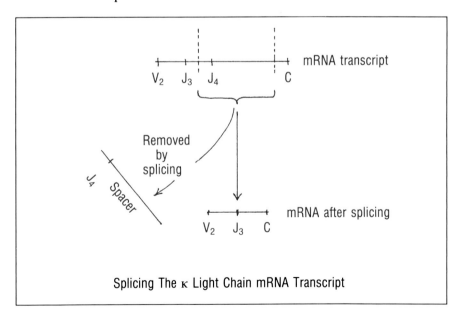

Splicing The κ Light Chain mRNA Transcript

The translation of the spliced mRNA produces a light chain κ polypeptide which is a constitutent of the antibody molecule.

This example shows how differentiation in lymphocyte cells can be brought about by mobile genes and mRNA splicing. Specific antigens are the triggers for the immune response generated by the genetics of differentiation.

The Genetic Control of Heavy Chain Polypeptide Synthesis In Antibodies

There are four gene sets for controlling the synthesis of heavy chain, antibody polypeptides. In addition to the V, J and C genes there is also a D gene set. There are probably more than 10 D set genes, situated on chromosome pair 12. These D genes can be joined to any permuations of V, J and C genes, so increasing the variation among heavy chain polypeptides in antibodies by at least a factor of 10.

Summary of Differentiation in Antibody Gene Expression in Mice

The presence of V, J and C genes for light antibody polypeptides, and of V, J, C and D genes for heavy antibody polypeptides allows the synthesis of more than 10,000 different H polypeptides and of about 1,000 different light polypeptides. These polypeptides can associate as two identical, light polypeptides with two heavy polypeptides to make about ten million different antibodies.

Further variation in antibody molecules is caused when V genes join J genes at different places along their base sequences. Somatic recombination and somatic mutations further add to the variety of antibody polypeptides produced from a relatively small part of the genome. Recombination and mutation are different from differentiation, of course.

'Mother' Effect Demonstrated By The Right Handed or Left Handed Coiling in the Shells of the Snail Limnea pereger – An Example Showing that Cytoplasmic Chemicals Can Control the Expression of Genes

During the production of eggs in the ovaries of some species, specialised chemicals (perhaps polypeptides) are made under genetic control and these lie inactive in the cytoplasm until they are needed. If fertilisation takes place, the specialised chemicals take on the rôle of gene control. The direction of the coiling of shells made by the snail Limnea pereger, left handed or right handed, has been shown to be controlled by the chemicals in the cytoplasm of each egg which is fertilised. It seems that fertilisation allows the influence of the cytoplasmic chemicals, contributed by the egg, to be different in their control of the nuclear genes by comparison to their role in the unfertilised egg.

The way this works in Limnea pereger is as follows:-

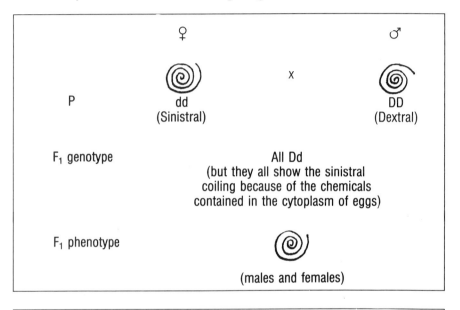

However all the F_2 generation have dextral shells because the cytoplasmic chemicals in the mother's eggs were all determined by the dominant D allele.

F_2 phenotype	(Males and Females)

Snails are hermaphroditic, so they can be self-fertilising. If snails of the F_2, containing the dominant allele, are selfed they will produce zygotes from eggs in which the cytoplasmic chemicals were produced under the influence of the dominant D allele.

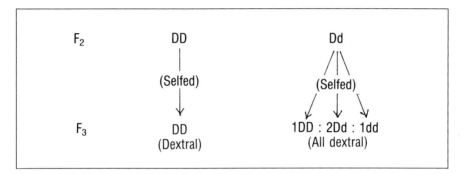

However when the F_3 double recessive is selfed the cytoplasmic chemicals in the eggs were produced under the control of dd, and this line will be sinistral if they continue to be selfed:-

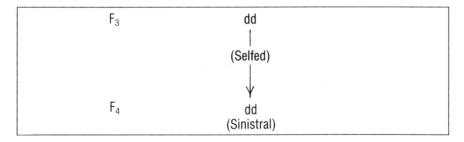

It can be seen from this example that the direction of coiling in Limnea pereger is controlled by the genotype of the mother, rather than the genotype of the individual.

2. The Modification of Gene Expression Caused by Hormones

In the case of hormones, chemicals are produced in small regions of animals and plants, and these seem to control gene expression in the cells of certain organs which act as targets for the mass transport systems that carry the hormones to them. Hormones therefore modify the potency of the chromatin of certain cells, rather than cause the reversion to the totipotency of the zygote.

The study of hormones is a large one and requires specialised reference books. Suffice to write here that hormones are broadly divided into protein hormones and steroid hormones.

It has already been seen that cells or clusters of cells can exert an influence on each other by chemical means to cause different sets of genes to be switched on or off in different cells. The endocrine glands of animals and small specialised regions of plants produce hormones which are transported throughout the entire organism. Target cells respond to hormones in different ways.

It has proved to be very difficult for investigators to study the molecular modifications of the expression of genes in animals and plants. The chromatin nature of DNA, repeating genes, introns and exons have all added to the difficulties of discovery. There are quite a large number of possible hormonal alterations which could change gene expression. Among these are:-

(i) Repression or derepression of single genes.

(ii) Alterations to the combinations of 'switched on' genes, the products from which interact with each other to cause phenotypic changes.

(iii) Alterations to gene transcription.

(iv) Alterations to gene translation.

(v) Modifications to the molecules specified by some genes may cause the chemicals produced by the translation of genes to associate together in ways different from those when the hormones are absent.

(vi) Modifications to the ability of membranes to transport molecules across them.

(vii) Releasing molecules from the inner surface of the outer cell membrane, which then move to the nucleus to modify the expression of genes.

(viii) Changing osmotic or pH conditions so that molecules in cells have their shapes and functions altered.

The Hormonal Control of Gene Expression in Animals

In animals the following are some examples of the expression of genes controlled by hormones:-

(i) Ecdysone in Insects

Ecdysone causes puffs in the salivary gland chromosomes of insect larvae. It may be that ecdysone removes the effects of gene repressors, so switching on some genes that would normally be suppressed. Ecdysone is known to be one of the hormonal causes of the metamorphic changes from grub to larva, to pupa, to adult. It stimulates the moults that allow metamorphosis to occur.

(ii) Thyroxin and Cortisone in Amphibians

There are considerable variations in the effects of thyroxin and cortisone, depending on the tissues studied. In some tissues thyroxin increases the nuclear RNA concentration, and this is followed by an increase in ribosomes close to the endoplasmic reticulum, and a subsequent increase in cytoplasmic polypeptide concentration. Thyroxin increases the activity of RNA polymerase, the enzyme used in gene transcription.
   Cortisone interacts with other cellular chemicals to cause an increase in RNA synthesis and thereby also an increased rate of polypeptide synthesis. Both the ability of ribosomes to synthesise polypeptides and the production of particular enzymes are affected by the cortisone concentration. Cortisone also alters the order of bases in some mRNA molecules.

(iii) Growth Hormone and Testosterone in Amphibians

Growth hormone increases the production of RNA polymerase. Testosterone also increases the catalytic effect of each RNA polymerase molecule.

(iv) Oestradiol and The Mammalian Uterus Wall

A radioactive (made purposely) sex hormone, oestradiol, was studied in its effects on the cells lining the mammalian uterus wall. An oestradiol/polypeptide complex was found in the uterus wall cells and this was proved to increase the rate of mRNA production. The polypeptide on its own was shown to repress mRNA production. The hormone in this case was acting as a gene derepressor.

(v) Cell Division in Endocrine Glands Caused by the Lack of Negative Feedback by A Hormone

The deficiency of a hormone can cause the production of new cells in the endocrine gland which should produce the hormone. Lack of iodine in the diet causes the thyroid gland to increase greatly in size (simple goitre) because the gland cannot make thyroxin. How cell division is derepressed in the absence of the hormone produced by this gland is not known.

(vi) The Production of Specific Chemicals in the Oviduct of Hens in Response to Hormones

Cells lining the oviducts of hens secrete chemicals of value to eggs as they pass along the passage. For example the hormone oestrogen causes the oviduct cells to make the egg-white polypeptide called ovalbumin. Progesterone on the other hand induces the production of a different egg-white polypeptide called avidin.

Oestrogens also greatly increase mRNA synthesis in these cells. It is possible that they do so by activating cell receptors of progesterone. Both these hormones increase rRNA and tRNA synthesis in cells lining the oviduct. Acid proteins in these cells, but not in cells from other tissues, are thought to bind the oestrogen/receptor and progesterone/receptor molecules to chromatin DNA to derepress genes.

(vii) Cell Steroid Hormone Receptors and The Action of Steroid Hormones After Transport on a Protein Carrier Molecule

Steroid hormones are hydrophobic, but they become soluble when attached to protein carrier molecules. They are transported in blood as a protein/steroid solute, which transfers the steroids to receptor sites on the outer cell membrane.

The cytoplasm of target cells have about ten thousand receptor molecules in the cytoplasm of each cell. The steriod/receptor/complex moves into the nucleus and binds to chromatin. It is possible that the same steroid/receptor complex affects different sets of genes in different types of cells. In those target cells that respond to steroid hormones, about fifty genes are controlled by each type of steroid. The steroid/receptor complex seems to bind to all regions of chromatin, but only alters a very small proportion of genes (there is no clear proof – except for cortisol – that each steroid/receptor complex has an increased preference for binding to those regions where the genes are found for those characters altered by the hormones). However, recombinant techniques have shown that cortisol/protein complexes bind specifically to certain receptors close to specific genes to increase their activity.

Steroid hormones may start a two step alteration to gene expression:-

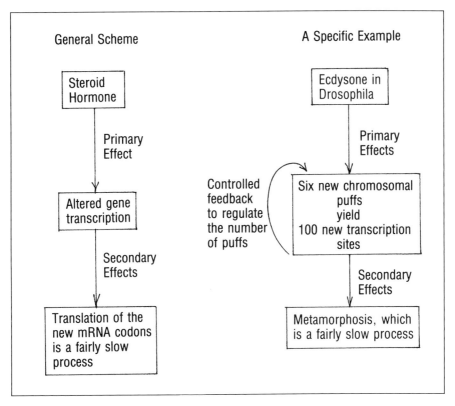

Defective cell receptors can prevent the modification to gene expression caused by a steroid hormone:-

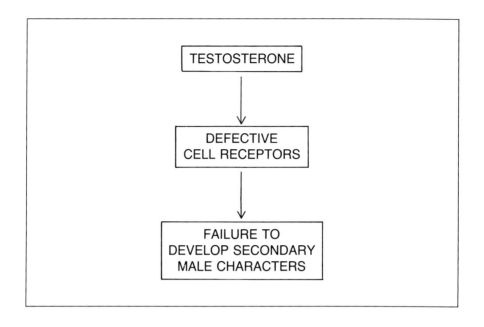

The Hormonal Control of Gene Expression in Plants

Plant hormones are organic substances (other than nutrients) needed in very small quantities, made in small localities in plants and transported to other sites of plants where they cause biochemical, physiological and morphological responses. They are growth regulators acting with other compounds such as nutrients and water.

Auxin (Indole Acetic Acid – IAA)

Only one naturally occurring auxin is known. Auxin causes an increase in mRNA synthesis. There is also a lesser increase in the number of tRNAs and rRNAs. It is possible that the gene specifically affected by auxin is the one that controls the synthesis of the enzyme RNA polymerase, which is needed for the synthesis of all three types of RNA. The exact mode of action is unknown, so the following tentative scheme, which shows auxin binding to a polypeptide to form a complex that modifies RNA transcription, is not yet proven:-

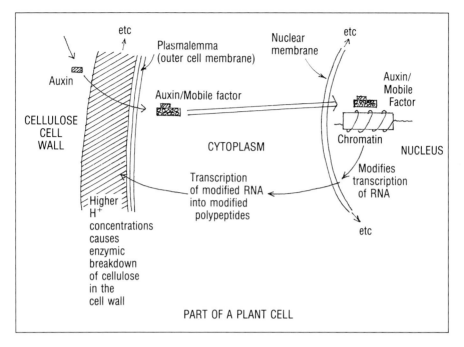

PART OF A PLANT CELL

Another theory put forward to explain the action of auxin on gene expression specifies that the modification of RNA polymerase leads to the release of H^+ from the endoplasmic reticulum or golgi channels into cell walls. H^+ may alter the shape of cellulases so that these enzymes loosen cellulose to allow cell elongation, using the internal turgor of the auxin affected cells. Why auxin should have this effect largely on longitudinal cell walls and then only in very limited part of the shoot and root is not known.

It should be noted that auxin operates in two stages:-

(i) At the biochemical level.
(ii) There is then a physiological consequence.

(i) can lead to several different (ii)s, depending on the species of plant and where cells are found in individual plants. Any general deductions about the influence of auxin on genes is likely to be very inadequate at present. It should be noted that IAA (auxin) is similar to a neurohormone in animals called serotonin. Serotonin is involved in the directional (polar) movement of nerve impulses across some synapses. Synapses are the small, fluid-filled regions between nerve endings and the next nerve in the nervous arc or between nerve endings and an effector such as a muscle.

The Modification of Gene Expression by Gibberellic Acids (GAs)

There are several naturally occurring gibberellic acids. Some aspects of gene transcription seems to be under the control of gibberellic acids and therefore the synthesis of some RNA molecules is modified by GAs (however GA_8 is known *not* to control transcription).

In general the addition of GAs causes more ribosomes to be made and the membranes of the endoplasmic reticulum also increase.

In particular those enzymes involved with the release of energy (hydrolases such as α–amylase) are produced in response to GAs.

The precise role of GAs in the increase of cell size and how they act synergistically (in unison) with auxin and cytokinin, and antagonistically (in opposition to) with abscisic acid is not known.

The Modification of Gene Expression by Cytokinin (K)

Cytokinin (K) operates with auxin and gibberellins to control the direction of plant cell growth, promoting lateral cell growth. Cytokinin also promotes cell division in growing regions of plants (it is of interest that cytokinin also promotes cell division in bacteria).

Cytokinins are known to exist in several forms in plant cells – free bases, ribonucleosides and ribonucleotides. They are also found as constituents of some of the tRNAs, near the position of the anticodon:-

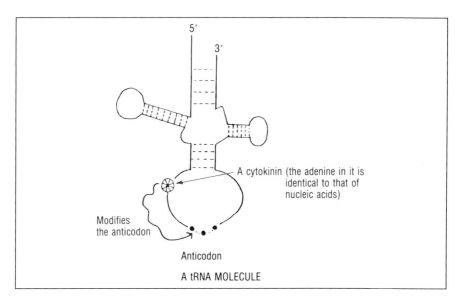

When cytokinins in the tRNA alter the ability of the tRNA to associate its codon with the anti-codon of mRNA, alterations are made to the amino-acids that are inserted at the positions of the codons. This can modify the function of polypeptides into which the amino-acids are inserted, but the details of how this causes changes to cell growth and cell division are not yet known.

The Modification of Gene Expression by Abscisic Acid (ABA)

Abscisic acid is a plant hormone that prevents growth and encourages senescence and dormancy. Some of the ways in which abscisic acid operates are:-

   i. The inhibition of the synthesis of some RNA molecules, especially those controlling hydrolytic enzyme synthesis.
   ii. ABA alters the electrostatic charge across cell membranes.
   iii. ABA causes the movement of K^+ out of guard cells, causing the stomata to close.
   iv. The creation of weakened cross-sections of leaves, near the stem end of petioles, creates the abscisic layer, which is the place where the petioles break at leaf fall.
   v. ABA acts as an antagonist to auxin, gibberellic acids and cytokinin.

The relationship between the genetic ways in which ABA decreases RNA transcription and the ways that growth promoter plant hormones increase RNA transcription is unknown.

The Modification of Gene Expression by Photoperiod

Light is known to alter the activity of a large number of enzymes. The synthesis of some enzymes, their degradation, activation and interconversion, as well as the transport of enzymes have all been shown to be controlled by light.

One plant enzyme that has been studied is phenylalanine ammonia lyase (PAL). This enzyme is synthesized when far red light is received by the phytochrome system. PAL is used to catalyse one of the stages of the biochemical pathway that converts phenylalanine into a part of the anthocyanin molecules and parts of other flavonoids.

Thus the activation of a phytochrome by far red light alters the expression of a gene involved in the synthesis of anthocyanins and other flavonoids.

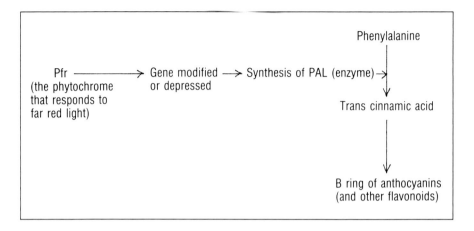

3. The Molecular Biology of Ageing

The repression and derepression of each cell's genes occur so that the 'switched on' set of genes maintains the cell in its differentiated form. Any damage or change to the set of genes needed by each differentiated cell could result in the malfunction or death of individual cells to the point that there are so many dead or malfunctioning cells that the whole organism dies.

DNA damage caused by mutations can be caused by a number of cellular, dietary or environmental factors:-

A. Cellular Factors

(a) Free Radicals

Some oxidases (enzymes such as cytochromes) that reduce oxygen in the inner mitochondrial membrane can make hydrogen peroxide, H_2O_2, and this chemical can, in the presence of variable valency metals such as iron, react to produce hydroxyl ions:-

$$H_2O_2 + 2Fe^{2+} \rightarrow 2OH^- + 2Fe^{3+}$$

(free hydroxyl radical)

Some other enzymes can add an electron to oxygen to make the superoxide free radical

$$O_2 + e^- \rightarrow O_2^-$$
(free superoxide radical)

The free radicals can be toxic in both these cases.

Free radicals are produced in the mitochondrial and chloroplast electron transport systems, and are therefore made in the inner mitochondrial membrane and in the thylakoids of chloroplasts. Ultra-violet light, X-rays, γ-rays, poisons and metal ions also create free radicals.

Membranes, enzymes and DNA are all damaged by free radicals.

Some naturally occurring molecules such as vitamin E, catalases and peroxidase enzymes as well as glutathione (a tripeptide) all reduce the toxic effects of oxygen. If the genes that control the synthesis of those molecules that give such protection from free radicals are altered, cell decay is accelerated.

Free radicals alter the number of DNA to DNA, and DNA to chromatin cross-linkages (page 483).

(b) Changes in Body Temperature

Body temperature changes can cause the loss of bases from one strand of DNA's double helix. This can be followed by the resulting single stranded region of DNA breaking more easily.

(c) Mutations during Mitosis or During DNA Repair

A number of cells receive DNA containing mitotic mutations. In addition, DNA may have incorrect sequences of bases inserted into it by enzymes to repair damaged regions. In either case a number of genes will operate incorrectly and, if there is a continuous increase in the number of incorrectly repaired regions, the death of the cell may follow.

B. Environmental Factors

Chemicals (including carcinogens), viral DNA, u/v radiation, γ-rays and X-rays all damage DNA in differentiated or dividing cells.

Note that in a differentiated cell much of the DNA is never expressed. Some 'genes' have multiple copies, so damage to one of the copies may have little effect on the polypeptide concentration resulting from the translation of the undamaged copies of the 'gene'. If the damage occurs in the recessive allele of a gene it may have no consequences for the cell, or for the cells that are further produced from it by mitosis. However, mutations to repressors or to derepressors, to hormonally controlled genes or to genes that control cell division could have more serious consequences. There is not yet any clear evidence that the degree of somatic mutation alters the rate of ageing of cells. It is not known how somatic mutations which change snRNAs, and thereby introns and exons, cause alterations to gene expression.

The Effects of Cross-Linkages of DNA to DNA and of DNA to Chromatin on The Synthesis of the Enzyme RNA Polymerase

It is estimated that between 6% and 9% of Drosophila DNA/chromatin is joined by cross-linked strands and that this percentage does not change with age. However the ability of the DNA/chromatin complexes to make RNA polymerases does decrease with ageing. By contrast, if DNA/chromatin is extracted from old cells and purified its capacity to make RNA polymerase is returned to that of a young cell. It may be that alterations to the positions of cross-strands with age alter the ability of certain genes to transcribe their mRNAs.

Age-related Changes in Cells

Although there are a few observed, regular features of ageing cells, it has so far not been possible to link the biochemical and morphological ageing changes in cells with changes in DNA/chromatin.

(i) Lysosomal Enzymes

Lysosomes contain enzymes that can digest the contents of a dying cell. In some studies the concentration of lysosomal enzymes increases with age, but whether these are free in the cytoplasm or still contained within the membranes of the lysosomes of old cells is still uncertain.

(ii) DNA to DNA, and DNA to Chromatin Cross-linkages

Cross-links may reduce the ability of a cell to repair damage to DNA. Cross-linkages may increase and alter with age, so reducing the transcribing ability of DNA.

(iii) The Efficiency of DNA Repair

The life span differences of different species may be related to differences in the ability of their cells to repair DNA damage with increasing age. In particular the error frequencies of some cytoplasmic DNA polymerases in human senescent cells have been shown to be three and a half times that of young human cells. Mispairing between G and T is the most frequent mistake in test tube conditions.

(iv) Membrane Fragility

Changes in the lipid/protein ratio of cell membranes cause greater fragility to membranes with age.

(v) The Movement of Molecules and Ions Across Membranes

Changes in the lipid/protein ratios in membranes cause changes in the ability of membranes to move molecules and ions across them and there do seem to be changes in both of these characteristics with age.

(vi) The Attack of Damaged Cells by Phagocytes

If cells rupture more easily when they are older by having less resilient membranes, they may be attacked by white blood cells which ingest them.

(vii) Disintegration of Cellular Components

There is an increased disintegration of the rough endoplasmic reticulum, a faster degeneration of muscle myofibrils and a more rapid deterioration in mitochondria with increasing age.

(viii) The Reduction of the Specificity of tRNAs for their Amino-acids

tRNAs seem to become less able to be specific for their particular amino-acid in ageing cells. There is the consequent insertion of the 'wrong' amino-acid when the tRNA brings it to the codon and this can have a damaging effect on the efficiency of polypeptides.

Conclusion

The modification of gene expression, caused by fundamental genes and their interactions with many other genes, makes the simplistic genetic crosses of earlier chapters inadequate in describing the cellular processes which lead to character expression. When hormones, differentiating chemicals, fundamental genes, ageing, suppressors and promoters are all considered together, the complexities of gene expression in higher organisms are still very imperfectly understood.

It will not be long, however, before the processes of cellular degeneration can be slowed down or even stopped altogether by the ingestion of drugs that prevent the ageing processes. The 'immortality' of man may one day have two alternatives, one rather less ethereal than the other!

Chapter 14 Questions

1. If mitosis provides identical genomes in cells arising from the zygote, why is gene expression different in different cells?
2. What is totipotency?
3. What causes differentiation?
4. What is the homeobox and what is its function?
5. How are spatial orientation, segmentation and differentiation controlled in the embryos of Drosophila melanogaster?
6. How do mobile genes control the differentiation of mammalian antibody production?
7. Give examples of 'mother-effect' in Drosophila melanogaster and in Limnea pereger.
8. How is gene expression modified by hormones:-
   a) in animals?
   b) in plants?
9. What is the damaging effect of oxygen in cells?
10. Apart from oxygen, what environmental factors accelerate the processes of ageing?

15

CANCER

Cancer cells are those cells which divide out of control to produce growths (also called tumours) made of new cancer cells which fail to differentiate. In laboratory conditions an animal cancer cell divides roughly once every day to produce a thousand million undifferentiated cells in a month. However, in living animals, cancer cells probably reproduce less rapidly than this, but there are nonetheless an enormous number of unwanted cells produced in cancerous tumours in a short time.

Cancerous growths occur in plants as well as in animals.

In animals many cancerous growths are detected in places quite far removed from the site where the original cell was transformed into a cancer cell. Cancerous cells can be carried around animal bodies by the mass flow systems of blood and lymph to be deposited in or on tissues to cause growths. The growths are made of cells having a different shape to the cells of the local tissue around them. The spreading process is called to METASTASISE and the growths are METASTASES.

In some cases cells which are transformed, to divide out of control, remain in a small locality, close to the original transformed cell. These growths are said to be benign and offer a negligible problem by comparison to metastases because they can be removed without any further uncontrolled cell division anywhere else in the animal's body.

Metastases

If a growth *is* malignant, the cancer cells in it do not all cling tightly to each other, and some break away and pass into the mass flow systems to spread widely through the body. The white cells of blood regard cancerous cells as 'foreign' invaders and there is some ingestion of cancerous cells by them. By occupying many white blood cells, the presence of cancerous cells in the blood stream lowers the resistance of the body to other invading diseases. Cancerous cells move into the lymph system where they collect in the lymph nodes. The detection and analysis of cancers are often made by draining lymph nodes and examining the cells in the lymph.

Cancer cells have, on average, about the same degree of independent mobility as an amoeba, so they can move into tissues after transportation to them in blood or lymph. Those cancer cells that lodge on, or move into body tissues begin to multiply locally. Blood vessels extend into the tumour and feed the cancerous cells, to the detriment of the normal cells around the tumour.

Cancer cells absorb sugar more rapidly and respire anaerobically to a greater extent than normal cells. They are also more immune to bacterial or viral infections than normal cells. All these properties give cancer cells a competitive advantage over normal cells. The degree of malignancy depends on the rate of tumour cell division, the degree of failure to differentiate, the rate of metastases and, to some extent, the proportion of the tumour provided with an adequate blood supply.

Some regular metastases sequences have been observed:-

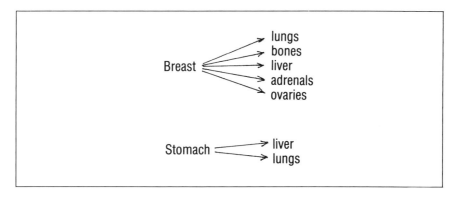

The Genetic Basis of Cancer

The word used for those genes that cause cancers is ONCOGENES. Several oncognes have been identified, some of them in man. It has proved to be difficult to identify the oncogenes, but there are signs of more rapid progress being made now that basic techniques for the identification of oncogenes have been established.

If care is taken to prevent the invalidation of results by ensuring the exclusion of cancer inducing chemicals (CARCINOGENS) from food, drink, air or surfaces that come into contact with animals, experiments with animal cells (including human cells) can provide some information about the genetic causes of cancer. The molecular study of cancer in humans is, of course, restricted to those cells that are removed from a person without causing serious damage – cheek cells from the mouth, lymph cells and blood cells can be studied.

At the present state of knowledge the following genetic facts about cancer cells have been established:-

(i) Cancer cells usually arise in those cells in which there is the replication of chromosomes. There are many cells in the body that have the capacity to divide either as a matter of routine (white blood cells and cells lining the gut) or in response to damage. There are 10^{12} new cells produced in an undamaged, adult human each day, so cancer can arise in mitotically dividing cells throughout a lifetime.

(ii) Mutations can occur in replicating cells to create oncogenes at mitosis or meiosis.

(iii) Mutations can occur in non-dividing cells if the cells are subjected to carcinogens, or penetrating radiation such as γ-rays.

(iv) Some individuals inherit a set of genes in which cancerous mutations are more likely to arise than in other individuals. Laboratory animals have been selectively bred to be susceptible or resistant to cancer. Human cancers of particular types often run in families.

(v) DNA from viruses can transform mammal cells into cancer cells, proving that viruses are the carriers of some oncogenes.

(vi) RNA from those viruses with an RNA genome, in the presence of the enzyme reverse transcriptidase, can be used to make DNA that transforms dividing, normal mammal cells into cancer cells.

(vii) In some cases it has been proved that cancer can be transmitted from one animal to another by way of viruses passed between them.

Cancerous Mutations That Have Been Identified

(i) A Point Mutation

R. Dhar, of the American National Cancer Institute, has identified a single base mutation that can change a normal gene into an oncogene (page 506).

(ii) Translocation of DNA between Chromosomes

The genes at one end of a chromosome can sometimes break away from their normal position on their chromosome and move to become attached at one end of another chromosome. In its new position, one of the genes (or the set of genes) in the translocated segment or in the chromosome to which it attaches is converted into an oncogene. The reason for transformation caused by translocation is not known. In humans, translocations between chromosomes 8 and 14 can switch on an oncogene.

(iii) Gene Amplification

If a normal gene is made in large numbers by amplification so that the repeated genes occur along a continuous length of DNA, the repeated genes can specify different phenotypic consequences compared to those of the single gene operating in the usual diploid condition. Gene amplification can convert some genes into oncogenes.

Viruses That Cause Cancer

Viruses were suspected of causing cancer as long ago as 1908, but it was not proved that some viruses can cause the transformation of normal cells into cancer cells until 1951. Viral DNA can be incorporated into a host cell genome by crossings-over when the host cell DNA replicates during cell division. A viral oncogene usually has to be incorporated into part of the host cell genome before it can exert its influence to cause cancerous cell division.

In the case of RNA viruses, the viral RNA has to be transcribed into the DNA copy by the enzyme reverse transcriptidase (RNA viruses have a gene for this enzyme), which is then made double-stranded by DNA base pairing, using enzymes produced by the host cell. Transformation of the host cell can then take place if replication of the host genome occurs and if an oncogene occurs in the double-stranded DNA copy that is incorporated into the host cell genome. In a similar way, a copy DNA fragment from reverse transcription of the virus RNA may contain the gene for reverse transcriptidase and, if this is incorporated into a host cell genome, the host cell encourages the transcription of RNA from viruses into DNA, and some of the DNA may contain oncogenes.

Viruses that depend on reverse transcriptidase for the synthesis of their DNA analogue are called retroviruses. All known cancer-inducing RNA viruses are retroviruses. Retroviruses are broadly divided into those that spread to a high proportion of cells very quickly (acute transforming retroviruses – ATRs) and those that spread from cell to

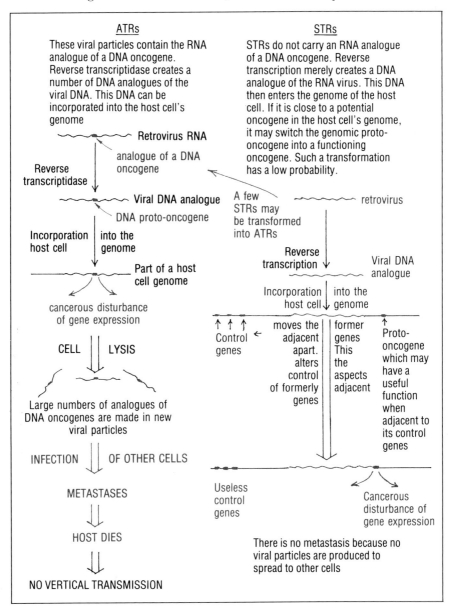

cell quite slowly (slow transforming retroviruses – STRs). Although it has yet to be proven, it is thought that some STRs can be changed genetically so that they become ATRs. An example of an ATR can be found in cancerous chicken cells. The diagram on page 490 summarises the mode of action of both ATRs and STRs.

It is likely that among the criteria that allow retroviruses to create oncogenic expression in host cells are:-

i) DNA sequences in the genome of the host cell must be altered by DNA analogues of the virus so that proto-oncogenes are switched into harmful oncogenes when DNA analogues of viruses deregulate the expression of the oncogene.

ii) Multiple copies of proto-oncogenes are derepressed by the viral DNA analogues that enter the genome.

iii) The viral DNA analogue has an oncogene in it, and multiple copies of the DNA analogue are incorporated into the genome of some host cells.

iv) That the DNA analogues of some retroviruses do not necessarily have to become incorporated into the host cell genome in order that the expression of the DNA analogues' oncogene occurs. Human T-cell lymphotropic virus type I (HTLV - I) may be able to cause leukemia without becoming incorporated into the human genome.

v) The general genetic inheritance of an individual, in which potentially cancerous retroviruses are found, will either resist or assist cancerous expression of viral base sequences.

Certain hormones have a controlling influence on the expression of virally induced cancers.

Of course, some proto-oncogenes in host cell DNA can be switched into oncogenes by several carcinogens other than retroviral particles. Some of these, including DNA viral particles, are mentioned elsewhere in this chapter.

It has not yet been clearly shown that DNA viruses become incorporated into the genome of their hosts before their oncogenes can induce cancerous cell divisions. It seems therefore that oncogenes in DNA viruses can be used to make polypeptides in their host cells that deregulate the mitotic cell division mechanism.

It is possible that DNA viruses contain some DNA equivalents to the oncogene analogues found in the RNA of retroviruses. It has yet to be established that there are proto-oncogenes in the genomes of host cells that are the same as oncogenes found in DNA viruses.

The Origins of Viral Oncogenes

Viral oncogenes were probably lifted out of their host cells, incorporated into viral base sequences, and have been exerting their damaging cancerous effects ever since. How long ago each oncogene was taken into viral DNA analogues or DNA is not known. Another uncertainty is whether or not oncogenes, as we know them now in viruses, were oncogenes at the time they became incorporated into viral base sequences. It is possible that former, harmless base sequences in viruses became cancerous later, either by mutations or by picking up or losing base sequences during the processes of entering and leaving their host cell's genome.

ATRs can pick up oncogenic analogues from contemporary host cell DNA. About two dozen host cell proto-oncogenes are known to be able to transform harmless viruses into ATRs. A very small number have been found in the human genome.

ATRs can only multiply in their host cells if otherwise harmless viral particles of a similar strain invade the cell. These otherwise harmless viruses can make polypeptides that ATRs cannot make for themselves, so allowing ATRs to multiply and to transform other cells into cancer cells.

Some new mutations in cells can, very infrequently, create new oncogenes or proto-oncogenes. Such mutations can be picked up as viral analogues at any time, and later become very numerous in replicating viruses.

The Lack of Any Known Connection between Specifically Identified Polypeptide Products Derived from Oncogenes that Lead to Uncontrolled Mitotic Cell Divisions

One of the less successful aspects of cancer research has been the failure to connect oncogenes with those polypeptide products that cause uncontrolled mitotic cell divisions in cancer cells.

The complex nature of introns and exons, together with the inter-regulation of upstream and downstream genes, makes the detective work of tracking down the functional regions of proto-oncogenes very difficult. More difficult still is finding the connection between these proto-oncogenes and the mitotic mechanism. At present, very little is known to connect oncogenic polypeptide products and the unleashing of cancerous cell multiplication.

The synergistic (acting in concert) and antagonistic (acting in opposition) inter-relationships between polypeptides derived from oncogenes, other gene products and the inducement or repression of other genes are unknown. The complicated inter-relationships of inducers and repressors and their interwoven network of controls are a knowledge void at present. When more is known about ultimate upstream genes, it may then be possible for scientists to work out the inter-relationships of genes and gene products downstream of them. Some of these genes will have complex inter-relationships with oncogenes.

An example of an ultimate upstream gene, which has, as yet, no known link with cancer, is given in Chapter 16.

A small number of genes have been identified that produce polypeptides which seem to have a function in the regulation or deregulation of mitosis. Among these are the src, sis, fos, myc, Ha-ras, Ki-ras, N-ras, cdc 2, mos, and abl proto-oncogenes. It seems that different tissues respond in different ways if these genes are changed into oncogenes.

When these genes are changed from proto-oncogenes into true oncogenes they switch on the production of polypeptides such as kinases. Such polypeptides regulate the uptake of specific metabolites in cells and these metabolites can sometimes cause cell growth and cell division. How they operate is not known.

DNA Cancer Viruses in Species Other than Man

When a virus containing a DNA oncogene enters a cell it loses its protein coat and its genes are translated into polypeptides. One consequence of this can be the onset of the replication of DNA in the host cell genome and cell division follows. Transformation of some of the host cells converts them into cancer cells which continuously divide, if they are provided with nutrients and air (some cancer cells in laboratory conditions have divided continuously for 20 years). Once the transformation of a host cell has occurred, the viral particles do not always persist in the cell, so that cancerous viral particles are not always passed on to the subsequent generations of cancerous cells. Therefore the absence of virus particles from cancer cells does not necessarily mean that the origins of the cancerous growth was *not* viral.

The following DNA cancer viruses have been identified:-

(i) Papoviruses

Of these medium sized viruses studied so far, Polyoma and SV-

40 are the most cancer inducing. They multiply in the nuclei of their host cells. Polyoma can cause cancer in a variety of animal tissues whereas SV-40 causes only flesh cancers (sarcomas). Neither of these viruses have yet been identified in humans.

(ii) Adenoviruses

These viruses have been found in the human breathing tract, but the only evidence of their cancerous properties comes from laboratory mammals in which they cause glandular cancers. Again these viruses multiply in the nuclei of their host cells.

RNA Cancer Viruses in Species Other than Man

RNA viruses with the potential to cause cancerous growth have been found in plants, insects, arthropods, birds and mammals. Most animals have 'inactive' RNA cancer viruses in nearly all their cells. The transmission of these viral particles can be made from generation to generation in sperms or eggs (VERTICAL TRANSMISSION) or from individual to individual by way of faeces, saliva or milk, (HORIZONTAL TRANSMISSION). Once analogous DNA has been transcribed from the viral RNA in the host cell, the incorporation and action of the analogous DNA containing an oncogene is similar to that of DNA cancer viruses.

Some examples of RNA cancer inducing viruses are:-

(i) Mouse Mammary Virus – Bittner (B-type) Virus

This RNA virus is passed from mother's milk into the offspring to cause mammary cancer in the daughters only.

(ii) Leukemias, Lymphomas or Sarcomas (flesh cancers) – C-type Viruses

Inactive C-type viruses are found in nearly all the cells of animals bodies. They can be activated to cause cancerous growth by hormones, carcinogens (cancer inducing chemicals), penetrating radiation or other viruses. Leukemia and lymphoma have been proved to be caused by C-type viruses in cats, apes, cattle and chickens.

(iii) Cat Leukemia Virus

Cats are susceptible to an infectious viral cancer, especially when they are young. The virus invades lymphocytes to convert them into cancerous cells. It also transforms other white blood cells into cancerous cells to cause leukemia. This virus is different from the C-type.

Oncogenic Viruses in Humans

Although they are not always present, virus particles or the detection of reverse transcriptidase in cells can point to cancer of viral origins. However, some normal cells transformed into cancer cells by viruses do not pass on either viruses or the gene for reverse transcriptidase to the subsequent generations of cells. Both virus type DNA and virus type RNA have been observed in human cells. Some oncogenic viruses can be passed from person to person.

Human DNA Cancer Viruses

(i) Herpes Viruses

These large viruses multiply in the nuclei of human cells and Herpes simplex type I and Herpes simplex II are known to cause cancers in humans. Type II is the major genital venereal disease of the last decade. The herpes virus Cytomegalovirus causes Kaposi's sarcoma, a skin cancer. This cancer is often found as one of the consequences of AIDS.

(ii) Epstein Barr Virus

This virus causes the uncontrolled division of mononuclear white blood cells. There are two types –

> infectious lymphoma, which disappears after a few weeks
> and
> lymphoma, which is continuous and causes death.

Lymphoma is common in Africa following malaria in children or young adults – called Burkitt's lymphona in Africa. Epstein Barr Virus has a different cancerous effect in South East Asia, where it causes cancers of the nasal cavity and the pharynx. Diet may be a factor in the expression of the cancer.

Epstein-Barr virus also causes nasopharangeal sarcoma.

(iii) Hepatitis B Virus

Viral DNA analogues have been found, corresponding to the base sequences in hepatitis B virus, incorporated into human cancer cells that arise in the liver.

(iv) Papillomaviruses

These small DNA viruses cause benign tumours in warts, and may also cause cancers of the skin, genitals and cervix.

Human RNA Cancer Viruses

The virus HTLV - I causes the cancerous multiplication of T-cells in human blood. This retrovirus seems to be associated with the place of birth of a person who develops this type of cancer. It acts as a very slow transforming retrovirus, and many people who suffer from this type of retrovirally induced cancer do so quite late in life.

The Relationship Between a Human Gene That Causes Controlled Cell Division When in Blood Platelets Repairing Wounds, and an Oncogene in Simian Sarcoma Virus

Three teams of scientists, one working at the Imperial Cancer Research Fund Laboratories, London (M.D. Waterfield, G.T. Scrace, N. Whittle and P. Stroobant), another at the University of Uppsala, Sweden (A. Johnsson, A. Wasteson and C.H. Heldin), and a third at The Jewish Hospital, St Louis, USA (J.S. Huang and T.F. Deuel) have found a genetic similarity between a human gene and an oncogene in a cancerous virus.

There is a sequence of 104 amino-acids common to the oncogenic virus Simian Sarcoma Virus, which is found in monkeys, and to a polypetide that causes cells on the sides of wounds to multiply in humans. The virus has not been found in humans.

The human protein is called 'Platelet Derived Growth Factor'

(PDGF), which is stored as granules in blood platelets and is only released to attach to specific sites on cell surfaces close to damaged blood vessels. When PDGF sits on the surface of a cell it attracts monocytes and neutrophils (both white blood cells), fibroblasts (cells which form and maintain the shape of connective tissues such as collagen), and smooth muscle cells, all of which are needed for tissue repair. PDGF also encourages mitotic cell division to form scar tissues in wounds.

The oncogene in Simian Sarcoma Virus is identified by the notation $p28^{SIS}$. The identical amino-acid sequences correspond to residues 67 to 171 in the $p28^{SIS}$ oncogene and to the codons derived from part of a gene in human chromosome pair 22. The cross-connection to identify the similarity of the two genes was made using the NEWAT protein data computer of Professor R.F. Doolittle at San Diego University, USA.

PDGF and the $p28^{SIS}$ polypeptides have their common amino-acid sequence near the carboxyl- end of the molecules. The amino-acid sequences at the amino- end of each molecule are different.

If it can be discovered why there is a difference between the expression of the $p28^{SIS}$ cancer inducing polypeptide and the controlled mitotic cell divisions caused by the PDGF, then a cure to negate the cancerous, transforming effects of the Simian Sarcoma Virus base sequence may be found.

About a year after $p28^{SIS}$ and PDGF were found in 1983, another viral oncogene was found by M.D. Waterfield and his team in London. The polypeptide encoded by the viral oncogene, V-erb-B, binds to a growth factor receptor on the surface of cells. The normal epidermal growth factor switches on the mechanism for cell replication.

Some Facts about Human Cancers

A fifth of the people who die each year in a developed country, such as the USA, have cancer as the ultimate cause of death. The statistics for underdeveloped countries cannot be directly compared with a country like the USA unless accurate surveys are made for people of comparable ages – with ageing there is an increasing proportion of deaths caused by cancer. From the data collected so far there is a general tendency to less cancer in underdeveloped countries, but local 'natural' environmental factors can increase the incidence of cancer – malaria or diet for example. Man-made environmental factors such as industrial pollution, or cigarette smoke greatly increase the number of cancerous mutations in dividing and, perhaps, in non-dividing cells.

About half of human cancer deaths are caused by the ineffectiveness of white blood cells in combating disease. A quarter of cancer sufferers are killed by the malfunction of an essential organ. 10% die from starvation and, in 7% of cases, blood platelets are not formed, so causing extensive internal bleeding.

Genetic inheritance and environmental factors influence the eruption of human cancers.

It is believed that many of the deaths at present attributed to cardiovascular problems should more correctly be attributed to cancerous growths blocking blood vessels.

Carcinogens and Human Cancer

The associations of particular cancers with occupations have been recognised for two centuries. Lists, which grow continuously longer, are being compiled of those chemicals in drinking water, food, medicines, and in industrial or domestic products that have proved to increase specific cancers in animals or humans.

The smoke from tobacco contains about 50 known carcinogens and cancers of the larynx, mouth, bladder and kidneys, as well as lung cancers, are higher in smokers than in non-smokers. Statistical correlations allow no doubt at all about smoking being a cause of cancer.

Preventive precautions for people to observe are that they should live in clean air, drink pure water, eat a carcinogen free diet and avoid smoking or radiation as far as possible.

Radiation and Human Cancers

There is always some radiation from outer space and from the radioactive elements in rocks or soil. Usually the levels from these sources are not very high. Ultra-violet radiation and weakly penetrating α particles can cause mutations to give rise to skin cancers. More penetrating particles, such as X-rays or γ-rays, can cause cancerous mutations deep inside the body.

Radiation may fragment viral DNA into segments that readily enter the host cells' genome and, if an oncogene is included in a viral DNA fragment, there may be the transformation of a normal cell into a cancerous cell.

Viruses and Human Cancers

Viral DNA and viral RNA (reverse transcriptidase has been found in human breast cancer cells) can transform normal human cells into cancer cells. In some cases (Herpes II) the viruses are transmitted from one person to another.

The Prevention or Cure of Cancers

Sensible preventive precautions by legislation, by industry, by the water authorities and by individuals can greatly reduce the risk of cancers caused by mutations.

A reduced sexual promiscuity reduces the chance of contracting venereal cancers such as Herpes II.

There is some successful surgery and radiation treatment to destroy or remove cancerous growths, but successful treatment depends on the degree of metastases. An example of genetic engineering being used in the fight against bone marrow cancer is given on page 556.

Experiments Designed to Identify Some of The Genetic Causes of Cancer

Background

There have been two different approaches to the genetics of cancer. Most of the best information obtained so far has come from the English speaking scientists of the USA and Britain. Their basic hypothesis is that quite a small number of oncogenes cause cancer and they have tried to identify them in animal cells and in viruses. The second approach has been that of French scientists who have made the assumption that there are quite a large number of genes (perhaps 2,000 or so) that are switched on in mammalian cancer cells but which are not translated in normal cells, and they are trying to identify the gene products and the genes that cause cancer.

It is known that the DNA of many, if not all, eukaryotic species is transcribed as interrupted sequences of bases:-

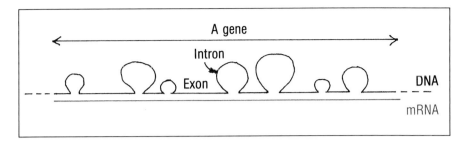

If the locality of one intron is examined in more detail it is seen that the loop of the intron is stabilised by its association, using base pairing, with snRNA. It may turn out that a similar, if not identical, molecule maintains the shape of all the introns and directs their positions between the exons of genes:-

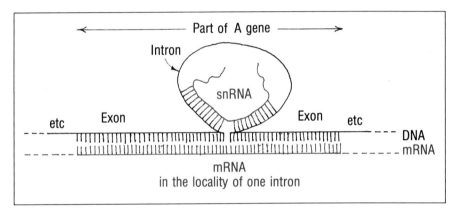

If there is a mutation in the gene where the snRNA is transcribed, then the effect of such a mutation will be widespread among many genes. This is because the boundaries between introns and exons will be changed by the snRNA gene in a large number of genes.

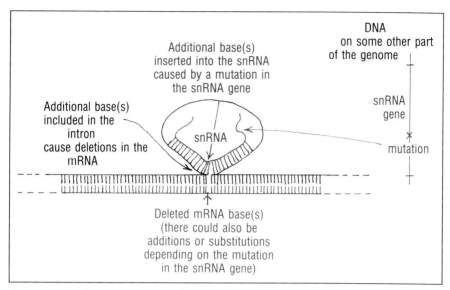

A change of the number of bases in the intron by $(1 \pm 3n)$ or by $(2 \pm 3n)$ will have more effect than a change of $(0 \pm 3n)$ bases (when n is a small number) on the amino-acid codon sequences in the mRNAs transcribed at a large number (if not all) the genes. If this hypothesis is right then a single mutation, or a few mutations in the snRNA genes could create a large number of changes to other genes, the combined consequences of which lead to uncontrolled, cancerous cell division. This hypothesis allows both the English speaking and French scientists to be correct and the next decade should prove the case one way or the other.

The hope is that a simple cure can be found for cancer. If there are only a small number of oncogenes, some of which control snRNA synthesis, there may prove to be ways of defeating the oncogenic polypeptides synthesized out of the genes. If there are a large number of genes that have the potential to cause cancer, any one of them may mutate to become an oncogene so making the prevention and cure for cancers very difficult to find. If the majority of cancers turn out to be induced by viruses, vaccines, or polypeptides with the properties of vaccines, will eventually be found to make human bodies immune to them.

A few of the classic experiments leading to the present state of knowledge about cancer are described very briefly below.

(i) The Proof That Viruses Can Cause Cancer

R. Vogt and R. Dulbecco at The California Institute of Technology introduced viruses into hamster embryo cells which were transformed into cancer cells. It was further shown that a control batch of cells without any viruses added to them, and cells that lose their viruses are not transformed into cancerous cells.

(ii) The Proof That Mutations in Chromosomal DNA Can Cause Cancer

(a) C. Shih at the Massachussetts Institute of Technology used the chemical METHYLCHOLANTHRENE to cause mutations in normal cells and created cancerous tumours. DNA extracted from tumour cells was purified and put into a medium in which there were mouse fibroblast cells (used for the shape maintenance of connective tissue like collagen). Transformed cancerous fibroblast cells were identified within fourteen days of introducing the DNA from tumour cells into the medium where the fibroblast cells were placed. When the transformed cells were introduced into mice, cancerous tumours formed. By contrast DNA extracted from normal cells caused no transformation of the fibroblast cells, which also failed to cause tumours when injected into mice.

(b) It has also been shown that DNA extracted from human cancer cells can transform normal mouse cells into cancerous cells and the introduction of these cells into mice causes tumours. This not only demonstrated the fact that human DNA contains cancerous genes, but also that the genes in the DNA of one mammalian species can cause cancer in another species.

(iii) Locating The Transforming DNA Segment That Causes Cancerous Tumours in Humans

The whole of the human genome has about 6,000,000,000 base pairs in it and contains genes that are each mostly made of 5,000 to 10,000 base pairs. Crossings-over occur in both sexes in humans so it is not possible to establish the position of autosomal genes by recombination frequency.

C. Shih carried out a second classic experiment at Massachussetts Institute of Technology in which he transformed mouse cells using human cancer cell DNA at the start of the experiment. When mouse cells divide they shed some human DNA at each cell division so that, at the end of several mouse cell divisions, there are only small segments of human DNA left in the mouse cell genome. Human

DNA is unique in having the three bases that specify the codon for inserting the amino-acid alanine occurring as adjacent, repeated, triple bases in quite long segments of human DNA. An indentifying probe was used to establish the fact that human DNA was incorporated into the cancerous mouse cell genome by recognising the presence of the repeated, alanine-determining DNA base sequences in the mouse cell genome.

Shih established that it was the human DNA segment that caused the cancerous growth in mouse cells by transferring small human DNA fragments from cancerous mouse cells into normal mouse cells. He found that the normal cells were transformed by the human DNA into cancer cells.

R.A. Weinberg and C.J. Tobin at the Massachussetts Institute of Technology and M. Wigler at the Cold Harbour Laboratory, New York State, USA followed up the experiments carried out by Shih by working out the base sequence of a human oncogene. They started their experimental work by making hybrid DNA, some taken from a human genome and the remainder coming from a virus called Bacteriophage λ. The reconstitution of λ viruses containing hybrid DNA was followed by placing the viruses in isolated localities on E. coli growing on agar.

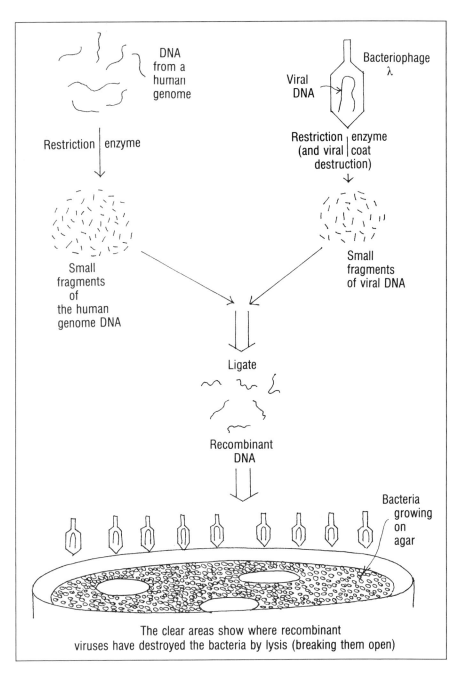

Each region of bacterial lysis contains a very large number of different hybrid phage particles. Following the technique invented by G.M. Cooper of The Harvard Medical School, USA it was necessary to divide the hybrid phage particles into ten groups.

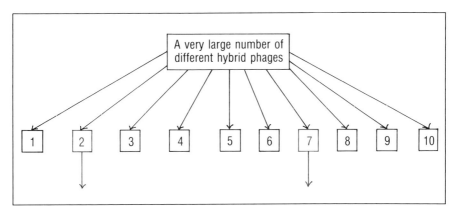

Cancer inducing phage particles were shown to be in a small proportion of the groups. Recombination caused transformation of the cells in these groups and thereby caused the cancerous growth of the cells. Each group containing oncogenic virus hybrid DNA was divided into ten once more:-

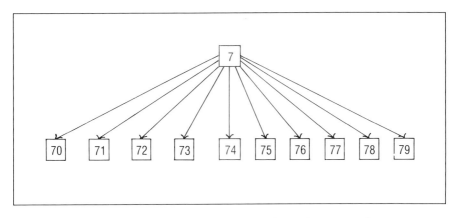

If 74 was the only sub-division showing the capacity to cause cancerous transformation, this group was sub-divided again. Sub-divisions continue in this pattern until the type of virus particle present is only that hybrid containing the transforming human DNA. This is evident when all the sub-divisions show cancerous growth.

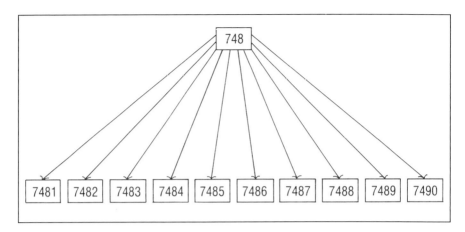

Having reduced the human DNA to a small fragment, and having established the certainty of a human oncogene in the hybrid virus DNA, it was possible to identify the human DNA fragment from its alanine codes.

Techniques to specify the exact sequence of bases in the transforming human oncogene have identified a sequence of 350 base pairs that always appear in the transforming human DNA fragment.

R. Dhar of The American National Cancer Institute found that in normal human cells the 350 base pair sequence of a gene is different from the cancerous 350 base pair sequence by a single base (point) mutation that substitutes T for G in one of the codes of the gene:-

Normal code: GGC   glycine   (via GGC in mRNA)
Oncogene code: GTC   valine   (via GUC in mRNA)

The substitution of one amino-acid for another in response to the substitution of T for G in DNA seems to make a polypeptide with cancerous properties, and the mutation therefore converts a normal gene into an oncogene.

Conclusion

Some progress has been made in identifying the genetic causes of some cancers. There are still many discoveries to be made before all the connections between proto-oncogenes, oncogenes, polypeptides and uncontrolled cell division are fully worked out. No doubt the combined efforts made by scientists who specialise in the genetics of cancer and those scientists involved in the world-wide Human Genome Project will unravel most of the mysteries of cancer by the year 2006.

Chapter 15 Questions

1. What are the characteristics of cancerous cells?
2. What is the difference between a benign growth and metastasis?
3. What are carcinogens?
4. What are oncogenes?
5. What categories of genetic change are known to cause cancer?
6. What are the connections between viruses and cancer?
7. Are there genes in humans with similar base sequences to known viral oncogenes?
8. How were oncogenes identified and located?

16

HUMAN GENETICS

It is appropriate to begin a description of human genetics by making a comparison between some genetically inherited characters in humans, including those of the chromosomes themselves, and the same characters in other primates. A number of scientists met in Italy, under the co-ordination of Professor A.B. Chiarelli, and their papers comparing genetic inheritance in primates were printed together in a book 'Comparative Genetics in Monkeys, Apes and Man'. A summary of the information presented at that time is very briefly described here. The species Homo sapiens is often included when the term 'primate' is used in the text.

Primate Chromosomes

A comparative study of the chromosomes in primates has thrown up some slightly confusing evidence concerning the evolutionary pathways of those primate species that are thought to be most closely related to man. Some chromosomal characters seem to point to one species being a closer relation to man than another; yet the reverse relationship of evolutionary distance seems to be indicated by other chromosomal characters.

(a) Chromosome Number

Species	Chromosome Number	Fluorescence of chromosome 13 and of the Y chromosome	Chromosomes 5 and 12
1. Man	23 pairs	Similar to gorillas	Same as primates*
2. Chimpanzees	24 pairs	Similar to primates*	Different from Primates*
3. Gorillas	24 pairs	Similar to humans	Different from Primates*
4. Primates*	21 to 36 pairs	Similar to chimpanzees	Same as man

Primates* are primates other than 1, 2 and 3 above

Human DNA contains approximately 6×10^9 base pairs in each somatic cell. It is not known yet what proportions of the human genome are introns, exons, spacer regions, mobile or repeated genes. It is possible that as little as 5% of human DNA in each cell is used for transcription.

The bar chart overleaf shows that the range of chromosome numbers for the old world primates (from which the ancestral forms of man are thought to have evolved) varies from 21 pairs to 36 pairs.

Man is the only primate species to have 23 pairs of chromosomes. The evidence collected so far points to the chimpanzee as being man's closest contemporary relation and a comparison between the chromosomes of man and chimpanzees suggests that both species evolved from a common ancestor with 24 pairs of chromosomes. In the ancestors of man there may have been the fusion of chromosomes 13 and 15 (as they appear in chimpanzees) to form what is now defined as human chromosome 1, thereby reducing the ancestral chromosomes by one pair in man's direct ancestors.

The bar chart also shows that 21 pairs of chromosomes is a widespread character of catarrhine (old world) primates, and that there are some numbers of chromosome pairs for which there are no known species:-

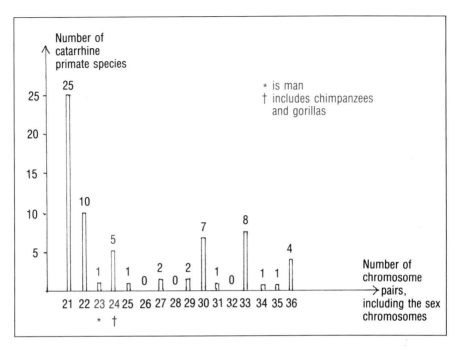

Although not directly related to human genetics, it is of interest that the shape and size of chromosomes in several species of primates that have 21 pairs of chromosomes, are very similar. There have been cases of matings between individuals of 'different species', and even of 'different genera', that have produced fertile, hybrid offspring. It is not clear if there is permanent fitness of the fertile hybrids, but the definition of 'species' in chapter 12 seems to be contradicted by the fertility of these hybrids. Another feature of the '21 pair' primate species is that their chromosomes show great similarities of behaviour at meiosis and this characteristic, together with the similarities of chromosomal shapes and sizes in several species, allows the possibility of viable hybrids. There is not much information about the 'fitness' of such hybrids in their natural habitat.

(b) Human Chromosomes Compared to Chimpanzee Chromosomes

The numerical description of human and chimpanzee chromosomes is given to those chromosomes that are thought to be equivalents in the modern species. The numbers are therefore in some respects different from those in the ancestral equivalents. In particular the chromosomes 13 and 15 mentioned here are different from those ancestral chromosomes that fused to make human chromosome 1.

Very similar   Pairs 3, 7, 11, 12, 13, 18, 19 and 22
Some similarity   Pairs 2, 9, 14, 15, 16, 20 and 21
Different   Pairs 4, 5, 6, 8, 10 and 17

(c) Down's Syndrome in Chimpanzees Caused by Trisomy (having a triplicate instead of a pair of chromosomes) of Chimpanzee Chromosome 22

The fusion of ancestral chromosomes 13 and 17 to form chromosome 1 in man makes chromosome 22 in chimpanzees the equivalent of chromosome 21 in man – the trisomy of chromosome 21 in man causes Down's Syndrome (pages 519 to 521). Chimpanzees with trisomy in chromosome 22 exhibit similar mental and physical characters as humans suffering from trisomy in their chromosome 21. Down's syndrome is easily recognised by the slightly mongoloid appearance of human individuals, even if they have a European mother and father. This mongoloid appearance is also shown by trisomic chimpanzees.

(d) The Quantity of DNA in The Chromosomes

Man has 12.4 (arbitrary units)

Gorillas have 12.6 (arbitrary units)

Macaque monkeys have 11.1 (arbitrary units)

The Comparison of Amino-acid Sequences in Primate Polypeptides

The way amino-acid sequences are used to infer the distances of common ancestral relations from two contemporary species was described in Chapter 11. The polypeptides α-haemoglobin, β-haemoglobin, the GM factor in blood, fibrinopeptides A and B, cytochrome c, lactate dehydrogenase, malate dehydrogenase, phosphoglucomutase, phosphohexose isomerase and five isomers of lactate dehydrogenase have been analysed to compare the differences in amino-acid sequences between some of the species of primates. Each analysis adds a further contribution to the reliability of ancestral trees inferred by differences in amino-acid sequences. Much of the evidence from these analyses points to the chimpanzee as our closest relative, with the gorilla second. The analyses of the amino-acid sequences in the α-haemoglobin of primates have revealed the following comparisons with the sequence in human α-haemoglobin:-

Chimpanzees — identical

Gorillas — only one amino-acid difference (lysine occupies position 104 which contains arginine in humans)

Primates — several differences. The macaque monkey has eight differences; lemurs have thirty differences.

Blood Groups

Tests are available to identify the blood groups A, B, AB, O, M, N and the Rhesus factor. It is also possible to identify the presence of the chemical H, which is the chemical that is thought to start all the biochemical pathways leading to the synthesis of the major blood group substances. The degree of the agglutination reactions of primates' blood to anti-A and Anti-B or anti-Rhesus, and the results of the inhibition saliva test (this test is used when there are no detectable results from the anti-A or anti-B tests) allows an estimate of the degree of similarity or difference between human blood and the blood of other primate species. There are sub-groups of some of the major blood groups, and these can be distinguished by the blood's reaction to agglutinogens. For example the degree of clumping with anti-A is:-

$$A_1(\text{human}) > A_1(\text{chimpanzee}) > A_2(\text{chimpanzee}) > A_2(\text{human})$$

In several primate species there is no agglutination reaction to human anti-A or anti-B, even though they have some A, B, AB or O properties, so other tests, such as the inhibition saliva tests, have been necessary to find out if a primate species has any of the blood properties of the human blood groups.

Once the A, B, AB, O, M, N, Rhesus factor and H allele frequencies have been estimated for each species, an attempt can be made to use the similarities and differences of the blood group allele frequencies to estimate the distances of the relationships between the primate species. Blood groups analyses along these lines have already been used to compare the blood groups of human races, and to infer from them the geographical origins of the races. It seems that European gypsies had their origins in Northern India. The origins of the races living in the East Indies and in Australia have been estimated using the allele frequencies of their blood groups and comparing them with people living on the Asian land mass. Aborigines are thought to be descendants of migrants from Southern India where the population has blood group proportions similar to the Aborigines.

The results of the various types of blood tests have yielded slightly confusing evidence for relating man to other contemporary primates. However the number of individuals used in non-human samples has been small and the gathering of information to complete a study of this sort is necessarily slow, if wild populations are not to suffer from interference from scientists which alters their patterns of behaviour.

Chimpanzee Blood Groups

Blood groups are O and A. A is sub-divided into A_1 and A_2, as in humans. The blood of chimpanzees shows all the properties of human blood of these two groups. There are similar racial differences in the blood group allele frequencies, as in humans. However chimpanzees have two fewer major blood groups than humans, in those individuals tested so far.

Orang-Utangs and Gibbons

Although not having as many similarities to the properties of human blood as chimpanzees, there are three of the human blood groups in these species – A (common), B (rare) and AB (quite rare), but the degree of A, B and AB differs from humans.

Gorillas

Lowland gorillas have only blood group B (their anti-A is different from that of man), while mountain gorillas have only blood group A (in those few individuals tested). Gorilla's blood shows more similarities to the blood of monkeys than to the blood of man.

Old World Monkeys

Their blood shows no significant reaction to anti-A or anti-B, so the inhibition saliva test has to be used. Although the properties of the blood are different in degree to that of humans, there have, nonetheless, been identified all the human blood groups among the range of old world monkey species. Baboons have all four major blood groups. Other species, including the Rhesus monkey, have only one of the major blood groups – some species have only A, others have only B.

New World Monkeys

The blood group tests on new world monkeys and analyses of the agglutinogens in their blood, show that the new world monkeys are more distantly related to man than are old world monkeys.

Lines of descent are best followed by the examination of widespread genes found in organelles donated by gametes. For example, mitochondria are contributed only by eggs. The examination of mitochondrial genes can therefore reveal inheritance, and therefore degrees of divergence, through females. Comparisons of races and species can be made in this way. The ancestry of human races can be compared by examining mitochondrial genes in human females.

This brief introduction to human genetics necessarily omits the evidence gathered from the morphology of primate bones or fossils which shows the evolution of form in the hard parts of the primate species over a period of time. There is an excellent, permanent exhibit at The Natural History Museum, Kensington, London which shows the best of the evidence gathered so far on the evolution of primates, eventually leading to the emergence of Homo sapiens. Although there still remain several unanswered questions, there is much evidence to support Charles Darwin's hypothesis that man has evolved under the same sort of environmental selective pressures as the other species found on earth. Man's capacity to modify the resources of the earth to his own advantage using an ever increasing sophistication of technology is a very recent phenomenon, perfectly 'natural' in the evolution of species (since man is one of them) and it is impossible to make any predictions about how the 'selfish genes' of 'technological' man will alter the selective pressures upon his own species, as well as those acting on all the others. Perhaps there is a long and painful process that occurs in a few isolated places in the universe that leads to the triumphal climax of producing a rational species. Then the flowering of the creative achievements of the 'rationals' is, perhaps, always followed very quickly by their discoveries about the nature of matter, by their misuse of its energy, and leads to the extinguishing of the assemblies of atoms that maintain 'life' on those few spaceships where life emerges from cosmic débris. The processes of natural selection have put into man many qualities, inherited from fiercely competitive ancestors, which make his species very badly qualified to control the use of his increasing technological powers. On a more hopeful note, the rationality of man has also striven for good, either through material idealism or religion. The capacity of the species to give birth to Mozart, Verdi, Turner, Picasso and Shakespeare, to build the great cathedrals of Europe, to make the pottery of China, to eradicate smallpox and to discover antibiotics, gives us the small glimmer of hope that the creative side of the psyche of our species will dominate over the destructive one.

A Mathematical Introduction to Genetic Inheritance in Humans – The Hardy Weinberg Equations

A brief mathematical introduction is needed if an understanding of the significance of allele frequencies in human (or any other) populations is to be reached. If a gene has two, and only two alleles, one of the alternative alleles may have the potential for producing damaging effects. Usually the damaging alleles are recessive so they are only expressed if they are inherited in the gametes from both parents to produce a homozygous, double recessive individual. It is only possible to detect the recessive allele when it is expressed in the double recessive because it is suppressed by the dominant allele in those heterozygous carriers of the allele.

Provided there are only two alleles for the gene there is a simple mathematical method for working out the allele frequencies for a gene, using the percentage of individuals which express the recessive character as a starting point in the calculations.

For example the approximate proportion of people in the United Kingdom suffering from fybrocystic disease is 0.05%. The recessive allele for this disease is expressed in homozygous recessive individuals by the deposition of unwanted material in mucus excreting tissues, such as the pancreas, the lungs and the intestines. There are therefore 99.95% of the population in whom the wild-type allele is expressed for the correct functioning of these organs. The question then is 'can the proportion of detected homozygous recessives lead to the calculation of:-

(i) the allele frequencies in the population?

(ii) the proportion of the population that are homozygous dominants and the proportion that are heterozygous?'

If p represents the dominant allele and q represents the recessive allele, when there are only two alternative alleles in a population, two mathematical statements can be made. Percentages are rewritten in the equational statement using a decimal point:-

$$4\% = 0.04$$
$$0.05\% = 0.0005$$

(i) The Equation for Working Out The Allele Frequency in a Population

$p + q = 1$ (all the wild type alleles added to all the recessive alleles add up to all the alleles in the population for one gene).

Using the fibrocystic disease example we know that $q^2 = 0.0005$. Therefore $q = 0.0224$ (note that when a square root of a number less than 1 is calculated the square root is larger than the square).

So:-

$$p + 0.0224 = 1.0000$$
$$p = 0.9776$$

The allele frequencies are therefore:-

wild type	:	fibrocystic disease
97.76%	:	2.24%

(ii) **The equation for Working Out the Proportions of Homozygous Wild-Types, Heterozygotes and Homozygous Recessives in the Population.**

$$p^2 + 2pq + q^2 = 1$$

(the proportion of homozygous wild type individuals plus the proportion of heterozygotes plus the proportion of recessive homozygotes add up to all the individuals in the population).

In our example, the calculation of the values of p and q have already been made in (i). It is now possible to calculate:-

$$p^2 = 0.9776^2$$
$$= 0.9557$$

and $2pq = 2 \times 0.9776 \times 0.0224$
$$= 0.0438$$

Homozygous wild type	:	heterozygous carriers:		double recessive
p p	:	2 p q	:	q q
95·57%	:	4.38%	:	0·05%

This example shows that, although quite a large proportion of the human population is made up of heterozygous carriers of a harmful recessive allele, the proportion of individuals who are unlucky enough to inherit the double recessive condition is small. This lack of genetic fitness in a small proportion of human individuals often prevents their development as far as breeding offspring of their own. In some cases death occurs early, in others later, but the human species as a whole can afford to lose some genetically defective individuals without any damage to the viability of the species. Of course, the medical profession tries to find solutions for genetic defects and this may, in the short term, increase the proportion of defective alleles in the breeding, human population. However, genetic engineering may eventually lead to preventive precautions being taken by substituting harmful alleles inherited from carrier parents and replacing them with wild-type alleles in eggs or sperms. The genetic screening of embryos may also persuade some mothers to abort embryos which have serious genetic deficiencies, or which are carriers of genetic deficiencies, so preventing the passing on of harmful alleles to a new generation. Genetic engineering and genetic screening may eventually lead to the removal of harmful alleles from the human species.

Of course, the Hardy Weinberg equations will apply to all those genes, no matter in what species, in which there are two, and only two alternative alleles, one of which dominates over the other.

The Identification of Human Chromosomes and Some Techniques for Discovering Aberrations in Them

Each of the 22 pairs of autosomes and each of the sex chromosomes has a characteristic position of the centromere that identifies each chromosome. The relative sizes of the chromosomes at a particular stage of cell division, when they respond to stains, has also been established. There are also distinguishing bands on the chromosomes that can be seen:-

(i) After pre-treatment with trypsin followed by the application of the Giemsa stain.

(ii) Using an ultra-violet illuminated microscope (care has to be taken about how this is done as u/v radiation is *very damaging* to the cells of the retina in the human eyeball) to show up distinguishing bands on each chromosome.

Large aberrations in human chromosomes can be detected because the chromosomes in which they occur show morphological differences of size, centromere position, or patterns of bands produced by staining or u/v radiation.

It is also possible to distinguish between cells taken from males and females because a small capsule (the Barr body) appears just inside the nuclear membrane in female cells, and this capsule is thought to be the second X chromosome. It is of less genetic activity than the other X chromosome which lies free in the nucleoplasm. Barr bodies do not appear in male cell nuclei. It has been demonstrated that some of the genes in the Barr body X chromosome are expressed in the phenotype – the Xg^a allele for one of the red blood cell antigens has been shown to be expressed when it is in the Barr body X chromosome.

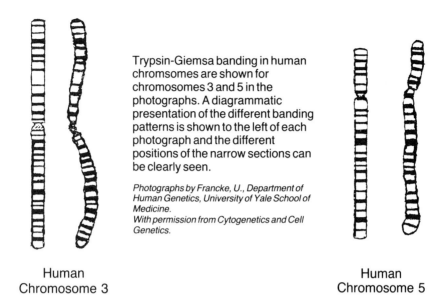

Trypsin-Giemsa banding in human chromsomes are shown for chromosomes 3 and 5 in the photographs. A diagrammatic presentation of the different banding patterns is shown to the left of each photograph and the different positions of the narrow sections can be clearly seen.

Photographs by Francke, U., Department of Human Genetics, University of Yale School of Medicine.
With permission from Cytogenetics and Cell Genetics.

Human Chromosome 3

Human Chromosome 5

Some Genetic Human Defects Caused by Aberrations in Chromosomes either as Large Mutations or as Alterations to Chromosome Numbers

1. Down's Syndrome (Mongolism)

(a) In most people suffering from Down's Syndrome there is an extra chromosome for chromosome 21, making chromosome 21 triploid. It is not yet clear if there is unequal migration of Chromosome 21 in the first or second meiotic division of egg production, or if the unequal migration of the chromosomes occurs in one of the earliest mitotic cell divisions after fertilisation. The increasing chance of inheriting Down's Syndrome with the increasing age of the mother but not of the father points to the egg, or the fertilised egg, rather than the sperm as being the origin of this genetic defect.

(b) Down's Syndrome is also caused by the translocation of a small fragment from one of the pair 21 chromosomes to one of the pair 15 (or 14) chromosomes. It is not yet established whether this occurs in the production of an egg or in the production of the fertilising sperm.

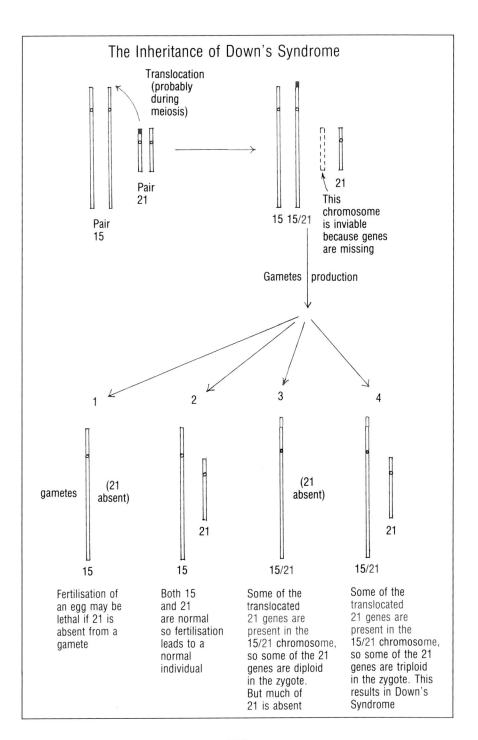

The expressions of Down's Syndrome are:-

   (i) Facial changes that appear Asiatic (mongoloid) to a European.
   (ii) Mental retardation.
   (iii) Stunted growth.
   (iv) Short, incurving little fingers.
   (v) Grey and white flecks in the iris of the eye.

2. An Extra Chromosome for Chromosomes 14, 15, 16, 17 and 18

Triploidy in chromosomes 14, 15, 16, 17 and 18 causes mental defects in humans.

3. Deletion of Part of Chromosome 5 – Cri du Chat Syndrome

A deletion of part of chromosome 5 causes mental deficiency and under-development of the larynx, the latter causing the high-pitched mewing emitted from an individual suffering from this syndrome.

4. A Translocation From Chromosome 22 to Chromosome 9

This translocation has been shown to result in the type of leukemia called Philadelphia chromosome leukemia.

5. An Extra X Chromosome in 'Males'

A small proportion of 'males' receive an extra X chromosome to give them a total of 47 chromosomes, the sex chromosomes being XXY (the incorrect migration of X chromosomes in meiosis or early mitosis after fertilisation causes the cells of the male to have an extra chromosome in all of them). XXY males cannot produce spermatozoa, but otherwise their male functions seem normal. There is a tendency to mental defects and sometimes there is the over-development of breasts (Klinefelter's syndrome).

6. An Extra Y Chromosome in Males

47, XYY males are usually tall, but otherwise lead normal lives.

7. The Deletion of One X Chromosome in Females

45 XO females are usually short, lack secondary female characteristics and do not show the normal symptoms of the oestrous cycle. A Barr body is usually absent from the nuclei of their cells. This condition is usually caused by the omission of an X chromosome in the fertilising sperm. It can also occur if a break occurs within the centromere during early mitotic division in the female zygote.

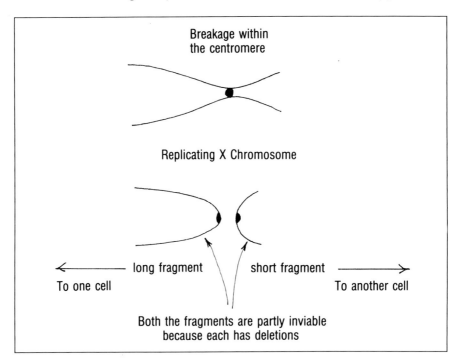

Whenever there is the deletion of an X chromosome or part of an X chromosome the condition is known as Turner's syndrome. Some women are therefore XX, but there are duplications and deletions on one of their X chromosomes. Both complete X chromosomes are needed to produce a normal woman, but the inheritance of the long fragment can allow the normal development of ovaries (even if the woman is unusually short in stature). The inheritance of the short fragment causes incomplete ovarian development, but the woman will have normal stature. Women with a normal X and one or other of the long or short X fragments have Barr Bodies in their cells.

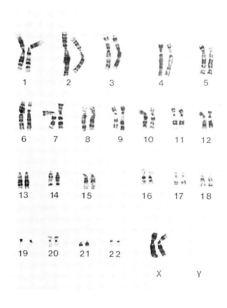

Karyotpe showing the arrangments of chromosomes in a human female with Cri du Chat syndrome. A deletion is shown at the top of the right hand chromosome of pair 5. This individual is written as 46, XX, del (5).

Photograph by L. Willatt..
With permission from East Anglian Genetic Service and The Science Photo Library, London.

The Karyotype, as shown by trypsin/giemsa staining band patterns of a human male showing trisomy (three chromosomes instead of two) in chromosomes 16. Each chromosome of the 22 homologous pairs of chromosomes can be identified by its band pattern and, to some extent, by its size. The X and the Y sex determining chromosomes also have characteristic band patterns and sizes.
Trisomy of chromosome 16 is the most common genetic cause of the spontaneous abortion of human babies. This aborted baby is described as 47, XY, +16.

Photograph by L. Willatt.
With permission from East Anglian Genetics Service and The Science Photo Library, London.

The Inheritance of Human Genetic Defects

1. Autosomal Allelic Defects

a. Dominant Defective Alleles on Autosomes

(i) Huntingdon's Chorea

The (dominant) allele for Huntingdon's Chorea is rare – about 1 in 2,000 people are heterozygous for the gene. The homozygous occurrence of the gene in an outbreeding species such as humans is therefore very rare indeed. It would probably be lethal if ever the homozygous condition occurred.

An individual unlucky enough to inherit the HC allele usually develops the symptoms quite late in life, often after having a family, so there is no means of forewarning an HC heterozygous individual of the likelihood of producing some children with Huntingdon's Chorea. Once the symptoms start there is a steady loss of mental powers and involuntary muscular spasms.

The Huntingdon's Chorea allele obeys the usual rules of the monohybrid cross between a heterozygote and a homozygote, except that the HC allele dominates over the wild-type allele.

	♀		♂
P	HC +	×	+ +
gametes	HC or +	×	+
F_1	HC +	:	+ +
F_1 genotype ratio	1	:	1
F_1 phenotype ratio	Huntingdon's Chorea (develops later in life)	:	normal
	1	:	1

(ii) Dwarfism

The dominant autosomal allele for Chondrodystrophic dwarfism arises as a spontaneous mutation in one of the autosomes during gamete production. Four out of five dwarfs die within a year, but some live to produce offspring of their own. Once the mutation has occurred the same monohybrid cross applies as for the dominant Huntingdon's Chorea allele.

DNA 'fingerprinting' can make use of characteristic unique band sequences that show on an autoradiograph for each individual. However, some bands are common to people that are closely related, such as children of common parents. In the autoradiograph shown in the picture it can be seen that some bands are common to both children and to their parents.
M = mother; F = father; C = a child. The pen points to an unrelated individual.

Photograph by David Parker.
With permission from The Science Photo Library, London.

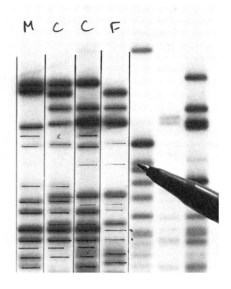

(b) Recessive Defective Alleles on Autosomes

(i) Fibrocystic Disease

Because the allele for fibrocystic disease is recessive it is only expressed in the double recessive. Most sufferers die of pneumonia which is encouraged by the deposition of unwanted material in mucus secreting tissues, including the lungs. The expression of the disease occurs in one in four of the offspring of heterozygous parents:

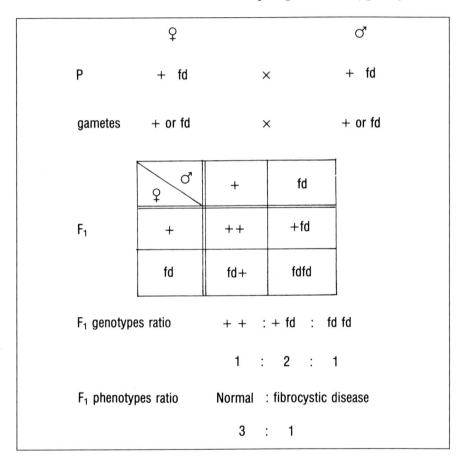

(ii) Albinism

The recessive allele for albinism is rarer than that for fibrocystic disease. It is expressed in only 0.005% of the population and is inherited in the same way as fibrocystic disease in one in four of the offspring of heterozygous parents.

(iii) Phenylketonuria

This is expressed in 0.004% of the population and the genetics of the defective recessive allele is the same as (i) and (ii). Sufferers have mental deficiencies.

(c) Autosomal Super Genes

Four adjacent loci on human chromosome 6 control (at least in part) the acceptance or rejection of transplants. Each locus has several genes in it, and there are several alleles for each gene. Each gene controls the production of an antigen. In modern surgery it is possible to make approximate matches between donors and recipients for these antigen genes so that rejection is minimised.

2. Sex-linked Allelic Defects

(a) X Chromosome Genes

(i) Colourblindness – A Recessive Allele on the X Chromosome

This recessive allele is expressed more often in males than in females because, when it occurs in one of the X chromosomes in a heterozygous female, it is usually recessive to the wild type allele on the second female X chromosome.

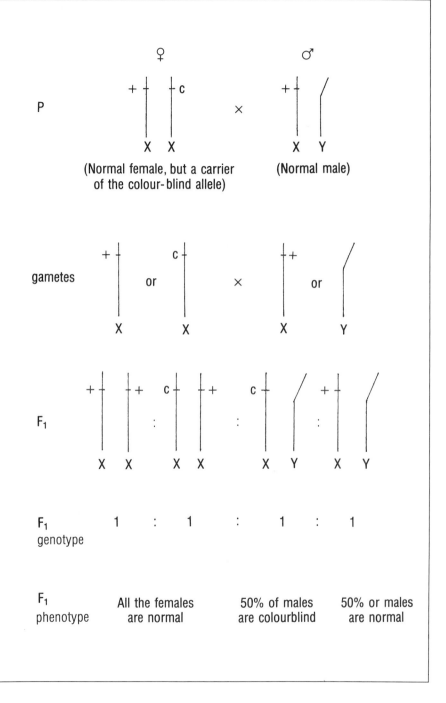

If a colour blind male marries a homozygous normal female:-

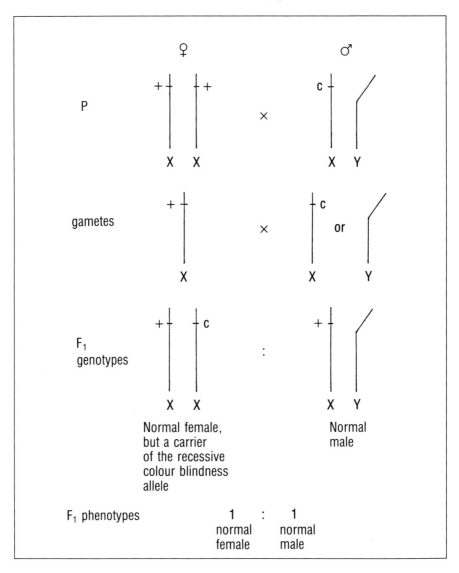

However if a colour blind male produces children with a heterozygous female carrier:-

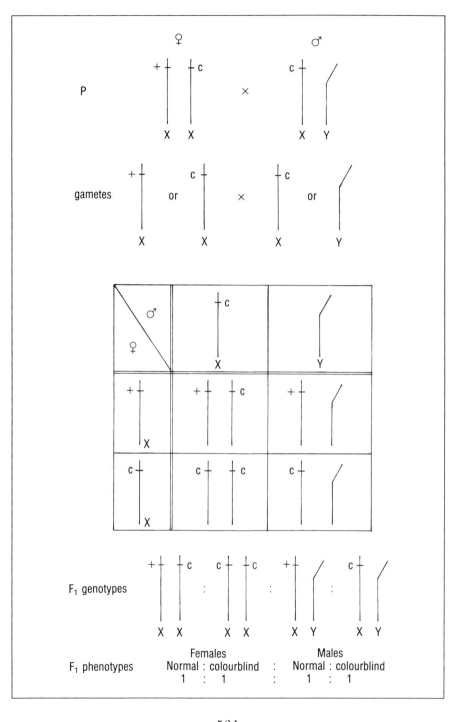

(ii) Haemophilia – a Recessive Allele on the X Chromosome

The inheritance of haemophilia occurs by the same genetic inheritance pattern as colour-blindness, once again making males more vulnerable than females. The expression of the mutant allele is caused by the failure to make polypeptides called anti-haemophilia globins, which are needed to cause clotting when blood vessels are damaged internally or at the skin surface. There are degrees of the severity of the disease, indicating that several sex-linked genes are involved in the synthesis of blood clotting polypeptides. It is possible that the sex-linked haemophilia genes are 'control' genes for the synthesis of the polypeptides encoded by clotting genes on the autosomes.

(iii) Glucose-6-phosphate Dehydrogenase Deficiency – A Recessive Allele on the X Chromosome

The mutant gene for glucose-6-phosphate dehydrogenase (an enzyme) deficiency is on the X chromosome and is found in up to 10% of some low latitude populations. This deficiency disease is linked to malaria resistance and individuals having the enzyme deficiency allele compete more favourably as heterozygotes, having better malaria resistance than the pure bred wild-types for the enzyme. Double recessives die.

This example of the selective advantage of the heterozygote is similar to that of the high incidence of the sickle-cell allele where again the heterozygote has a degree of competitive advantage, also in low latitudes, where malaria is prevalent.

(b) Y Chromosome genes

(i) Baldness – a Gene that can only be expressed in Males

The allele for baldness can be passed from a bald father into his daughters as well as his sons, showing that the gene for baldness is on an autosome. However, as it is only expressed in males, there must be a baldness controlling gene in the Y chromosome.

Genes For Human Sex Determination

The general discussion on gene regulation by inducers and repressors appeared in Chapter 4. Their function as controllers of differentiation was briefly mentioned in Chapter 14.

In particular, it was discovered in 1987 that 'zinc-fingers' from certain gene-regulating polypeptides associate with specific loci on DNA to cause the differentiated expression of downstream genes controlled by zinc-finger polypeptides.

The analysis of base sequences in the X and Y chromosomes of several placental mammal species has shown that sex-determining loci exist in specific regions of both X and Y chromosomes. The base sequence in some parts of the sex-determining gene on the X chromosome is fairly similar to the corresponding parts of the sex-determining gene on the Y chromosome. This suggests that the X and Y chromosome both evolved from a common ancestral gene that became modified by mutation, genetic drift and natural selection during the evolution of sexual differentiation.

Some species of placental mammals have more than two sex-determining loci, some on the X chromosome, others on the Y.

The Loci of Ultimate Upsteam Genes That Determine Sex in Humans

The human Y chromosome has a gene – the 'ZFY' gene – that encodes the production of a 'zinc-finger' polypeptide. The X chromosome contains a 'ZFX' gene that also encodes a zinc-finger, sex-determining polypeptide. The ZFX polypeptide is, of course, different from that encoded by the ZFY gene.

Whenever the ZFX gene is homozygous, as in XX humans, the person is female. Whenever the ZFY gene is present it modifies the expression of the ZFX gene. XY individuals are therefore male.

The analysis of DNA deletions in males (ZFY absent) and DNA additions in females (ZFY present in addition to ZFX) has revealed the approximate loci of the ZFY and ZFX genes on the Y and X chromosomes respectively. The analysis was carried out using the DNA taken from the cells of individuals known to have sexual abnormalities.

The ZFY Gene Locus

Whenever 300,000 bases from one particular locus were absent from the Y chromosome of abnormal males, the males failed to develop male gonads. Whenever the same 300k base sequence was observed in the DNA of abnormal XX individuals they had male gonads, despite the absence of a complete Y chromosome. This 300k region of the Y chromosome was given the name 'Testicular Determining Factor' – the TDF.

The TDF base sequence has been further analysed. It was found that about 140k encodes a polypeptide that regulates the transcription of DNA by binding to it at specific loci. The polypeptide also binds to some RNA molecules. The 140k gene was thought to be the ZFY gene within the TDF fragment. The ZFY gene is in the short arm of the Y chromosome and constitutes about 0.2% of the Y chromosome DNA.

The ZFX Gene Locus

Using a similar system of deletions and additions in analysing the DNA of sexually abnormal humans, the ZFX gene locus has been found in the short arm of the human X chromosome.

Y Chromosome Analysis

A team of investigators at the Whitehead Institute for Biomedical Research in the United States of America, under the direction of Dr. D.C. Page, analysed the consequences of Y chromosome aberrations. They concentrated their studies on the short arm, the centromere and that part of the long arm closest to the centromere. There are between 30 and 40 million base pairs in this part of the Y chromosome.

It was possible to divide the examined part of the chromosome into 155 loci. Of these, only two showed male determination and both came from the short arm of the Y chromosome.

The Further Analysis of the Two Male Determining Loci Using Probes for Genetic Mapping

When the base sequences of the two TDF male-determining loci had been established in human DNA, base sequences in the two loci were compared with the base sequences on the TDF regions found in frogs, three species of Drosophila and in yeast ADR 1. The TDF human base sequence was also compared with the same Y chromosome regions of mammals. Significant similarities of base sequences were found in one part of the 300k TDF fragments taken from several species. The locus within which similar base sequences were found is about 140k bases in length. It has been ascribed as the '1A2' region.

A number of DNA base sequences within the 1A2 region of the species listed above conform with the codons needed for the production of a multiple finger polypeptide. The function of the polypeptide is known to be downstream gene regulation. One particular 1A2 exon sequence of bases, uninterrupted by introns, has been of particular value in revealing the properties of a zinc-finger polypeptide used for downstream gene control.

It seems that 'finger' production is dependent on zinc cations and on pairs of cysteine and pairs of histidine molecules inserted at regular intervals into the polypeptide. Each protruding finger is made when a zinc cation forms a chemical complex with the amino-acids of the polypeptide. The pairs of cysteines and pairs of histidines are necessary for forming the zinc complex and therefore for ensuring the finger formation and functional properties of the polypeptide.

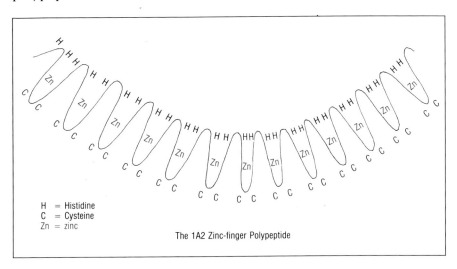

The 1A2 Zinc-finger Polypeptide

This simplistic presentation makes no attempt to show the three dimensional aspects of the polypeptide. It shows thirteen fingers dependent upon the bending of the molecule when pairs of cysteines and pairs of histidines, forming a complex with zinc, create the necessary folds.

Lys, Arg, His, Asn, Gln and Thr are the amino-acids that most readily associate with specific sites on DNA to cause downstream gene control. These amino-acids are found in abundance at the exposed ends of the zinc-fingers.

It is probable that there is a large, uninterrupted exon on the 1A2 gene that is largely responsible for the zinc-finger polypeptide. This 1A2 derived polypeptide also contains large hydrophobic groups at specific loci. These are associated with tyrosine, phenylalanine and leucine. These amino-acids also occur at regular intervals along the length of the polypeptide. The hydrophobic groups encourage specific folding of the fingers for aligning them with specific sites on downstream genes.

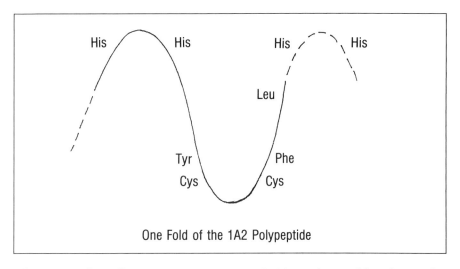

One Fold of the 1A2 Polypeptide

The repeating fingers occur every 27-32 amino-acids along the molecule, so making the fingers vary slightly in length. For any one finger the positions of the important amino-acids are:-

Sequence Number				10	12	15	19	25	28	32
Amino-acid				Tyr	Cys	Cys	Phe	Leu	His	His

Similarities of Exons in the X and Y Chromosomes

Single stranded nucleotide strands can be made in vitro with a base sequence that pairs with an unblocked strand of DNA in a cellular gene. If large numbers of radioactive single-stranded probes are released in the presence of DNA containing the complementary sequence of bases, one of the probes is likely to pair up with the DNA.

In examining both the X and Y human chromosomes using probes, it has been found that there are exon regions in the 1A2 region of the Y chromosome that correspond quite closely to exons in the apparent sex-determining region of the X chromosome.

The following exons on the 1A2 gene of the Y chromosome showed considerable similarities to exons in the ZFX locus of the X chromosome:-

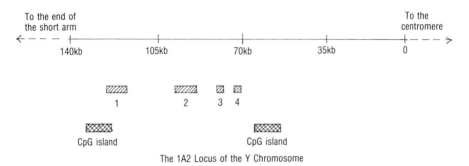

The 1A2 Locus of the Y Chromosome

The 'CpG' islands are clusters of triple C-G unmethylated paired bases. Most mammalian 'start' codons are contained within a CpG island. In the case of the 1A2 exons, the CpG island at the 125k position probably contains the 'start' codon, which also lies within exon 1.

The CpG island at about the 50k position very likely contains the termination codon. The implication of the CpG islands, and their relationships with the exon sequences that are conserved in both the X and Y human chromosomes, is that transcription is from left to right in the above diagram.

The TDF Region of the Y Chromosome

The 1A2 locus of the TDF region on the Y chromosome seemed, at first, to produce the most significant genetic information so far about sex-determination in humans. However the whole region of the two (out of 155) male determining DNA loci has yet to be evaluated fully. From the studies made so far, the TDF region of the Y chromosome has been divided as follows:-

The TDF Region of the Y Chromosome

It is not yet known what influence, if any, the 1A1, 1B and 1C loci have on sex determination in humans.

More Recent Discoveries About the ZFY

After Page and his colleagues had put forward the ZFY gene locus as a candidate for the male determining gene, more recent work suggests that the ZFY may control some aspects of spermatogenesis, but that another gene locus is responsible for the ultimate male determining gene.

A Better Candidate for the TDF (Testis-Determining Factor) in Humans

The need to modify man's understanding of biological inheritance with the emergence of new knowledge is obvious. This is especially true when new information arrives at the rapid rate, typical of the late 20th century. Readers of this book may be somewhat surprised to learn now that the ZFY gene, identified by Page and his colleagues, is not in fact the testis-determining factor (TDF). Page's work is included in this chapter as an example of seemingly very good experimental work that was viewed with considerable reservations by another group of scientists.

In many ways the ZFY gene put forward by Page and his colleagues, with a start codon in the 125kb position, had many of the qualities of a gene that controls transcription. In addition, the zinc fingers in the gene's polypeptide product seemed to corroborate the discovery of an important gene that could control other genes downstream of it. The identification of a CpG island containing a start codon added further weight to the hypothesis that the ZFY gene was the TDF gene.

A second group of scientists to search for the TDF carried out their experimental work at the Imperial Cancer Research Fund Laboratories in London. They disbelieved the Page findings and the London team made detailed studies of DNA taken from several XX males. Among these were four individuals of special significance from France and Algeria. The four males have very small fragments of Y-derived translocated DNA in one of their two X chromosomes. Detailed analysis of each fragment of Y-derived DNA in these four individuals has shown that the fragments lack the ZFY gene. In the knowledge that individuals lacking the ZFY genes can exhibit a considerable degree of maleness, especially in relation to testes development, the Y-derived fragments from the four males were searched for a better candidate for the TDF gene. The London research found a gene even closer to the Y chromosome boundary than the Page ZFY gene.

It was the possibility that the TDF gene is closer to the 'boundary' than the ZFY gene that caused Page and his team in the United States of America to miss what seems to be the correct gene for TDF.

The 'Boundary' between the Distal and Proximal parts of the X and Y Chromosomes

In essence, the short arm of the Y chromosome is the proximal part of the chromosome, and the long arm is the distal part. The distal part is autosomal in function.

The X chromosome also has proximal and distal parts that are separated in exactly the same position, in relation to the distal end of the chromosome, as the separation position in the Y chromosome. The separation position in both X and Y chromosomes is called the 'boundary'.

The boundary of the Y chromosome in man is occupied by a sequence of 303 base pairs that encode an Alu sequence. This sequence is absent in the X human chromosome, even though a boundary (of less clear definition) also exists in the X chromosome. The distal side of the boundary in both X and Y chromosomes has been ascribed the name 'pseudo-autosomal'. This name implies the seeming lack of sex determination. Proximal to the boundary of the X and Y chromosomes are their sex-determining regions.

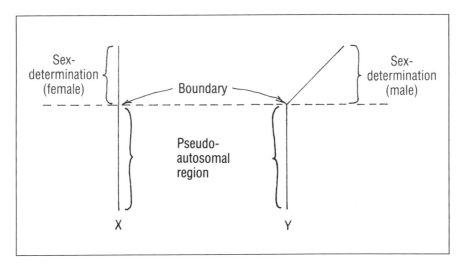

An interesting comparison has been made between the 224 base pair sequence immediately proximal to the boundary in X and Y human chromosomes. The study revealed a 23% base-pair divergence. This small divergence suggests that, in the primate ancestors of man, the boundary was further towards the proximal end of the sex-determining chromosomes than it is today.

The distal side of the boundary shows 99% base homology. By contrast, those base pairs further than 225bp from the boundary on the proximal side exhibit very low base sequence homology. The implications are that where there is 99% homology, on the distal side of the boundary, crossings-over at meiosis are usually undamaging wherever they occur. However it is not uncommon that a proportion of crossings-over in the 77% homologous region can cause inviability. Crossings-over in the non-homologous region on the proximal side are almost certain to have damaging effects. The boundary as it exists now, is therefore a likely position, in both X and Y chromosomes, that separates crossing-over viability in the distal, pseudoautosomal region from the proximal sex-determining region in which crossings-over must not occur.

The 225bp sequence which shows 77% homology in X and Y chromosomes can make a contribution to the study of the evolutionary divergence of man and primate species from their common ancestor.

One method of attempting to work out how long ago the boundary shifted from the ancestral primate boundary to its present position is to estimate the average number of base pair substitutions at meiosis between one generation of primates and the next. The estimated number can be recalculated as the average number of base substitutions per base pair per year. The present divergence between the X and Y chromosomes of pair 23 can be used to extrapolate backwards in time to chromosomes that were identical for this 225bp sequence in the ancestral primate in which divergence began. This method has the disadvantage of not knowing the average generation time, nor the average point mutation rate during the whole time span of divergence. One 'guesstimate' is that the rate of base pair point substitution is 1.5×10^{-9} per base pair per year. This would put the beginning of X and Y chromosome divergence for this base sequence at about 45 million years ago. However, the average generation time used as one of the determinants of this date was 33 years, and this seems rather long for the generation time of man's ancestral primate ancestors.

A second method for calculating the divergence date of origin is to make comparative studies of this 225bp sequence in existing primates and use the (open to criticism) calculation methods of Chapter 11. Perhaps this second method of calculation will lead to a figure rather less distant in time than the 'guesstimated' 45 million years go of method 1.

It is of interest that the great apes (gorillas, chimpanzees and orang-utangs) have their X and Y chromosome boundaries in the same positions as those in man.

This discussion about the boundary has relevance to the search for the TDF, because the repeated Alu sequence allows probes to identify where the sex-determining region (the proximal region) of a Y chromosome starts. A translocated segment taken from a Y chromosome and inserted into an X chromosome can be identified quite easily by the repeated Alu sequence. Having found this sequence in an X chromosome, it is then necessary to find out how much of the proximal section of the Y chromosome has been inserted into the X chromosome.

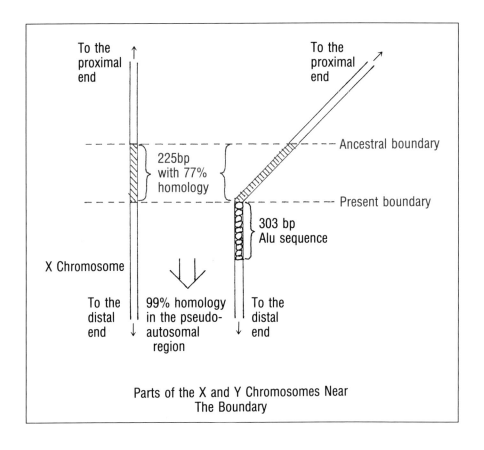

Parts of the X and Y Chromosomes Near The Boundary

The Discovery of the pY53.3 – A Strong Candidate for an Exon in the TDF Gene in Humans

The study and analysis of DNA taken from the four French and Algerian individuals mentioned earlier gave particularly valuable information about the location of a gene that determines testicular development.

All four were unable to make spermatogonia and one had intersex bilateral ovotestes. Nonetheless they all had testicular development, to some degree at least, and all four lacked the region of DNA where Page had located the ZFY gene. ZFY was therefore eliminated as the TDF.

The X boundary was identified in the normal position in one of the pair of X chromosomes in all four individuals. The Y boundary was also found in the second X chromosomes and exactly matched the normal boundary position.

Fragments of DNA were cut from the proximal side of the Y boundary and analysed. The analysis revealed that in these four individuals the greatest number of Y-derived, proximal base pairs inserted into an X chromosome is 60kb. The ZFY identified by Page was between 152kb and 202kb from the Y boundary. Having eliminated ZFY as the TDF, the search in London now concentrated on those conserved (invariant or conservative) base sequences in a narrowed region of DNA, within 60kb of the boundary. One fairly long conserved base pair sequence was found. This was ascribed pY53.3.

The next problem was to untangle the intron-exon regions within the open reading frames of the DNA. The pY53.3 contains two overlapping reading frames, one encoding 99 amino-acids and the second encoding 223. Within the longer reading frame there is a sequence of 80 codons that matches an amino-acid sequence found in polypeptides taken from different species, some of the polypeptides being found only in the nuclei of cells. What the exact significance is of having this common sequence, in species as diverse as the fission yeast Schizosaccharomyces pombe, a wide range of animals and man, is not known. Codons here are DNA triplicate codes.

It is of interest that some of the polypeptides encoded by the conserved 80 codons are nuclear, non-histone proteins related to high mobility polypeptides such as HMG 1 and HMG 2. The high mobility polypeptides have a function in controlling chromosome structure and in switching genes on and off. The implication here is that the conserved pY53.3 section of DNA encodes polypeptides that exercise translation control over gene expression elsewhere in the genome.

The conserved 80 codon sequence is Y-specific for a wide range of animals. A comparison between conserved segments taken from rabbits and man show that 64 codons are identical. A further eight are conservative (code for the same amino-acid). There is therefore a 90% overall similarity in the Y-specific conserved pY53.3 sequence in humans and rabbits. Conserved sequences in genomes are rare and are likely to have very important functions. Far reaching gene control consequences caused by the polypeptide products of such a conserved sequence is therefore a real possibility. It is not yet known which of the 80 conserved codons are critical for the correct functioning of the transcribed polypeptide product.

The Testes Test

It was, of course, necessary to show that pY53.3 is associated with testicular development in humans. A probe, corresponding to 0.9

kb of the pY53.3 region in human Y chromosomes, matches the sequence of bases found in an RNA molecule produced by the nucleus of adult human testes cells. No other human tissues tested so far produce RNA molecules with this base sequence.

Some Further Facts About pY53.3

The largest open reading frame of the pY53.3 is probably the last exon of the sex-determining region-Y gene (SRY). This suggested that the SRY gene may very well turn out to be the testis determining factor gene (TDF gene). Further tests needed to be carried out before the TDF gene could be identified for certain. The absence of the Y-derived SRY gene in XY females would corroborate the positive information supplied by the four French and Algerian XX males.

The open reading frame of the pY53.3 has a 3' stop codon nearby together with a poly-A section typical of stop codon regions. At the 5' end of the conserved sequence there is a potential splice site (splicing – Chapter 14), but the splice acceptor site is not necessarily an intron-exon boundary.

If there are small conserved exons that are part of the SRY gene and which are vital for the correct function of this gene's encoded polypeptides, they may prove quite hard to find by comparison to the finding of the 80 codon large conserved sequence exon. If the functional region of the SRY polypeptide is all encoded in the large exon pY53.3, there need not be any further conservation elsewhere in the gene. Further work will no doubt follow the 1990 results obtained by Dr P.N. Goodfellow and his team at the Imperial Cancer Research Fund in London. The exon sequence of the entire gene needs to be discovered together with details of splicing, if splicing exists in the gene. Further work on the functions of the polypeptide encoded by the SRY gene is an obvious line to follow. Cross-references with the translated functions of similar, conserved sequences at the carboxyl- end of S. pombe genes and to the functions of non-histone nuclear polypeptides with similar conserved sequences elsewhere also need to be made.

The Function of the ZFY Gene Discovered by Page and His Colleagues

Whenever the ZFY gene is absent there is the failure to produce spermatogonia, even if the SRY gene is present. The ZFY gene may therefore have a function in spermatogenesis. It is now clear that it is not the ultimate male-determining TDF.

The zinc fingers in the polypeptide encoded by the ZFY gene probably confer transcriptional control qualities to the polypeptide.

Within a year or so of the postulation that ZFY is the testis-determining factor gene (TDF), a much better candidate has emerged. The rapidity of world-wide research is likely to create an enormous quantity of new and improving information about the human genome. This is especially true at a time when techniques for analyses of DNA, stop codons, start codons, splicer regions and the recognition of open reading frames are improving very quickly. Page found out that the TDF gene is close to the boundary and steered the London team towards the right region of the Y chromosome to search for the TDF. There is the hope that each successive improvement to our understanding of the human genome will cause it to lose much of its mystery. An understanding of ultimate upstream genes and their encoded polypeptide products will lead, one day, to the reasons why cancer cells start to divide out of control.

Experiments Using Sry, The Mouse Equivalent of The Human SRY Gene, That Have Transformed an XX Mouse into a Male

Scientists have followed up the finding of the pY53.3 exon with researches into the Y chromosome TDF region in several eutherian mammal species. In experiments carried out at The Laboratory of Eukaryotic Molecular Genetics at Mill Hill, London and at The Human Genetics Laboratory, Imperial Cancer Research Fund, London, a DNA sequence corresponding to the pY53.3 exon in humans has been found in the Y chromosome of mice. The gene that contains the pY53.3 exon in humans has been ascribed SRY. The equivalent gene in mice is ascribed Sry. Both SRY and Sry are very close to the pseudo-autosomal boundary of the Y chromosome in both species. The London scientists have been successful in using the mouse Sry gene to transform XX embryos into mice with some male characters. They have also produced, in 1991, an adult XX male using the same gene.

The Sry gene, and control genes upstream and downstream of it, can be synthesised in a pure state by way of a viral phage – f741. Large numbers of such DNA fragments can be made. The full DNA sequence of the purified fragments is about 14kb in length, 8kb upstream and 5kb downstream of the Sry gene itself.

The Sry gene is known to encode a polypeptide in which the configuration of the molecule allows chemical binding with specific regions of DNA. The Sry derived polypeptide therefore has the properties of a gene regulator for downstream genes.

Male Characters Expressed in Transgenic XX Mouse Embryos

The 14kb DNA sequence containing the pY53.3 exon was injected in large numbers into 158 XX mouse zygotes. After implantation into surrogate mothers, two of the injected embryos showed testis chords 14 days after implantation. These two foetuses were shown to have multiple copies of the Sry gene incorporated into their genome. The precise locations of these insertions were not defined.

In normal XY mouse embryos, extracts taken from proto-gonad tissue shows that the Sry derived regulatory polypeptide is produced after about 10½-12 days. This 36 hour period coincides with the switch in proto-gonad cells which causes them to become testes rather than ovaries. In such normal XY mice, testis chord formation also begins after 14 days.

Male Characters Expressed in Transgenic XX Mouse Adults

In further experiments, zygotes were again injected with the 14kb DNA fragments in large numbers. Of the 49 males (out of 93 animals) that were born, three were transgenic.

Two of these three mice showed no male characters, even though they contained incorporated copies of the Sry gene. Indeed, these two mice developed full female characters and were able to give birth to viable offspring. There are two possible reasons why the Sry gene was not expressed as maleness in these two individuals:

  i) The location of the Sry genes in the genome of the XX mouse did not allow adjacent control genes to promote the production of Sry's polypeptide.

  ii) The insertion of the Sry gene only occurred in a proportion of somatically dividing proto-gonad cells during the early stages of embryo development. Only a small proportion of gonad tissues would receive the Sry gene in this case. When there is only a low proportion of gonad cells that contain the Sry gene, the predominance of the hormones produced from untransformed cells allows the development of a normal XX female.

A very great degree of success in transforming an XX zygote, so that an adult male grew and developed from it, was shown in the third mouse. This individual contained multiple copies of the Sry gene. The following adult male characters were evident in him:-

a) He was externally male in his genitals.
b) He behaved normally in his sexual behaviour towards females and achieved coitus with them.
c) His testis contained seminiferous tubules, Leydig cells (which must have produced testosterone for the development of secondary male characters – see e) below), peritubular myoid cells and Sertoli cells. However, at a similar age to normal XY males, his testes were less than a quarter the mass of testes taken from a brother of the same litter.
d) He had a vas deferens, accessory glands and seminal vesicles.
e) He had secondary male characters.

However, none of his sexual behaviour with females resulted in progeny. This was because his testes were unable to make sperms. Essentially, therefore, the experiments produced a single, sterile, adult male.

A search for female tissues in this XX mouse revealed none. The development of female organs in males is suppressed by anti-Müllerian hormone, so this male must have been able to produce AMH in his Sertoli cells.

Secondary male characters are considerably promoted by the hormone testosterone. This male must have been able to make this hormone.

The Failure of Human SRY to Transform XX Mice Into Males

Mouse Sry and human SRY differ in 29 of the 79 exon codons that encode the binding part of the regulatory polypeptide produced by the Sry or by the SRY gene. Although they vary from species to species, these 79 codons are functionally essential for switching on maleness. Single base mutations in this 79 codon sequence can lead to the failure of testicular growth in humans.

Elsewhere in the Sry and SRY genes, the codons differ considerably.

The codons encode the non-binding part of the polypeptide.

With so many differences in the Sry and SRY genes, it is not altogether surprising that injection of human SRY into mouse zygotes has so far failed to transform any XX mice into males. This is true, even when multiple copies of SRY are incorporated into an XX mouse's genome, efficiently transcribed and then translated at critical phases of embryonic and adult development. No evidence of male tissues nor of male organs have been found in mice that grow and develop from SRY injected zygotes.

Conclusions

The successful transformation of an XX mouse into a male, albeit sterile, using multiple copies of a purified 14kb mouse DNA sequence is a strong pointer to the notion that the ultimate upstream male-determining gene is contained within this section of the mouse Y chromosome. A comparison of this 14kb region from very close to the pseudo-autosomal boundary in other mammal species, including man, shows some fairly strong similarities to the Sry in mice. In man, the SRY was identified in the Algerian and French XX male individuals, so adding weight to the probability that mouse Sry and human SRY are the ultimate upstream male determining genes.

The spermatogenesis gene in mice may be found to be a mouse equivalent of the Page ZFY gene. The two genes, one Sry or SRY for general maleness through testis formation, the other Zfy or ZFY for spermatogenesis, would benefit from being close to each other on the genome. Page's ZFY is not included in the 14kb fragments used in the London experiments.

It may now be possible to examine the downstream series of gene regulation that starts with SRY in humans and Sry in mice for sex determination and other characters.

3. The Selective Advantage or Disadvantage of the Heterozygote

For a new mutant allele to be succesful it is usual that those heterozygotes that inherit a new allele have to compete at least as successfully as the wild-type homozygote in the early generations after the mutation is formed.

(i) Sickle-cell Anaemia – An Example of Polymorphism

It would appear, at first inspection, that the inheritance of the sickle cell allele, which is expressed strongly in the homozygote and is partially expressed in the heterozygote, would always be at a selective disadvantage compared with wild-type individuals who have normal, red blood cells. However the sickle cell condition also confers a resistance to malaria, so there is a disproportionate success of sickle cell heterozygotes by comparison to homozygous wild-types in mosquito infested areas of low latitudes. The frequency of the sickle cell allele is as much as 20% in some African populations, producing about 4% of homozygous sickle cell children. Homozygotes for this condition die before reproducing.

HB^A = normal blood cells allele.
HB^S = sickle cell anaemia allele.

$HB^A HB^A$ = normal blood cells; no malaria resistance
$HB^A HB^S$ = some sickle cell symptoms; malaria resistance
$HB^S HB^S$ = death from sickle cell anaemia

(ii) The Unexplained Selection Against the Heterozygote Foetus for the Rhesus Factor

The persistence of two Rhesus inheritance alleles is difficult to explain, as the heterozygous foetus carried in an Rh- woman is always selected against. There may have been eras when genes interdependent with Rh+ gave rise to individuals which were at a selective advantage, while at other times genes interdependent with Rh- were at a selective advantage. In the same way that cancer is always genetically damaging, so too the damaging Rh+Rh- association persists, causing damage to individuals but not to the species.

The Association of Genetic Inheritance With Disorders that do not Seem to be Directly Caused by the Genes

Statistical correlations, after allowing for racial effects, have been applied to the occurrence of some disorders and the genetic inheritance of some other characters:-

(i) Duodenal ulcers are more common in people with blood group O than in people with other blood groups.

(ii) Duodenal ulcers are more common in people with CDe/cDE Rhesus inheritance, than in people with other Rhesus combinations.

(iii) Cancer of the oesophagus is associated with a genetic disorder called tylosis, a disorder controlled by a rare, autosomal dominant allele that is detected as thickening of the skin on the palms of the hands and on the soles of the feet.

Linked Genes

(i) Nail-Patella Syndrome

The very rare dominant allele for nail patella syndrome (skeletal disorders such as underdeveloped knee caps, elbow joint deformities and abnormal finger nail growth) seems to be linked on the same autosome with the ABO blood group gene.

The Inheritance of Human Blood Groups

(a) The Major Blood Groups

i is a recessive allele to I^A or I^B
I^A is dominant over i
I^B is dominant over i
I^A and I^B are codominant for each other.

ii = Blood Group O

$\left.\begin{matrix} I^A i \\ I^A I^A \end{matrix}\right\}$ = Blood Group A

$\left.\begin{matrix} I^B i \\ I^B I^B \end{matrix}\right\}$ = Blood Group B

$I^A I^B$ = Blood Group AB

(b) Rhesus-factor Inheritance

The Rhesus-factor inheritance is more complex than the inheritance of the major blood groups. There are thought to be three closely linked genes on one of the pairs of autosomes, and the combinations of alleles in the three genes determine which Rhesus-factor is expressed:-

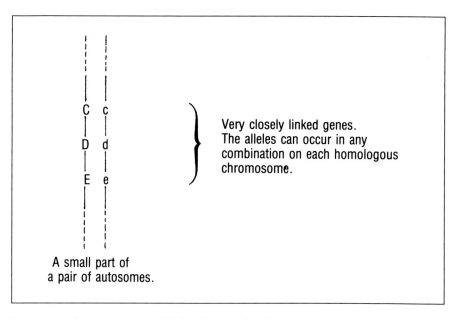

It seems that two out of the three dominant genes can be enough to give a Rhesus-positive phenotype.

(i) Common Rhesus-positive combinations

CDe/cde   (R_1r)   - 35%
cDE/cde   (R_2r)   - 14%
CDe/cDE   (R_1R_2) - 13½%

(ii) Rhesus-negative combinations

Full Rhesus-negative is cde/cde. However cdE/cde and Cde/cde also produce mother/offspring complications.

The Rhesus-Factor – Factor D

About 85% of the human population have the Rhesus Factor, an antigen, in the red cells of their blood. Thus:-

85% of the population are Rh+
15% of the population are Rh−

If Rh+ blood enters the bloodstream of an Rh− individual there is the slow production of Rhesus antibodies which cause the agglutination of the red cells from the donor in the recipient's bloodstream. But the first introduction of Rh+ blood into a Rh− recipient does not cause a large production of Rhesus antibodies, so that there are produced only a small number of antibodies which wait in case they are needed for a second invasion of the Rh− person by Rh+ red cells. On the second invasion there is the immune response when large quantities of the Rhesus antibodies are produced.

The Rhesus Factor and Pregnancy

♀ 15% Rh−	×	♂ 85% Rh+
Children (F_1) are likely to be Rh+ (but see page 551 for the three Rhesus genes and the allele combinations that produce Rh+ or Rh−)		

A developing Rh+ foetus in the womb of its mother has its bloodstream flowing very close to that of the mother in her placenta, but separated by placental membranes. However at parturition (birth) both the mother and baby bleed at the placenta and some of the baby's blood may enter the mother's bloodstream.

For the first child, therefore, there are usually no problems, because the Rhesus antibodies are not produced by the mother until after it is born. However the persistence of the Rhesus antibodies in the mother's bloodstream after the birth of the first child can cause considerable damage to the red cells of subsequent children, particularly if there is much bleeding at parturition. There may also be some seepage of the anti-Rhesus antigen across the placenta during development of the foetus in the womb.

Immediate complete transfusions of blood into the baby after birth, or while it is still in the womb, prevent the Rhesus antibodies causing Erythroblastosis Foetalis (haemolytic disease of the newborn, in which there is the massive destruction of red blood cells).

Another treatment is that of giving the mother anti-Rhesus globulins which coat the foetal cells in a way that blocks the Rhesus factor, so preventing the production of Rhesus antibodies in the mother.

About 1 baby in 200 is affected by Rhesus incompatibility instead of the 12.75% expected, partly because a number of the babies are first children and partly because there are clean breaks at the birth of the first child. There are also interactions of the ABO blood groups that prevent incompatibility (see next section).

The 1 in 200 affected babies may be stillborn, suffer heart failure because of severe anaemia, or severe jaundice may develop in them after birth (broken down blood pigments = bile pigment = jaundice) or there may be severe and permanent brain damage (spastic paralysis).

Interactions of the ABO Blood Groups with the Rhesus Factor

The different blood group antibodies can destroy the red cells of incompatible blood. The following table shows that in some cases the invasion of the mothers bloodstream with red cells from a Rh+ foetus leads to the production of Rhesus antibodies, whereas in other cases it does not. The deciding factors are the blood groups of both foetus and mother.

	FOETUS	MOTHER		
	+	−		
	A	A	(i) & (ii)	
(i)	β β β β β (○)	β β β β (○)	The mother makes anti-Rhesus antibodies slowly. These then persist so that a second baby of similar blood group would be worse affected if there were bleeding at the first parturition.	UNSAFE
	B	B		
(ii)	α α α α α (○)	α α α α α (○)		
	A	0	(iii) & (iv)	
(iii)	β β β β (○)	α β α β ⚹ β α β α → Breaks down in response to α	Although there are Rh+ antigens in the baby's red cells that pass into the mother's bloodstream the mother's antibodies (inherited with her blood group) break the cells down before the mother makes any anti-Rhesus antibodies.	SAFE
	B	0		
(iv)	α α α α (○)	α β α β ⚹ β α β α → Breaks down in response to β		

The symbols of the table are:-

A = blood group A, which contains Anti B = β

B = blood group B, which contains Anti A = α

O = blood group O, which contains both Anti A (α) and Anti B (β)

+ = Rhesus antigen in the red blood cells.

− = No Rhesus antigen in the red blood cells.

○ = a red blood cell

⚹ = a broken down red blood cell.

The Geographical Distribution of Human Blood Groups Related to Antigens Carried in Disease Bearing Viruses and Bacteria

Although the smallpox virus is thought to have been eradicated in all places, except for a few medical laboratories where vaccines against it are produced, it has been established that it contains an antigen A. It is therefore less likely to be attacked by white blood cells in people of blood group A or AB than if it invaded someone of blood groups O or B. Observations made in India during the last epidemics to occur there showed this to be true.

It is possible that the differences in the proportions of individuals of each blood group in different parts of the world were caused, at least in part, by the antigens in the disease bearing organisms that caused the worst plagues in each geographical region.

Genetic Engineering in Humans

It must be emphasised that there are at present too many risks involved in most of the techniques discussed in Chapter 8 for their use in the genetic engineering of human cells. The inability to insert genes at predetermined loci next to control genes leads to deregulation and to unacceptable risks of damage. In addition there are problems if bacterial or synthesised genes are used for human gene therapy. This is because there are differences in polypeptides (mostly sugar residues added to eukaryotic polypeptides) in prokaryotic and eukaryotic cells even though the base sequence in a transforming gene is the same in the bacterial cell and in the human cell. The safety level towards which genetic engineers must aim is similar to that of drugs or other medical treatment. There is much to be done before this safety level can be reached in human genetic engineering.

It has, however, been possible to use genetic engineering in culture for mapping the positions of some human genes. This is done either by cotransformation frequency or by the use of radioactive probes produced by genetic engineering.

Cotransformation experiments are carried out by fusing mouse cells with human cells. Human chromosomes are mostly shed and mouse cells' chromosomes are largely retained. The few human chromosomes that are retained in mouse cells are often fragmented. In a very small proportion of cases there is the incorporation of human DNA fragments into the mouse cell replicating genome. The expression of dominant human genes can be observed in a proportion of mouse cells. The cotransformation frequency of linked human genes allows the construction of a genetic map for human genes. The closer together linked genes are on a human transforming DNA fragment, the higher the likelihood that two crossings-over will incorporate both genes into the recipient cell's genome.

The genetic mapping of human genes is also carried out using radioactive probes. The probes are small fragments of DNA in which there is the order of bases which corresponds with an mRNA molecule. The mRNA molecule is identified as having the order of bases that are translated into a gene's polypeptide. A large number of synthesised probe molecules are injected into human cells *in vitro*. Probe DNA and the homologous sequence of bases in the gene unpair their bases in such a way that probe DNA and gene DNA pair up. The positions of the radioactive probes on the genome has allowed the assignment of genes for particular polypeptides to particular chromosomes, using this technique.

A further method for estimating the distance apart of genes on human chromosomes is to irradiate DNA fragments in which there are known to be linked genes. Irradiation breaks the DNA into smaller fragments. The ratio between cotransformation and independent transformation of the two genes is then calculated. The higher the rate of independent transformation the further apart the genes lie on the chromosome fragment.

The Potential Uses of Human Genetic Engineering

It would be of immense benefit to the human species if mutant, damaging alleles could be replaced by wild-type alleles in the eggs, sperms or 'four cell' embryos of those families which carry or express damaging mutant alleles. Ten in every thousand European babies come into this world with an inborn error of metabolism under single gene, Mendelian control. Many other ailments also have genetic causes. In the case of 'four cell embryo' genetic engineering it may become possible to freeze three of the four cells in case the genetic engineering of the first one is not entirely successful.

One line of investigation that shows much hope for the future is the genetic engineering of human bone marrow taken from cancer patients. Some drugs that kill cancer cells also kill bone marrow cells. It may become possible to engineer bone marrow cells genetically so that drug resistant genes are inserted into the replicating part of the marrow cell's genome. Bone marrow cells can be removed from a patient by surgery and replaced by genetically engineered, drug resistant bone marrow cells. The undesirable, marrow-killing effects of anti-cancer drugs may be much reduced in this way, so allowing larger and more effective doses of anti-cancer drugs, provided there are no side-effects elsewhere.

The book ends on this hopeful note.

Chapter 16 Questions

1. What is the evidence that man is quite close to other primate species in his inheritance?
2. It is known that only two alleles exist for one of the human genes. Suppose that 4% of the population show the double recessive character. What is the proportion of heterozygous carriers in the population? What proportion of the population is homozygous dominant?
3. How can human chromosomes be distinguished from each other? How are aberrations detected in them?
4. What are the causes of Down's Syndrome?
5. What well known human genetic defects are caused by large mutations or by changes to chromosome numbers?
6. How is Huntingdon's Chorea inherited in humans?
7. What causes dwarfism?
8. Why are Huntingdon's Chorea and dwarfism unusual genetic defects?
9. How are fibrocystic disease, albinism and phenylketonuria inherited in humans? What is surprising about the proportion of carriers of these diseases in the human population?
10. What proportion of the human population is likely to be the carrier of at least one damaging recessive allele?
11. How is colourblindness inherited in humans?
12. How is haemophilia inherited in humans?

13. How is baldness inherited in humans?
14. Give a detailed account of the gene that is the ultimate upsteam genetic control for maleness in humans?
15. What is polymorphism?
16. How is sickle-cell anaemia inherited in humans?
17. Give examples of human disorders that ultimately depend on genetic inheritance, but whose expression depend on factors other than the genes themselves.
18. Give an example of linked gene disorder in humans.
19. How are blood groups inherited in humans?
20. Give a detailed account of the inheritance of the Rhesus factor in humans. What problems can be caused for mother/baby incompatibility by the Rhesus factor?
21. What is the potential for genetic engineering in humans?
22. How far should the manipulation of genes in humans be allowed to go?
23. What might be the consequences for mankind of finding a chemical that very greatly diminishes the ageing effects of oxygen in cells?

POSTSCRIPT TO CHAPTER 16

Inevitably, in a subject where new information is arriving very quickly, some recent discoveries were made when the main body of the book was not able to be changed. Some of these recent discoveries illustrate further ways in which there can be deviations from simple Mendelian inheritance. Since they apply to the human species, these post-script topics should be read in conjunction with chapter 16.

Differential Expression of Human Genes, Depending Upon the Sex of the Parent from Whom The Mutation Was Inherited

Seemingly indentical mutations in homologous chromosomes can lead to differential expression of the mutations. The sex of the parent from whom the mutation was received can alter the expression of mutations. Diabetes can be expressed differently from indentical mutations received from mothers by comparison to fathers. Some cancers are only expressed if the (proto) oncogene(s) come from one sex of parents but not from the other sex. How upstream and downstream differential gene expression is controlled in these cases is not known.

Repeating Genes Caused by Mutational Gene Amplification in Humans

When genes are repeated excessively by mutational gene amplification, the effects can be diverse and widespread. Many cell lines can give rise to several defective organs. Cancer can also result from gene amplification. Mental and psychiatric problems can be caused by excessive mutational gene amplification.

Extra-chromosomal Inheritance in Humans

Some germ cells that are fertilised contain mutant plastid genes such as those genes in mitochondrial DNA. If only a proportion of the plastids in the fertilised germ cell contain the mutation, only a proportion of cell lines that come from the zygote will contain the mutation. From an evolutionary point of view, and from the point of view of the next generations arising from the individual containing mutant cell lines, only cell lines that develop and grow into the gonads have any chance of having any further significance for individuals and for the species.

3n Inheritance in Humans

If there is the failure to halve the number of chromosomes in a germ cell after meiosis and a 2n germ cell is fertilised, there is often the shedding of one set of the 3n chromosomes. Often such an individual is inviable and is aborted as an early foetus. Sometimes, however, a complete set of chromosomes is shed to leave a viable 2n individual in all its important cell lines. If the 2n germ cell is the egg, the chromosomes that are shed are usually those from the sperm. Such an individual will inherit all their chromomsomes from their mother. The daughter of a heterozygous mother could, as a result of chromosome meiotic replication and the production of a 2n germ cell, be a double recessive for a damaging allele, even though the father was a pure bred wild-type for the gene. Clearly the Laws of Mendelian Inheritance are flouted in cases of abnormal meiosis followed by fertilisation.

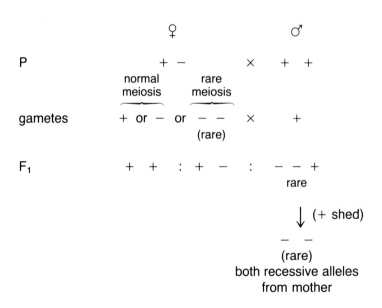

APPENDIX 1

THE ORGANISATION OF NON-DIVIDING CELLS

Introduction

It may, at first, seem that the organelles listed in the table of this appendix have little value for the study of genetics. However, as techniques for analysis are constantly improving, the information provided by the genes of those organelles that have DNA of their own is likely to lead to a profound understanding of cell biology, as well as to more and more evolution comparison studies.

So far as those organelles that have their own DNA are concerned, there will be some polypeptides found in them that are under the control of organelle genes and other polypeptides that are under the control of chromosomal genes. The cooperative actions of organelle and chromosomal genes in different species is likely to be a further source of comparative evolutionary information.

Of course, many organelles have no DNA of their own. Nonetheless comparative polypeptide studies of those organelle polypeptides that are widely distributed are also likely to broaden man's systematic understanding of organelles, cells and species.

Cell organelle genetic studies are already being applied to human races. For example, mitochondria are inherited only through those found in eggs. The study of mitochondria in the different human races may lead to some useful pointers regarding the migrations of human females, perhaps back to their points of origin.

Before looking at the tabular summary of cell organelles, it is as well to remind ourselves of the fact that cells are broadly divided into two groups – PROKARYOTIC and EUKARYOTIC.

Prokaryotes are much simpler in structure than eukaryotes. They have only one chromosome, which is circular. They have no nucleus and very few membrane-bound organelles. Bacteria and 'blue/greens' are prokaryotic cells.

By contrast, eukaryotic cells have many specialised, membrane-bound organelles, including a nucleus (red blood cells have no nucleus and are an exception to this general rule; some cells, such as the fertilised eggs of D. melanogaster, have more than one nucleus in them). Eukaryotic cells are conveniently sub-divided into 'plant cells' and 'animal cells'.

Fungi don't quite fit into any of the three categories of cell mentioned so far. Although they are eukaryotic cells, and quite close to plants in many of their organelles, they have some very important differences from plants. For example, they lack chloroplasts and are therefore unable to photosynthesize. They derive much of their nutrition from rotting plant and animal material. The fungi are a specialist study.

Viruses belong to neither broad category of cell. Indeed, since viruses depend very greatly on the cells they invade for their nutrients, viruses are not true cells and are not mentioned in the table of this appendix.

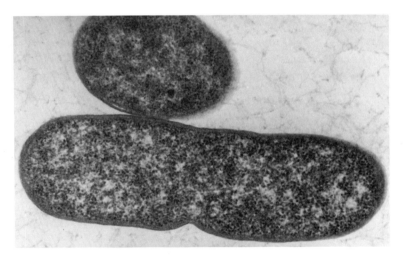

Transmission electron micrograph of Escherichia coli.
The upper oval shape is a non-dividing E. coli. The lower, more elongated shape shows the beginnings of the formation of a new cell wall during cell replication. There is the absence of cell organelles, but the presence of many ribosomes widespread in the protoplasm.

Magnification x 41,000

Photograph by Dr. Tony Brain.
With permission from The Science Photo Library, London.

The following table divides cells into their appropriate categories.

Non-Dividing Cells

Organelle	Eukaryotic		Prokaryotic	Function
	Animal	Plant		
1. **Nucleus**	✓	✓	✗	The principal site of genes which are found on chromosomes. Most inheritance characters are controlled by the genes in nuclear chromosomes.
2. **DNA + histones**	✓	✓	✗	Histones are protein molecules that attach to DNA. It is very difficult to separate eukaryotic DNA from histones.
3. **DNA without histones**	✗	✗	✓	In bacteria and blue/greens the DNA is found in a single circular chromosome. The DNA of prokaryotes is quite easy to purify. Prokaryotic DNA was used to prove that DNA is the material of inheritance in most living things.
4. **The Nucleolus**	✓	✓	✗	Part of one pair of chromosomes is engulfed in the nucleolus. Repeating genes in this part of the chromosomes called the 'nucleolus organiser' are used to make ribosomal RNA (see Chap.3).
5. **Nucleoplasm**	✓	✓	✗	The watery medium in which the chromosomes, chromosome components, mRNA and mRNA components float. The nucleoli float in the nucleoplasm.

Organelle	Eukaryotic		Prokaryotic	Function
	Animal	Plant		
6. **Nuclear Membrane**	✓	✓	✗	It separates the nucleoplasm from the rest of the cell. There are pores in this double membrane and the nuclear membrane is continuous with the membrane of the nuclear body and the endoplasmic reticulum. Ribosomes are often found on the outer membrane; Filaments are often found on the inner nuclear membrane.
7. **Cytoplasm**	✓	✓	✗	The watery medium between the nuclear membrane and the outer plasma membrane. In the cytoplasm float the metabolites needed by the cell, especially those needed for polypeptide synthesis. Many cell organelles bounded by membranes float in the cytoplasm.
8. **Protoplasm**	✗	✗	✓	The prokaryotic equivalent of cytoplasm.
9. **Plasmids**	✓	✓	✓	Plasmids are small circular DNA fragments which float in cytoplasm or protoplasm. They are very useful for genetic engineering.
10. **Endoplasmic reticulum**	(Rough) ✓	(Rough) ✓	✗	The E.R. is a membrane-bound passage connecting the nucleus, the golgi body and the outer cell membrane. Materials can pass from the surface of the cell to the nucleus and to the

Organelle	Eukaryotic		Prokaryotic	Function
	Animal	Plant		
10 (cont.)				golgi without having to pass through the cytoplasm. Materials produced in the nucleus can be passed through the E.R. to ribosomes on the cytoplasmic side of the E.R. – rRNA, tRNA and mRNA. Rough E.R. has ribosomes on its large surface area. The packing together of cisternae made of E.R. is subject to rapid shape changes.
11. **Ribosomes**	✓	✓	✓	Ribosomes are used together with mRNA and tRNA for polypeptide synthesis. (See separate chapters on The Nucleolus and on Polypeptide Synthesis). In eukaryotic cells ribosomes are found on the cytoplasmic side of the endoplasmic reticulum and of the nuclear membrane. In prokaryotic cells ribosomes are widely distributed in the protoplasm (the living watery medium in which the metabolites of bacteria and blue/greens float).
12. **Golgi Body**	✓	✓	✗	Specialised secretory cells produce and store their secretory products in the golgi body. The golgi complex also absorbs chemicals in cells that are not secretory cells. By having a direct link with the outside of the cell and with the nucleus through the endoplasmic reticulum, these organelles are well placed for both absorption and secretion.

Organelle	Eukaryotic Animal	Eukaryotic Plant	Prokaryotic	Function
13. **Lysosomes**	√	√	×	Lysosomes contain the enzymes needed to digest the contents of cells after death. The lysosome membrane separates the enzymes from the cytoplasm so long as the cell is alive.
14. **Mitochondria**	√	√	×	Mitochondria are the respiration centres of eukaryotic cells. They contain their own DNA which governs some, but not all the proteins needed for the working of the mitochondria. Their membrane is double, the inner being essential for energy release.
15. **Kinetoplasts**	√	×	×	Energy transmission; extensions of mitochondria close to the cilia of ciliated protozoa.
16. **Chloroplasts**	×	√	×	Chloroplasts are used for converting light energy into chemical energy. They contain their own DNA which govern some, but not all the polypeptides they need.
17. **Blue/Green pigments**	×	×	√	Similar functions to chloroplasts, but contain no DNA.
18. **Leucoplasts**	×	√	×	The sites in plant cells where glucose is converted into starch.
19. **Chromoplasts**	×	√	×	The sites where plant pigments are made.
20. **Elaioplasts**	×	√	×	Oil storing colourless plastids in plants, mostly found in monocotyledons.

Organelle	Eukaryotic		Prokaryotic	Function
	Animal	Plant		
21. **Cytoplasmic filaments**	✓	✓	×	These protein threads fulfil many specialised functions in non-dividing cells. Muscle fibres are one example. Some filaments of cells are used for cell division.
22. **Microbodies**	✓	✓	×	Microbodies store granular and crystalline materials.
23. **Large Vacuoles surrounded by a membrane**	×	✓	×	The contents of large vacuoles are 'dead'. In plant cells the large vacuoles are a reservoir of cell nutrients, mostly soluble sugar, in cell sap. The osmotic properties of cell sap suck water into plant cells so maintaining high turgor and pushing the outer cell membrane against cell walls.
24. **Microtubules**	✓	✓	×	These structures are found near the outer surface of cells. They assist in maintaining cell shape. They may be the point of origin of the spindles in dividing cells.
25. **Small Vacuoles surrounded by a membrane**	✓	×	×	These are usually the remnants of pinocytotic vesicles which pinch off fluid bathing a cell. The contents are 'dead'.
26. **Outer Plasma Membrane**	✓	✓	✓	The plasma membrane is partially permeable and acts as a molecular and ionic sieve. Molecules pass across it either by passive diffusion or by active (using energy) transport. The plasma membranes of eukaryotic cells are connected to the nucleus, golgi and lysosomes

Organelle	Eukaryotic Animal	Eukaryotic Plant	Prokaryotic	Function
26. (cont.)				by the endoplasmic reticulum. Bacterial plasma membranes can be made into bacterial cell walls at times.
27. **Mesosomes (generalised prokaryotic membrane)**	×	×	✓	An infolding of the plasma membrane. They enhance oxidative phosyphorylation. They may also enhance cell wall formation.
28. **Pinocytotic Vesicles**	✓	×	×	Invaginations of animal cell plasma membranes pinch off extra-cellular fluid so allowing bulk transport into small vacuoles. The small vacuoles gradually disappear as the contents of the pinocytotic vesicles move across the membrane into the cytoplasm.
29. **Microvilli**	✓	×	×	These projections increase the surface area of animal cells, so assisting the passage of materials into and out of cells. They are found most frequently in cells involving transport or secretion of large quantities of materials – cells of the intestinal epithelium.
30. **Desmosomes**	✓	×	×	Used as structures for cell to cell attachment. In the locality of attachment the plasma membrane thickens in both cells. Fine fibres stitch the thickened membranes together, but often leave a gap between the membranes. Fusion of membranes can occur to assist faster cell to cell transport of materials.

Organelle	Eukaryotic		Prokaryotic	Function
	Animal	Plant		
31. **Plasmadesmata**	√	√	×	These structures are similar to desmosomes with fused membranes. However it is necessary for holes to be made through cell walls before plasma streaming can occur between plant cells.
32. **Cell Walls**	×	√	√	Cell walls in plants are made of 'dead' cellulose. In plants the cell walls prevent turgor from bursting the outer cell membrane by retaining the membrane which is pushed against the cell walls. Plant cell walls offer very little resistance to the movements of water and metabolites and may be considered freely permeable and unselective pathways through which solutions can flow. Bacterial cell walls can be made under certain circumstances to reduce the permeability of the perimeter of the bacterium. Under these conditions the metabolism of the bacterium can be very much reduced. Encapsulated bacteria of this sort can exist for many years without erupting to reproduce.

Questions Appendix 1

1. How would you represent a typical prokaryotic cell pictorially? Label your diagram.
2. How would you represent a typical plant cell pictorially? Label your diagram.
3. How would you represent a typical animal cell pictorially? Label your diagram.
4. Of what value will be the detailed analyses of cell organelles?
5. Give an example of an organelle polypeptide that has already been compared in a number of species.
6. Compare prokaryotic and eukaryotic cells.

APPENDIX 2

THE MATHEMATICS OF POPULATION GENETICS

No book on a biological subject would be complete without a brief discussion of decision-making when faced with a set of experimental observations. 'Experience' has often been the criterion used by non-scientists for decision-making. In recent years scientists have collaborated with mathematicians to find mathematical methods of interpreting data. In particular, scientists have needed to know the degree of certainty that can be placed on the acceptance or rejection of their experimental hypotheses when the results of their experiments arrive.

The purpose of this chapter is to introduce students to simple statistical methods. Inevitably there are increasingly complex mathematical methods that can be applied to biological systems, but these are more appropriate to books writen by mathematicians. A list of reference books on statistical methods is given for further reading.

It must be emphasised that for many biological experiments, including those of genetics, the use of appropriate mathematics to interpret the results is absolutely essential. Biologists must increasingly become adept at using mathematics. Having made this assertion, it must then be stated straightaway that the mathematics of this chapter are very easy to understand. Do not be worried by mathematical symbols. They are no more nor less than the 'shorthand' learned by secretaries, and can be unravelled with the same ease.

Before any of the mathematics of statistics are explained it is, perhaps, appropriate that something should be mentioned about decision-making, and about 'The Scientific Method'.

Subjective and Objective Decisions

Subjective Decisions

Subjective decisions are those that are NOT absolutely certain and depend upon interpretations by individuals. Different individuals can and do interpret the qualities of objects, characteristics or events in different ways. Qualitative assessments are therefore not definitive and are often open to a variety of interpretations.

Philosophical decisions such as 'right' and 'wrong' are matters for subjective decisions.

A biological example of a subjective assessment might be that of the 'superiority' or 'inferiority' of the human races. Much has been written on the subject, but there are no clear-cut definitive answers, only subjective assessments largely based upon unquantified data.

Objective Decisions

Objective decisions are often those that are based upon measurable criteria. In such cases the data will be true without question, no matter who assesses it. For example, the lengths of individual units can be measured using a standard scale. Another 'objective' example would be the number of individual units in a sample.

Objective assessments can sometimes extend to other certainties, this time without the possibility of obtaining numerical certainties. For example humans belong to one main blood group or another. There are no intermediate blood groups, so the assessment of blood groups is made by objective decisions. So too are decisions such as 'alive' or 'dead'.

General Criteria For Carrying Out Scientific Experiments

There are two very important criteria for carrying out scientific experiments.

(i) Follow the 'scientific method'.

(ii) Quantify results whenever possible so that mathematical analysis can be applied to them.

The Scientific Method

A scientific investigation should be carried out in the following way:-

Problem
A question arises in the mind of a scientist and he is unable to find the answer to it without carrying out his own experiment(s).

Hypothesis
The scientist makes a sensible guess at the answer to the problem.

Experiment
An experiment is designed for providing data that leads to the answer to the problem.

Results
The results, quantified and with objective measurements whenever possible, are recorded.

Conclusion
The conclusion simply states whether the:-

> (i) HYPOTHESIS WAS RIGHT
> or
> (ii) HYPOTHESIS WAS WRONG

Further action is needed if the hypothesis was wrong:-

> **New Hypothesis**
> **New Experiment**
> **New Results**
> **New Conclusion**

The 'Scientific Method' is repeated, often tiresomely, until the hypothesis is shown to be correct.

Each hypothesis that is proved to be correct opens up new questions and the progress of scientific discovery continues in this orderly way.

The Mathematical Analysis of Quantified Experimental Results

For those experiments that lead to quantified results, it is necessary to state mathematically derived degrees of certainty about the conclusion.

The inference here is that, although objective results can often be obtained from experimental work, the interpretation of the results cannot be 100% certain. 100% certainty requires that an infinite number of results be used. Of course, limitations of man/woman hours and constraints of finance do not usually allow 100% certainty of decision making even to be approached. The requirement of infinite results for absolute certainty cannot of course, be reached.

Confidence Levels

Although objective results can be obtained from many experiments, as has been previously stated, the interpretation of the results cannot be 100% certain unless an infinite number of experiments could be carried out. Also constraints of time and finance may prevent the approach to 100% certainty.

For each experiment, therefore, a *degree* of certainty is chosen. For those experiments which have very important consequences, such as the safe use of medicinal drugs, the degree of certainty has to be very close to 100%. For those of less importance the 95% confidence level is often chosen.

CONFIDENCE LEVEL is the term used by mathematicians for the degree of certainty in making decisions using quantified data.

Two Ways of Expressing Confidence Levels

The confidence levels placed on decisions can either be written on the scale:-

        a)   0% to 100%

              or

        b)   0.0 to 1.0

a) 0% to 100% Scale

(i) An example might be the confidence level needed for the absence of side effects in a medicinal drug:-

$$99.9999\% \text{ confidence level}$$

Such a confidence level demands that 999,999 individuals out of 1,000,000 will show no side effects.

(ii) Another example might be the confidence level needed for the assertion that one variety of buttercup grows on humps in a meadow and another variety grows in the hollows:-

$$95\% \text{ confidence level}$$

Such a confidence level demands that 95 out of a hundred comparisons of the two varieties (by sampling) conforms to the expectation that there is a statistically significant difference in the habitats of the two varieties.

b) 0.0 to 1.0 Scale

The conversion of percentages into decimal point numbers is carried out as follows:-

$$99.9999\% = 0.999999$$
$$95\% = 0.95$$
$$5\% = 0.05$$

Statistically Significant Differences

In essence, confidence levels are those that *refute* the null hypothesis with the confidence level ordained by the person conducting the experiment.

For example, if the experimenter wishes to have a 99.9999% confidence that a drug is safe, one has, first to make the hypothesis that the drug is not safe. This is the null hypothesis in this case. If the experimental results give one or more damaging results out of 1,000,000 then the null hypothesis is accepted at the confidence level chosen.

Now, it may be that it is impossible to carry out the test on exactly

1,000,000 people. The experimenter therefore has to use a smaller number of people for the experiment and must then be able to interpret the results with the help of statistical tables. The chi-squared (χ^2) test would be appropriate in this case and this test is explained fully later in this chapter. The interpretation of the results of the experiment using statistical tables would lead to an acceptance or to a rejection of the null hypothesis with the stipulated degree of confidence, depending on the experimental results.

Although the statistics of the buttercup experiment samples (page 596) are different – the 'Standard Error of the Difference' is the appropriate statistical evaluation in this case (fully explained later) – the acceptance or rejection of the null hypothesis would again apply. Of course, the confidence level would be 95% in this case, rather than the much higher confidence level needed for the testing of medicinal drugs.

If the null hypothesis can be *rejected* at the confidence level chosen, then it can be said that there *is* a statistically significant difference between expectation and the actual results.

If the null hypothesis can be *accepted* at the confidence level chosen there will *not* be a statistically significant difference between them.

The use of statistical tables, using second-hand data, will demonstrate later in this chapter just how easy it is to accept or reject the null hypothesis at a chosen confidence level. An explanation of symbols appears wherever they are used in equations. Plain English is used wherever possible to describe the ways statistical problems can be resolved. Simple genetic problems of statistical evaluation have been chosen as examples.

How to Decide What Sort of Statistics to Use – Discontinuous Variation and Continuous Variation.

Many quantifiable properties of living things (or inanimate ones for that matter) can be divided into two broad categories:-

i) Discontinuous Variation – Chi- squared (χ^2) Statistics Apply.

Some characters of living things are clearly defined with distinct differences between individuals. For example, the common human blood groups are A, B, AB and O. No intermediates exist. Blood groups in humans give a very good example of discontinuous variation.

Many dominant and recessive characteristics that have alternative alleles exhibit discontinuous variation. For example the expression of the long wing allele in D. melanogaster is distinctly different from the expression of the double recessive vestigial wing allele. Here again is an example of discontinuous variation.

A large number of discontinuous characters under the control of alleles or super-alleles were demonstrated in the genetic cross tables of Chapter 5. Some of the questions at the end of this chapter refer back to the genetic crosses of Chapter 5, where 'expected' ratios of offspring can be found.

ii) Continuous Variation – 'Normal Distribution' Statistics Leading to 'Standard Error of the Difference' Statistics.

Some characters of living things vary very slightly from individual to individual. For example, even if a very large number of small equal differences between the heights of boys were chosen, all the boys in the world aged eleven would contribute some individuals for each height category between the smallest eleven-year-old boy and the tallest. Variation by very gradual degrees is called 'Continuous Variation'.

Continuous variation can be analysed in a number of ways, depending on the nature of the experiment.

If a continuously variable character is distributed in a population in an equal way about a mean value, the statistics to use are those of 'The Normal Distribution'.

If it is necessary to compare different, normally distributed samples the statistics to use are 'Standard Error of The Difference'.

Many population characters of continuous variation are *not* normally distributed. There are many mathematical ways of interpreting distribution that are *not* normal, but these are best approached through specialist statistical books. So far as populations of living things are concerned, it is best to start with the simple statistics that lead to 'χ^2' and to 'Standard error of the Difference'. These are now explained.

The Statistics of Chi-squared (χ^2).

χ^2 is the statistic appropriate to discontinuous variation. A very simple example is that of the mating of two heterozygous fruit flies.

Each fly contains the dominant allele for long wing and the recessive allele for vestigial wing:-

$$+ \quad vg$$

We wish to predict the ratio of the phenotypes of the F_1 flies and to find out if there is a statistically significant difference, at the 95% confidence level, between the expected ratio and the numbers actually produced in the experiment.

Predictions

	♀		♂
P	+ vg	×	+ vg
gametes	+ or vg	×	+ or vg
F_1 genotypes	+ + : + vg	:	vg + : vg vg
F_1 phenotypes	3 long wing	:	1 vestigial wing

The Experiment Results.

There were a total of 1064 F_1 fruit flies produced from the parent heterozygotes. These were:-

Long wing	787
Vestigial wing	277
Total	1,064

The next step is to compare the experimental results with the expected results.

Do *not* allow the expected results to cause you difficulty. Simply look at the total number of flies produced in the experiment and divide them into two groups according to your predicted ratios. There were 1,064 flies all told. Of these, three quarters ought to be long wing and one quarter will be vestigial wing, according to your prediction:-

¾ × 1,064 = 798 long wing

¼ × 1,064 = 266 vestigial wing

An inspection of the observed results and the predicted ones shows that there are indeed differences between 'observed' and 'expected' results:-

	Long wing	Vestigial wing
Observed	787	277
Expected	798	266

We now need a method for asserting a degree of confidence that the observed results differ from expectations by amounts that represent a 'statistically significant difference'. Mathematicians have devised a formula for producing a number. The number produced is then compared with a table. The comparison allows a decision to be made regarding a statistically significant difference at the chosen confidence-level. The number is provided by the formula for χ^2.

The Formula for χ^2

Mathematicians tell us that:-

$$\chi^2 = \Sigma \frac{(O-E)^2}{E}$$

where Σ = sum of
O = observed values
E = expected values

In our fruit fly example:-

	Long Wing	Vestigial Wing
Observed	787	277
Expected	798	266
O−E	−11	+11
(O−E)²	121	121

We have therefore obtained for both long wing flies and for vestigial wing flies the O−E values and from these the $(O-E)^2$ values. We already know the expected values for each type of offspring. We can now feed the appropriate figures into the χ^2 formula.

$$\chi^2 = \frac{(787-798)^2}{798} + \frac{(277-266)^2}{266}$$

$$= \frac{121}{798} + \frac{121}{266}$$

$$= \underline{\underline{0.607}}$$

Having obtained the χ^2 number (in this case 0.607) we now need to consult statistical tables, very appropriately supplied for us by the mathematicians. The χ^2 table is given at the end of this appendix.

Interpreting the Calculated χ^2 Using Tables

The formula for χ^2 gives the 'sum of' a number of values. The larger the number of values, the larger the number produced by χ^2 is likely to be. For example the sum of six $\frac{(O-E)^2}{E}$ is likely to be larger than the sum of only two of them:-

$$\chi^2 = \frac{(O_1-E_1)^2}{E_1} + \frac{(O_2-E_2)^2}{E_2} + \frac{(O_3-E_3)^2}{E_3} + \frac{(O_4-E_4)^2}{E_4} + \frac{(O_5-E_5)^2}{E_5} + \frac{(O_6-E_6)^2}{E_6}$$

is likely to produce a larger number than

$$\chi^2 = \frac{(O_1-E_1)^2}{E_1} + \frac{(O_2-E_2)^2}{E_2}$$

The mathematicians had to take account of the number of

$$\frac{(O-E)^2}{E}$$

values that are added together before making the χ^2 statistical table. They did so by inventing a term called 'THE DEGREE OF FREEDOM', the symbol of which is N.

Luckily the degrees of freedom is very easy to evaluate, it is simply:-

The number of $\frac{(O-E)^2}{E}$ values minus one.

In our example the degrees of freedom are given by:-

$$N = 2-1$$
$$= \underline{1}$$

because there is one $\frac{(O-E)^2}{E}$ calculation for the long wing flies and a second for the vestigial wing flies, making two $\frac{(O-E)^2}{E}$ values altogether.

Once the value of χ^2 has been calculated and when the 'degrees of freedom' is known, it is then a matter of consulting χ^2 tables to decide whether or not there is a statistically significant difference between the assertion of the null hypothesis (which says there is no statistically significant difference between the expected numbers of flies and the numbers produced in the experiment) and what actually happened.

Using The χ^2 Table

The χ^2 table shows 'Degrees of Freedom' in a column on the left side of the table. Across the top of the table is the probability. The way the probability is expressed uses the decimal point convention.

Some explanation of how the mathematicians made the table is necessary. It must be remembered that the decimal point 'probability' numbers could also be represented as percentages. For example 0.99 represents a probability of 99.0% and 0.05 represents a probability of 5.0%.

The next question is – why was the table constructed with 'degrees of freedom' down one side and the 'probability' across the top?

Probability and Degrees of Freedom

For any particular 'degrees of freedom', which is obtained from the formula:-

$$D \text{ of } F = N-1$$

where N is the number of $\frac{(O-E)^2}{E}$ added together in the formula, the probability of the numbers given in the table occurring by chance alone is given across the top of the table. For example there is the probability that χ^2 will be equal to or greater than 7.851, when the degrees of freedom is 3, in only 5%(0.05) of cases by chance alone. If, therefore, the calculation of χ^2, for D of F=3, was 7.851 or greater, then it could be stated with 95% confidence that there was a statistically significant difference between the expected numbers and the observed numbers.

Returning to the fruit flies example, we know that the D of F = 1 and that we have chosen the 95% confidence level. Turning to page 607 of the χ^2 table, the table gives us a number. In this case the number is 3.841.

Probability → 0.99.... ....0.60.........0.10.........0.05.........0.01.........0.001

D of F
↓
1   3.841
2
3

All we need to do now is to compare our calculation of χ^2 with the tabulated χ^2 for D of F = 1 and probability 0.05. If our *calculated* χ^2 is *greater* than the tabulated figure, then it can be stated with 95% confidence that a statistically significant difference exists between expected and observed numbers. The reason that this assertion can be made is that the table tells us that the number 3.841 occurs by chance alone for D of F = 1 in only 0.05 (5%) of cases. Therefore, if we obtain a χ^2 calculated using the experimental results, equal to or greater than 3.841, we can say with 95% (100% − 5%) confidence that a statistically significant difference does exist between expected and observed numbers of fruit flies.

In fact our calculation of χ^2 gave:-

calculated χ^2 = 0.607

The calculated χ^2 is now compared with the tabulated χ^2 of 3.841. Since calculated χ^2 is smaller than tabulated χ^2 we have to say that there is NOT a statistically significant difference between expectation and observation at the 95% confidence level. The null hypothesis therefore has to be accepted.

The null hypothesis stated that there was NO statistically significant difference between expected numbers and observed numbers. We therefore accept with 95% confidence the experiment's hypothesis that the ratio of progeny produced by the crossing of two heterozygous fruit flies is:

3 long wing : 1 vestigial wing

An Extract From the χ^2 Table

In general, confidence levels below 95% are unacceptable. The extract from the χ^2 table shown below gives tabulated χ^2 numbers for confidence levels between 95% and 99.9%.

Probability	0.05	0.01	0.001
D of F			
1	3.841	6.635	10.83
2	5.991	9.210	13.82
3	7.851	11.340	16.27
4	9.488	13.28	18.47
5	11.070	15.09	20.51

An examination of the table shows that, as the D of F increases, it is necessary to have a larger tabulated χ^2 for each probability level. This conforms with the earlier discussion on the necessity to take account of the number of $\frac{(O-E)^2}{E}$ values fed into the χ^2 formula.

In order to familiarize students with the interpretation of the χ^2 table, some questions appear at the end of this appendix. It is appropriate to give a second example here.

A Second χ^2 Example

Suppose we cross F_1 heterozygous 'walnut' chickens, as in the Chapter 5 examples:-

F_1 ♀ Rr/Pp   ×   ♂ Rr/Pp

The genetic cross table in Chapter 5 predicts the following ratio of phenotypes in the F_2 generation:-

walnut	:	rose	:	pea	:	single
9		3		3		1

Quite a large number of hens and cocks of the F_1 generation produced the following offspring:-

walnut	:	rose	:	pea	:	single
846		375		235		128

The total number of offspring was 1,584

With what level of confidence can it be asserted that there is a statistically significant difference between expected numbers of chickens and observed numbers?

The Calculation of χ^2

Since the expected ratio of chickens is:-

walnut	:	rose	:	pea	:	single
9		3		3		1

the expected numbers of each type are:-

walnut : rose : pea : single
$9/16 \times 1,584 = 891$: $3/16 \times 1,584 = 297$: $3/16 \times 1,584 = 297$: $1/16 \times 1,584 = 99$

We now make a table of observed and expected values in order to calculate O−E and thereby $(O-E)^2$:-

	walnut	rose	pea	single
Observed	846	375	235	128
Expected	891	297	297	99
O−E	−45	+78	−62	+29
$(O-E)^2$	2025	6084	3844	841

In this example:-

$$D \text{ of } F = N - 1$$
$$\text{In this case } N = 4$$
$$\text{Therefore the D of } F = 4 - 1$$
$$= \underline{3}$$

The tabulated χ^2 for D of F = 3 and for probability 0.05 is 7.851. A comparison of our experimental χ^2, which is 44.196 and tabulated χ^2 allows us to reject the null hypothesis very firmly at the 95% confidence level. Quite simply the table tells us that a χ^2 of 44.196 is likely to occur by chance alone in far fewer than 5 cases out of 100.

At the 95% confidence level therefore, it can be stated that there is a statistically significant difference between expectation and experimental observations.

The results cast doubt on the genotypes of the chickens used in the experiment. An experimenter would have to carry out further experiments to find out whether or not the F_1 chickens were all heterozygous for both genes.

The Application of Statistics to 'Normally' Distributed Continuous Variation

As was briefly mentioned earlier in this chapter, continuous variation implies that there can be very small quantifiable differences between individuals in a population.

Suppose that a sample was taken of boys aged 11 in a small town. It was decided that, between the shortest boy and the tallest, there should be categories of height, each category having an upper and lower limit. Each category of height has the same height interval between their upper and lower limits.

Having found out how many boys there are within the limits of each category, it is very useful to represent the frequencies of each height category on a 'bar chart'. The vertical units for each bar are 'numbers of boys'. The horizontal units represent the height of boys. Each bar represents the number of boys found in each height category.

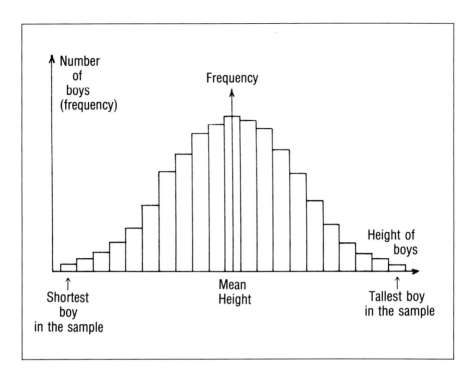

Normal Distributions

The statistics of a sample are 'normally' distributed if a quantifiable character has:-

   i) A mean value

   ii) An equal distribution of the sample either side of the mean, the numbers in each category diminishing with distance from the mean.

Adjacent bars presented as in the diagram above are collectively called a *histogram*. In normally distributed histograms it can be seen that most individuals fall into categories quite close to the mean value. Fewer and fewer individuals are found in categories as they become increasingly separated from the mean.

Although the mean height in the above histogram is found in this case in the longest bar (i.e. the one containing the largest number of boys), the mean value of a normally distributed histogram does not necessarily coincide with the longest bar.

The statistics of a normally distributed sample can be used to give very useful mathematical evaluations such as the *mean*, the *standard deviation*, the *standard error* and the *standard error of the difference*. These will all be discussed in detail later.

Large Samples

In the same way that χ^2 depends on time and economy of resources, so too do sample sizes for normally distributed characters. However, suppose that all governments in the world collaborated to make a world-wide definitive answer to the height of boys aged eleven. With such a large number of boys, it would be possible to divide height categories into a large number of them, each category having very, very small differences between upper and lower category limits. Each bar of the histogram would be very thin indeed, so much so that it would be possible to represent each bar with a thin line. If the midpoints at the top of each line (the top of each thin line represents the number of boys in each category; i.e. the 'frequency' in each category) were joined together, it would be possible to draw a smooth curve with a normal distribution:-

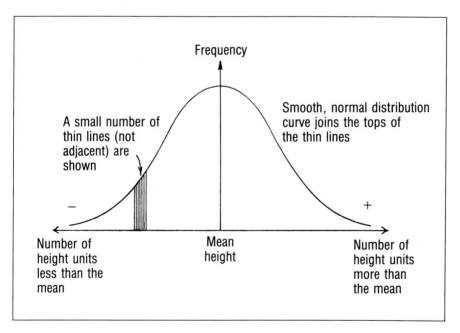

Different population studies will, of course, provide different normally distributed curves:-

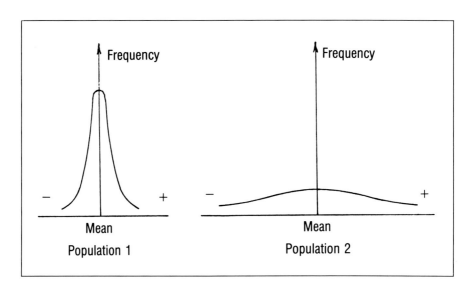

Population 1 clearly concentrates the bulk of its individuals close to the mean, whereas population 2 has quite a high proportion of its individuals separated from the mean by an amount that is considerable. The spread of the population is therefore different for each population.

Much of the discussion in this 'normal distribution' section of statistics concentrates on mathematical ways of describing the spread of the population.

The Standard Deviation of Infinitely Large Populations

It is convenient to introduce the mathematical concepts of normally distributed statistics using the theoretical assumption, never actually achieved of course, of having an infinitely large population. Smooth curves represent the distribution of such populations and the pictorial presentation of smooth curves makes the understanding of concepts easy.

The 'spread' of any normally distributed population can be calculated mathematically using the data collected from a sample. Once the 'spread' has been calculated, the use of statistical tables reveals considerable information about the population. There are two steps needed to evaluate the 'spread'.

1. Calculate the Mean Value

For infinitely large samples it is possible to calculate the TRUE MEAN of the sample. The mean value is simply the sum of all the individual values divided by the total number of measurements. In mathematical terms this can be represented by the following equation:-

$$m = \frac{\Sigma x}{n}$$

where m = the mean

Σ = the sum of

x = any one individual measurement

n = the number of individuals measured

2. Calculate The Standard Deviation

For infinitely large samples it is possible to calculate the *true* standard deviation of the sample. The deviation, 'd' of any one measurement is:-

d = the difference between the individual value and the mean value m.

This can be best understood with reference to a normal distribution curve. Each individual falls into a category represented by a thin line bar. The frequency of each category is represented by the top of the thin line. The normal distribution curve joins the tops of the thin lines:-

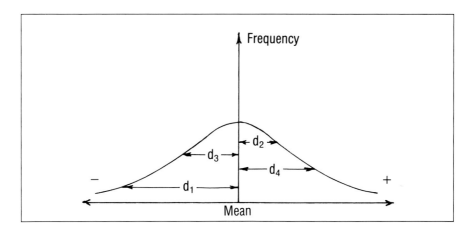

d_1, d_2, d_3 and d_4 show the deviations of four individuals from the mean.

The standard deviation of the sample is a mathematically calculated number which describes the spread of the normal distribution, taking the data provided by the sample into account:-

$$\sigma = \pm \sqrt{\frac{\Sigma d^2}{n}}$$

Where σ = the standard deviation
Σ = the sum of
d = the difference of each individual observation from the mean
$\sqrt{}$ = the square root of
n = the number of individuals in the sample

Straightaway the concept of infinite samples poses problems. This is partly because we never achieve anything remotely approaching huge numbers as a sample size. It is also because the formula has ∞ (infinity) as a number to be fed into it. Luckily for statisticians, the mathematicians discovered that the evaluations of samples of medium size do not differ very much from the perfect results obtained from infinitely large samples. Samples having more than sixty individuals in them give statistical numbers that do not differ very much from the numbers that apply to infinitely large samples. This becomes clear when the statistical table applicable to the normal distribution is examined (it is a table at the end of this appendix).

It is nonetheless appropriate to demonstrate the usefulness of the theoretically perfect calculations of standard deviations. We can turn our attention to small samples later.

The Usefulness of σ – The True Standard Deviation

The statistical table appropriate to normal distributions is designed to tell us what proportion of a population is found between limits defined by:

The Mean $\pm$ Chosen multiples of the Standard Deviation

We know how to calculate the mean so we can now use the formula to calculate the Standard Deviation. As examples to illustrate how the information provided by the statistics of the sample can be used, the following diagram illustrates two pieces of information provided by the normal distribution table.

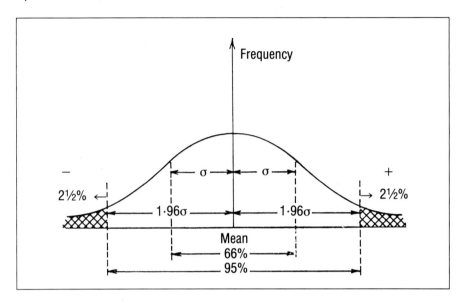

The normal distribution table, called the t-table, tells us that:-

i) within the limits of:
   (the mean ± σ)
   there are 66% of the population
ii) within the limits of:
   (the mean ± 1.96σ)
   there are 95% of the population.

By inspecting the above diagram we can also see that 2½% of the population lie beyond the limit of (mean + 1.96σ) and 2½% of the population are on the other side of the diagram, beyond the limit of (mean − 1.96σ).

What percentage lie beyond the limit (mean + σ)?

What percentage lie beyond the limit (mean − σ)?

The Variance

Mathematicians have meticulously tidy minds. They are not much in favour of allowing the standard deviation to present an untidy ± double signature. They have therefore invented a method for the removal of the ±.

By squaring any number, whether the number be positive or negative, the square is always a positive number. When mathematicians squared the ± Standard Deviation they called the square 'The Variance'.

$$\text{The Variance} = (\pm \text{ Standard Deviation})^2$$
$$= \sigma^2$$

The variance is therefore how mathematicians describe the spread of a sample. It is not, perhaps, as good as the Standard Deviation for visualising what defines the spread of samples that are normally distributed. It is nonetheless a term that has to be understood, because it fits into another formula very neatly later in this discussion.

Small Samples

When samples of sixty or fewer individual measurements are taken, the value of the standard deviation obtained from the formula is only an ESTIMATE of the true standard deviation of the whole population. As the size of samples becomes progressively smaller, the estimates of the standard deviation become progressively less reliable.

The values of the estimated standard deviations obtained from small samples are nearly always too low. An adjustment to the formula is therefore necessary.

1 must be subtracted from the number of individual observations to correct, rather imperfectly, for the tendency to underestimate the standard deviation. The formula for the ESTIMATED standard deviation in small samples is therefore:-

$$s = \pm \sqrt{\frac{\Sigma d^2}{n-1}}$$

where $s =$ the ESTIMATED standard deviation

d = the difference between each individual observation from the mean

$\sqrt{}$ = the square root of

n = the number of individuals measured

Many samples in early undergraduate work or in schools are necessarily small. This formula for 's' is therefore of considerable importance.

The Degrees of Freedom in Normally Distributed Small Populations

It is true to say that, if there are n values of a variable, only (n−1) of them can be selected randomly. This is because, having chosen (n−1) values out of n, the final value is automatically chosen and cannot be regarded as a random selection. The Degrees of Freedom in normally distributed samples is the term used to describe the number of units in a sample that can be randomly selected. For a sample of n individual observations:-

$$\text{Degrees of Freedom} = n-1$$

It has already been seen that for small samples:-

$$s = \pm \sqrt{\frac{\Sigma d^2}{n-1}}$$

This can now be rewritten:-

$$s = \pm \sqrt{\frac{\Sigma d^2}{D \text{ of } F}}$$

The Variance of Normally Distributed Small Samples

Again the variance is a mathematical tidying up process:-

The Variance = s^2

The ± again disappears.

It is more important in many ways to visualize the estimated standard deviation (s), as it appears on normal distribution diagrams, than the concept of the variance (s^2). However, it is also necessary to know how to obtain the variance, because this symbol appears in equations for 'Standard Error of the Difference'. The SE_{diff} is described later.

The Use of Statistical Tables To Evaluate The Probablity of An Individual Measurement Deviating From The Mean by Specific Multiples of The Standard Deviation

In the same way that the Degrees of Freedom create different numbers for the 'probabilities of occurrence by chance alone' in the χ^2 table, so too do the Degrees of Freedom cause changes to the numbers found in the probability columns of the normal distribution table (called the t-table).

Mathematicians have calculated the values to put into the t-table as being the probabilities of any one individual value deviating from the mean by tabulated multiples of the standard deviation. The table therefore takes into account the spread of the sample, which is indicated by the standard deviation. The t-table also takes into account the number of individual observations used in the sample. Before looking at the formula for the 't' of the t-table, try to visualize a normal distribution curve, with its estimated mean and its estimated standard deviation marked on it.

The formula for 't' is:-

$$t = \frac{\text{The difference of an individual observation from the mean}}{\text{The estimated standard deviation}}$$

The following diagram illustrates the meaning of 't' for a sample containing 55 individual observations. The Degrees of Freedom are therefore $(55-1) = 54$. The t-table for D of F = 54 gives the following information:-

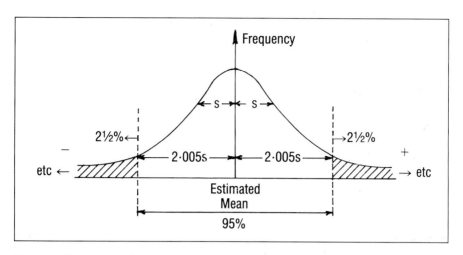

For D of F = 54, the t-table shows that the probability of finding an individual that differs from the estimated mean by a multiple of the standard deviation equal to 2.005s is 5% (0.05). Have a look at page 608 of the t-table to confirm that this is true.

Let us now look on page 608 at a second example which has 61 individual observations in it:-

    D of F = 61−1
            = 60

The following information is provided by the t-table for
D of F = 60 :-

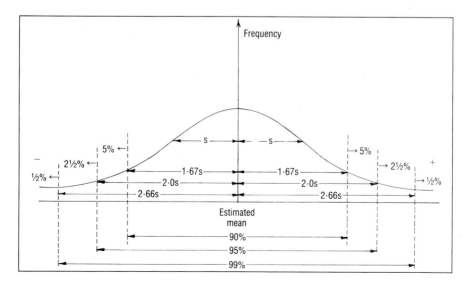

For a D of F = 60, can you tell by looking at the last diagram on page 595 the probability of finding an individual observation of:-

 i) Estimated mean + 2.66s or larger?

 ii) Estimated mean − 1.67s or less?

What else does the t-table tell you for degrees of freedom = 60?

Once the concepts of the normal distribution are understood, together with the ability to use the t-table, we can now move on to the last stages of this introduction to elementary population statistics. These are 'Standard Error' and 'Standard Error of the Difference'.

Standard Error

It is necessary to repeat that, when the number of individual observations in a sample is small, we can only *estimate* the statistics of the population as a whole. In small samples we therefore find the *estimated mean*. The reliability of the estimated mean decreases as the number of individual observations in the sample falls.

There is a mathematical relationship between the numbers in a sample, their estimated standard deviation and the reliability of the estimated mean. The name mathematicians have given to the reliability of the estimated mean is STANDARD ERROR:-

$$SE = \pm \frac{s}{\sqrt{n}}$$

where SE = standard error

  s = estimated standard deviation

  n = the number of individual observations in the sample

The following diagram shows that the collection of normally distributed, individual observations in a small sample allows the following properties to be calculated for the sample:-

 i) The estimated standard deviation.

 ii) The estimated mean.

 iii) The standard error.

All three properties listed above can be calculated simply by feeding the values of individual observations and the number of observations into appropriate formulae.

In other words the reliability of the mean can be calculated. The smaller the calculation of the standard error, the better the reliability of the mean. The diagram below shows that the mean has an unreliability measured by the standard error.

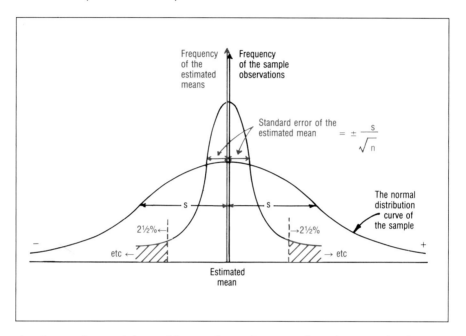

Again mathematicians tidy up the ± by squaring the standard error and calling it the variance:-

Variance of The Estimated Mean = (Standard Error)^2

It will be seen in the discussion that follows on 'The Standard Error of the Difference' that it is important to understand how to calculate the 'Variance of the Estimated Mean'. It is also useful to understand how it is derived.

The 'Standard Error of the Difference' is the statistical method used to compare two samples and can be used to test the *null hypothesis* that there is no statistical difference between the two samples.

Standard Error of the Difference

Suppose that the hypothesis is made by a population geneticist that there is a statistically significant difference between the number of pistils per flower in early and in late flowering buttercups, a common wild flower in England. A statistical method, which includes a stated confidence level (we will choose the 95% confidence level) has to be found which allows us to accept or to reject the hypothesis. As usual the null hypothesis will be used for the purposes of statistical application. The way this is done appears near the end of the calculations.

The Buttercup Samples' Data

	Early Variety	Late Variety
Number of observations in the sample	$n_1 = 294$	$n_2 = 375$
Mean number of pistils per flower (estimated)	$m_1 = 17.85$	$m_2 = 12.15$
The standard deviation (estimated)	$s_1 = 4.24$	$s_2 = 3.38$

When the statistics of the two samples are illustrated, two normally distributed curves can be seen:-

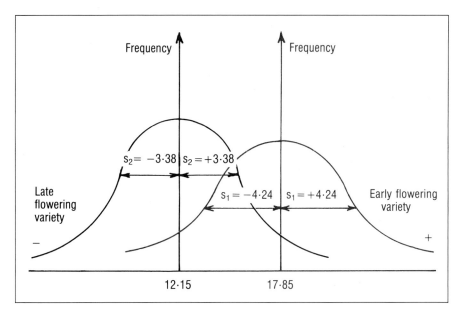

The numbers in the diagram were, of course, provided by the individual observations of the two samples. The usual formulae for calculating the estimated mean and the estimated standard deviation were used.

The two normally distributed curves have quite a considerable overlap. Their means are quite considerably different. It is impossible at a glance, therefore, to assert whether or not there is a statistically significant difference between the two samples. So we must use a statistical test to find out if there is a statistically significant difference, at the chosen confidence level (95%), between the 'early' and 'late' samples. The statistic to use when comparing normally distributed samples is:-

'The Standard Error of the Difference'

This statistic usually has its name abbreviated to:-
SE_{diff}

The Concept of Standard Error of the Difference

It is useful, but not absolutely essential, to have an understanding of the way 'the standard error of the difference' equation is formulated. In practice, it is far more important to be able to feed numbers, obtained from samples, into the SE_{diff} formula. The buttercup numbers are used later as an example of how to do this.

If the theory behind the formulation of the SE_{diff} equation becomes a little difficult, don't despair. Your understanding of the symbols and equations used for normal distribution calculations are perfectly adequate for applying the SE_{diff} formula and for interpreting the statistical table appropriate to SE_{diff}. But do give the theory of SE_{diff} a try. It does help.

In essence, the aim of the SE_{diff} formula is to give a statistical method for deciding whether the observations in each of two samples are such that a specific degree of confidence can be used for accepting or refuting the null hypothesis, which states that there is no statistical difference between the two samples.

The formula takes into account the following information:-

i) The number of individuals' observations in each sample.

ii) The estimated standard deviation of each sample.

iii) The variances of each sample (provided by squaring each estimated standard deviation).

iv) The standard error of each sample.

The formula for SE_{diff} (given later) provides a calculated number. This formula-derived number is then compared with the actual difference between the estimated means. The appropriate statistical table is then used to make decisions by comparing the actual difference between estimated means and the formula derived SE_{diff}.

In other words, mathematicians have provided a formula and some statistical tables that allow us to make assertions at specific confidence levels about different samples. The numbers we need are all provided by the individual observations of the two samples.

The argument that explains the normal distribution curve for the SE_{diff} is as follows:-

i) Suppose a large number of samples were taken during the seasons when each variety of buttercup was flowering. Each sample contains a number of flowers. Estimated means and estimated standard deviations can be calculated for each sample and the estimated normal distribution curve for each sample can be drawn:-

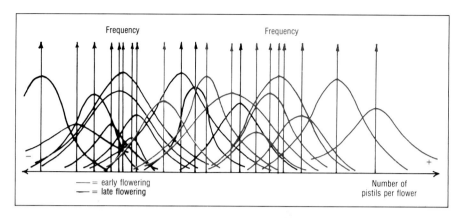

   ii) Write the data provided by each sample on separate pieces of paper.

   iii) Place the pieces of paper in a hat.

   iv) Shake the hat thoroughly.

   v) Randomly (without bias) remove one piece of paper at a time from the hat and record the sample data written upon it.

   vi) Each piece of paper must be put back into the hat after recording what is written upon it. This ensures that there is an equal probability of selection for every piece of paper each time one is pulled out of the hat.

   vii) Work out the differences between the estimated means of successively selected samples:-

   1st piece of paper sample mean = a
   2nd piece of paper sample mean = b
   3rd piece of paper sample mean = c
   4th piece of paper sample mean = d
               etc.

The differences between successive estimated means are given by the calculations:-

$$a - b = \pm z$$
$$b - c = \pm y$$
$$c - d = \pm x$$

viii) Draw a normally distributed graph of the randomly selected differences between successive estimated means:-

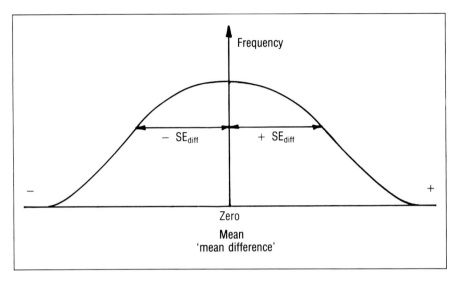

The mean mean difference is zero because half the differences between successive estimated means will be positive and half will be negative.

From the above diagram it can be seen that the SE_{diff} has some similarity with the estimated standard deviation. However, whereas the estimated standard deviation tells us something about the distribution of individuals in one sample, the SE_{diff} tells us something about the spread of 'differences between successive means', the mean differences being provided by several samples selected consecutively at random.

Mathematicians once again provide us with a formula for SE_{diff}.

The Formula For SE_{diff}

The formula for SE_{diff} depends on whether or not we are dealing with small samples. This is because, when small (fewer than sixty) samples are taken, 'n−1' replaces 'n' in the formula.

In the buttercup example, the number of individual observations in each sample is well above sixty. In this case, therefore, 'n' is the appropriate number to feed into the formula

The formula is:-

$$SE_{diff} = \pm \sqrt{\frac{s_1^2}{n_1} + \frac{s_2^2}{n_2}}$$

where SE_{diff} = standard error of the difference
 s_1 = estimated standard deviation for early flowering buttercups
 s_2 = estimated standard deviation for late flowering buttercups
 n_1 = number of 'early' individual observations
 n_2 = number of 'late' individual observations

It should be noticed that, for any sample having sixty or more individual observations in it, $\frac{s^2}{n}$ is the estimated variance of the estimated mean. And the variance of each sample is, of course, its SE^2.

The SE_{diff} can now be written:-

$$SE_{diff} = \pm \sqrt{\text{sum of the variances of each sample}}$$

And, since the variance of each sample is SE^2, the formula can be rewritten yet again:-

$$SE_{diff} = \pm \sqrt{SE_1^2 + SE_2^2}$$

The Buttercup Example

The information provided by the two samples is:-

Early Flowering   **Late Flowering**

$n_1 = 294$ $n_2 = 375$ (flowers)
$m_1 = 17.85$ $m_2 = 12.15$ (pistils)
$s_1 = 4.24$ $s_2 = 3.38$ (pistils)

We can now feed the data into the SE_{diff} formula:-

$$SE_{diff} = \pm \sqrt{\frac{s_1^2}{n_1} + \frac{s_2^2}{n_2}}$$

$$= \pm \sqrt{\frac{4.24^2}{294} + \frac{3.38^2}{375}}$$

$$= \pm \sqrt{0.0916}$$

$$= \pm \underline{\underline{0.303}}$$

The calculated SE_{diff} is now compared with the observed differences between estimated means of the two samples $(m_1 - m_2)$. The comparison can be shown in a diagram:-

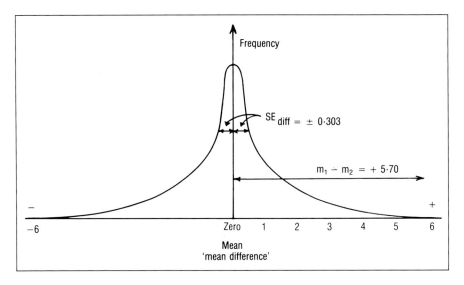

The observed difference between estimated means is:-

$$m_1 - m_2 = 17.85 - 12.15$$
$$= \underline{\underline{5.70}}$$

Comparing Two Samples for Statistical Significance

The statistical comparison of two samples is carried out by way of the NORMAL DEVIATE. This is given the abbreviation 'd'.

$$d = \frac{\text{observed difference between estimated sample means}}{SE_{diff}}$$

$$= \frac{m_1 - m_2}{SE_{diff}}$$

In the buttercup example:-

$$d = \frac{5.70}{0.303}$$

$$= 18.81$$

Mathematicians have provided us with a statistical table for the normal deviate. It is the 'd-table'. Again the table is constructed for different probabilities, and these allow confidence levels to be chosen by an experimenter. If the normal deviate, calculated from the observations of the experimenter, is larger than the tabulated number in the d-table at the chosen confidence level, then there is a statistically significant difference between the two samples.

However, for statistical analyses we always use the comparison between numbers obtained from the observations of the experiment and tabulated numbers to either accept or reject the NULL HYPOTHESIS at the chosen confidence level.

The Buttercup Null Hypothesis

The null hypothesis in this case states that there is NO statistically significant difference between the pistil number in early and late flowering varieties of buttercup.

Testing The Normal Deviate For Statistical Significance

In the buttercup example:

$$d = 18.81$$

An examination of the d-table shows that our calculated value of d is much larger than the tabulated value of d for 5% (0.05) probability. Once again the tabulated numbers for each probability level are those that occur by chance alone. If the calculated value of d is larger than the tabulated value of d, we can reject the null hypothesis.

In our buttercup comparison study we can firmly reject the null hypothesis at the 95% confidence level and state that there IS a statistically significant difference between the two varieties of buttercup. Of course we have only used pistil numbers for our observations. The conclusion to the experiment now refers back to the original hypothesis (pages 598 and 605) which predicted that there is a statistically significant difference between the pistil numbers in the two buttercup varieties. The conclusion simply states that, at the 95% confidence level, the hypothesis was right.

Questions Appendix 2

1. Distinguish between 'objective and 'subjective' decisions.
2. What is 'The Scientific Method' for conducting experiments?
3. Distinguish between continuous and discontinuous variation.
4. What statistical evaluation should be applied to:-
    i) discontinuous variation?
    ii) continuous variation?
5. What is the formula for χ^2 and what do the symbols represent?
6. Samples of the shrub Calluna were taken from five different areas and each plant was classified according to three types – erect; spreading; bushy.
    The following observations were made:-

Area	Erect	Spreading	Bushy
1	20	45	24
2	19	19	26
3	55	40	22
4	11	23	11
5	3	20	10

    a) What statistical evaluation should be applied?

    b) With what confidence level could the hypothesis – 'there is a statistically significant difference between the proportions of the plant types in the different areas' – be made?

7. Strains of carrots that are usually biennial (they flower and set seeds in their second year after germination) were compared. The proportions of plants which bolted (flowered and set seeds in their first year) by the normal harvest dates were:-

	Strain 1					Strain 2				
Samples	1	2	3	4	5	6	7	8	9	10
Number of roots examined	349	352	347	351	356	352	358	358	354	356
Number bolting	2	11	8	16	4	11	45	22	53	18

(i) Is there a significant difference between the bolting habits of each carrot strain within their own five samples?

(ii) With what confidence level can it be asserted that there is a statistically significant difference in the inherited bolting tendency of each variety?

8. The intelligent quotient (I.Q.) in a human population was found to be normally distributed about a mean value of 100. The standard deviation was found to be 16.
    a) What is the probability that an individual chosen at random will have an IQ greater than 120?
    b) What is the probability of the average I.Q of a sample of 10 individuals lying between 98 and 102?

9. The wing length of house flies is normally distributed. The mean length is 4.5mm with a standard deviation of 0.4mm. If the wing-length of 1000 flies were measured:-
    a) How many would be expected to be greater than 5.0mm?
    b) How close to the true mean would be the sample mean?

10. The lengths of the M_1 and M_2 lower molar teeth were measured (mm) in ten Phenacodus primaevus fossilised skulls:-

Skull	1	2	3	4	5	6	7	8	9	10
M_1 length	9.8	10.5	12.4	11.1	12.2	11.9	12.8	13.1	13.2	11.4
M_2 length	10.2	10.7	12.4	11.4	12.0	12.3	13.3	13.1	13.6	10.8

With what level of confidence can it be said that M_2 fossil molars are longer than M_1 molars?

TABLE of t.

Degrees of Freedom	Probability →				
	0.10	0.05	0.02	0.01	0.001
1	6.31	12.71	31.82	63.66	636.62
2	2.92	4.30	6.97	9.93	31.60
3	2.35	3.18	4.54	5.84	12.94
4	2.13	2.78	3.75	4.60	8.61
5	2.02	2.57	3.37	4.03	6.86
6	1.94	2.45	3.14	3.71	5.96
7	1.90	2.37	3.00	3.50	5.41
8	1.86	2.31	2.90	3.36	5.04
9	1.83	2.26	2.82	3.25	4.78
10	1.81	2.23	2.76	3.17	4.59
11	1.80	2.20	2.72	3.11	4.44
12	1.78	2.18	2.68	3.06	4.32
13	1.77	2.16	2.65	3.01	4.22
14	1.76	2.15	2.62	2.98	4.14
15	1.75	2.13	2.60	2.95	4.07
16	1.75	2.12	2.58	2.92	4.02
17	1.74	2.11	2.57	2.90	3.97
18	1.73	2.10	2.55	2.88	3.92
19	1.73	2.09	2.54	2.86	3.88
20	1.73	2.09	2.53	2.85	3.85
30	1.70	2.04	2.45	2.75	3.72
60	1.67	2.00	2.38	2.66	3.50
∞	1.65	1.96	2.33	2.58	3.29

THE χ^2 DISTRIBUTION.

Degrees of Freedom ↓	Probability →							
	0.50	0.25	0.10	0.05	0.025	0.01	0.005	0.001
1	0.455	1.32	2.71	3.84	5.02	6.63	7.88	10.8
2	1.39	2.77	4.61	5.99	7.38	9.21	10.6	13.8
3	2.37	4.11	6.25	7.81	9.35	11.3	12.8	16.3
4	3.36	5.39	7.78	9.49	11.1	13.3	14.9	18.5
5	4.35	6.63	9.24	11.1	12.8	15.1	16.7	20.5
6	5.35	7.84	10.6	12.6	14.4	16.8	18.5	22.5
7	6.35	9.04	12.0	14.1	16.0	18.5	20.3	24.3
8	7.34	10.2	13.4	15.5	17.5	20.1	22.0	26.1
9	8.34	11.4	14.7	16.9	19.0	21.7	23.6	27.9
10	9.34	12.5	16.0	18.3	20.5	23.2	25.2	29.6
11	10.3	13.7	17.3	19.7	21.9	24.7	26.8	31.3
12	11.3	14.8	18.5	21.0	23.3	26.2	28.3	32.9
13	12.3	16.0	19.8	22.4	24.7	27.7	29.8	34.5
14	13.3	17.1	21.1	23.7	26.1	29.1	31.3	36.1
15	14.3	18.2	22.3	25.0	27.5	30.6	32.8	37.7
16	15.3	19.4	23.5	26.3	28.8	32.0	34.3	39.3
17	16.3	20.5	24.8	27.6	30.2	33.4	35.7	40.8
18	17.3	21.6	26.0	28.9	31.5	34.8	37.2	42.3
19	18.3	22.7	27.2	30.1	32.9	36.2	38.6	43.8
20	19.3	23.8	28.4	31.4	34.2	37.6	40.0	45.3
21	20.3	24.9	29.6	32.7	35.5	38.9	41.4	46.8
22	21.3	26.0	30.8	33.9	36.8	40.3	42.8	48.3
23	22.3	27.1	32.0	35.2	38.1	41.6	44.2	49.7
24	23.3	28.2	33.2	36.4	39.4	43.0	45.6	51.2
25	24.3	29.3	34.4	37.7	40.6	44.3	46.9	52.6
26	25.3	30.4	35.6	38.9	41.9	45.6	48.3	54.1
27	26.3	31.5	36.7	40.1	43.2	47.0	49.6	55.5
28	27.3	32.6	37.9	41.3	44.5	48.3	51.0	56.9
29	28.3	33.7	39.1	42.6	45.7	49.6	52.3	58.3
30	29.3	34.8	40.3	43.8	47.6	50.9	53.7	59.7

Bibliography

The first good attempts to describe biological inheritance were written in England and France during the nineteenth century. It is with these papers that the earliest bibliographical refences of this book begin. Of course, much of what was written at that time has had to be revised as each step in understanding inheritance at the cellular level emerges. An attempt has been made to provide the bibliography with a mixture of original papers and text books which describe new knowledge and some of the experiments which provided it.

An intellectual break was made during the second half of the nineteenth century away from a perception of the world as an unchanging place. A world of stress in changing environments was presented and accepted by many scientists at that time. Although not widely known at the time, the bridge between observers of nature and the study of genetics in the modern sense was made in Austria in the 1860s.

The first half of the twentieth century saw the transition from visual observations of simple genetic crosses to important discoveries about the nature of the material of inheritance itself. The link between DNA and polypeptide synthesis was postulated.

Although the decades since the middle of the twentieth century are an artefact of little significance in the march of discovery, it is quite convenient to use the simple system of years with round number endings to embrace some important advances in genetics.

The 1950s saw the establishment of the structure of DNA. A complete analysis of the amino-acid sequence in a protein was made for the first time. Some of the processes of polypeptide synthesis were discovered. During the 1950s the best advances in knowledge came from the studies of bacteria and viruses.

An acceleration of discovery occurred during the 1960s. The entire sequence of processes involved in polypeptide synthesis was established. The degenerate genetic code was fully worked out. Ecological studies could be related to genes for the first time. The analysis of amino-acid sequences in widespread polypeptides began to give an insight into the history of genes and the relationships of species. Extra-chromosomal inheritance emerged as a new genetic study.

The discovery of the enzyme reverse-transcriptidase and the ways in which is can be used to synthesise genes revolutionised man's potential for advances in biotechnology. Genetic engineering was established as an immensely useful tool for use by humanity. By the end of the 1970s individual genes could be indentified, both from their positions on the genome of eukaryotic species and for their precise base sequences. As a result of these eukaryotic studies, introns, exons and the nature of chromatin were identified.

If the 1970s saw a switch from prokaryotic studies to eukaroyotes, the 1980s was the decade of specific gene identification. A few of the identified genes were found to be potentially cancerous – the proto-oncogenes. A small number of connections between proto-oncogenes, oncogenes and uncontrolled mitosis emerged. Machines for the rapid analysis of genes and polypeptides were established in several centres of research throughout the world. International cooperation set up the Human Genome Project, with the aim of making a comprehensive analysis of the entire human genome by the year 2006. The relationships between mobile genetic elements and their host cells began to be indentified.

The 1990s have already seen the positive identification of important genes. The human gene that determines maleness was found in 1990 and was proved to be the sex-determining gene in 1991. The inter-relationships of genes and gene controls is likely to be the main advance of the 1990s. Advances in our understanding of gene control, allied to better techniques in biotechnology should lead to more rapid advances in cancer research and cell differentiation studies. Biotechnology is becoming big business in several areas of science such as medicine, veterinary science and the agricultural sciences.

Most of the bibliography references were written in the twentieth century. An attempt has been made to include some of the best historical records both in books and in individual papers. Wherever possible, recent papers that are breaking new ground have been included too.

The names of several English language journals and a few written in other languages appear in the bibliography. The journal references include some of the famous papers in furthering our understanding of genetics.

Long bibliographical lists inevitably present problems to inexperienced students. The text of this book has been put together to remove the need for wide reading during the transition from school to university in the early months of a biology course that includes genetics. Your own particular requirements in genetics will depend upon the type of biology course you follow. Specialist courses in genetics will, of course, demand further reading. Each chapter in this book has a list of publications that could be used for further reading. However, tutors of universities also provide a list of publications for further reading and it would be wise to stick closely to those. The list of refences given for each chapter in this book would allow an extended general view of how a modern understanding of genetics has emerged during the history of discovery.

Chapter 1 bibliography is presented so that each section of the chapter has its own bibliography list. The bibliography for each section of this chapter is presented alphabetically according to the surname of the author(s) or editor(s).

Chapters and appendices other than chapter 1 have collective bibliographies presented alphabetically according to the surname of the author(s) or editors.

The complexities of evolution have made it necessary to have a rather long bibliography for chapter 12. There is also a supplementary reading list for chapter 12 which provides references for several rather narrow aspects of ecological genetics.

It is to be hoped that the text of this book, together with carefully chosen selections taken from the bibliographical references, will give students a satisfactory understanding of biological inheritance, according to their particular needs.

Librarians are very helpful in many matters regarding the provision of books and in explaining where to find them. Reference libraries that specialise in science usually keep several years of periodicals, and some have complete sets in sequence. It is not always easy to obtain passes to enter some of the prestigious reference libraries. A talk with your tutor, college librarian or departmental librarian will establish the libraries to which you have access by right.

Library arrangements sometimes exist whereby books not available at your local library can be obtained on request from libraries elsewhere. Such requests go through local library staff. National computer systems have been established in some countries so that the whereabouts of particular books and their availability can be very quickly discovered. Of course, there will be a limit to the number of books that can be borrowed on one library card at any one time. The effective use of library materials is an essential facet of student life.

It is to be hoped that you enjoy the fascinations of nature through reading this book and from your carefully selected further reading.

CHAPTER 1

1a) **Bateson, W.** *Mendel's Principles of Heredity*, Cambridge University Press, 1909.
Mendel, G. *Observations on Experiments in Plant Hybridisation*, Verh. naturf, Ver. in Brünn Abhandlungen IV, 1865.

1b) **Brooks, J.L.** *Just Before The Origin* (Alfred Russel Wallace's Theory of Evolution), Columbia University Press, 1984.
Darwin, C.R. *On The Origin of Species by Natural Selection*, John Murray, 1858.
Darwin, C.R. *Variation of Animals and Plants Under Domestication*, John Murray, 1871.
Darwin, C.R. (Edited by his son, F. Darwin) *Life and Letters of C.R. Darwin – An Autobiographical Chapter*, John Murray, 1887.
Jordanova, L.J. *Lamarck*, Oxford University Press, 1984.
Lack, D. *Darwin's Finches*, Peter Smith, 1968.
Lamarck, J-B. *Histoire Naturelle des Animaux sans Vertèbres* (The Introduction), Paris, 1815.
Lamarck, J-B. *Philosophe Zoologique*, Paris, 1809.
Lamarck, J-B. *Recherches sur L'Organisation des Corps Vivans*, Paris, 1802.
Wallace, A.R. *On the Law Which Has Regulated the Introduction of New Species*, Ann. Mag. Nat. Hist., 2nd Series, 16, 184-96, 1855.
Wallace, A.R. *On The Tendency of Varieties to Depart Indefinitely from the Original Type*, J. Proc. Linn. Soc. London, Zoology 3, 53-62, 1858.

1c) **Griffiths, F.** *The Significance of Pneumococcal Types*, Journal of Hygiene XXVII, 113-159, January 1928.

1d) **Franklin, R.E.** *Evidence for the 2-Chain Helix in Crystalline Structure of Sodium Deoxyribonucleate*, Nature 172, 1953.
Lewis, K.R., and John, B. *The Organisation of Heredity*, Arnold, 1971.
Watson, J.D. *The Double Helix*, Penguin, 1968.
Watson, J.D., and Crick, F.H.C. *The Structure of DNA*, Cold Spring Harbour Symposium of Quantitative Biology 18, 123-131, 1953.
Watson, J.D., and Crick, F.H.C. *Molecular Structure of Nucleic Acids – A Structure for DNA*, Nature Vol 171, No. 4356, 737-738, 1953.

1e) **Beadle, G.W., and Tatum, E.L.** *Genetic Control of Biochemical Reaction in Neurospora*. Proceedings of The National Academy of Science, USA, 27, 499-506, 1941.

Jacob, F., and Wollman, E.L. *Sexuality and the Genetics of Bacteria*, New York & Co., 1961.
Lederberg, J., and Tatum, E.L. *Recombination in Strain K12 of E.coli*, Nature 158, 558, 1946.
Watson, J.D., Tooze, J., and Kurtz, D.T. *Recombinant DNA – A Short Course*, Scientific American Book, W.H. Freeman, 1983.
Yudkin, M., and Offord, R. *Comprehensible Biochemistry*, Longman, 1973.

1f) Yudkin, B., and Offord, R. *Comprehensible Biochemistry*, Longman, 1973.

1g) Crick, F.H.S., Mrs. L. Barnett, Brenner, S., Watts-Tobin, R.J., and Shulman, L. *The Genetic Code*, Scientific American, October, 1962.
Nirenberg, M.W., and Mathei, J.H. *The Dependence of Cell-free Protein Synthesis in E.coli upon Naturally Occurring or Synthetic Polyribonucleotides*, Proceedings of The National Academy of Science, USA, 47, 1588-1602, 1961.

1h) Crick, F.H.S. *The Genetic Code*, Scientific American, October, 1962.
Streisinger, G., Okada, Y., Emrich, J., Newton, J., Tsugita, A., Terzaghi, E., and Inouye, M. *Frameshift Mutations and the Genetic Code*, Cold Spring Harbour Symposium Vol, XXXI, 1966.

General Chapter 1
Peters, J.A., (Editor), *Classic Papers in Genetics*, Prentice-Hall, 1959.
Various Authors, *Scientific American, October, 1985*.

CHAPTER 2

Alberts, B., Bray, D., Lewis, J., Raff, M., Roberts, K., and Watson, J.D. *Molecular Biology of the Cell*, (2nd Edition), Garland, 1989.
Asimov, I. *The Genetic Code*, John Murray, 1964.
Chothia, C. *Principles that Determine the Structure of Proteins*, Ann. Review of Biochemistry 53, 537-572, 1984.
Clarke, B.F.C. *The Genetic Code*, Studies in Biology No. 83, Arnold, 1979.
Clarke, B.F.C. (*Editor*), *Gene Expression*, Murksgaard, Copenhagen, 1984.
Lewis, K.R., and John, B. *The Organisation of Heredity*, Arnold, 1971.
MacLean, N. *Haemogoblin*, Studies in Biology No. 93, Arnold, 1978.
Szekely, M. *From DNA to Protein*, Macmilan, 1980.
Various Authors, *Scientific American*, October, 1985.
Woodhead-Galloway, J. *Collagen – The Anatomy of a Protein*, Studies in Biology No. 117, Arnold, 1980.
Wynn, C.H. *The Structure and Function of Enzymes*, (Studies in Biology No. 42,) Arnold, 1979.
Yudkin, M., and Offord, R. *Comprehensible Biochemistry*, Longman, 1973.

CHAPTER 3

Busch, H. (**Editor**), *The Cell Nucleus*, Volume I, Academic Press, 1974.
Hadjiolov, A.A. *The Nucleolus and Ribosome Biogenesis*, Springer-Verlag, 1985.
Jordan, E.G. *The Nucleolus*, Carolina Biology Reader No. 16, Carolina Biological Supply Company, 1980.

Scheer, U., Trendelberg, M.F., and Franke, W.W. *Regulation of Transcription of Genes of Ribosomal RNA during Amphibian Oögenesis. A Biochemical and Morphological Study,* Journal of Cell Biology 69, 465-489, 1976.

CHAPTER 4

Barish, N. *The Gene Concept,* Chapman and Hall, 1966.
Butler, J.A.V. *Gene Control in the Living Cell,* George Allen and Unwin, 1968.
Carlson, E.A. *The Gene – A Critical History,* W.B. Saunders, 1966.
Clark, B.F.C. *The Genetic Code,* Studies in Biology No. 83, Arnold, 1979.
Haskell, G. *Practical Heredity with Drosophila,* Oliver and Boyd, 1961.
Howard, B.A. *Vectors for Introducing Genes into the Cells of Higher Organisms,* Trends in Biomedical Sciences 8, 209-12, 1983.
Hughes, S.H. *Synthesis, Integration and Transcription of the Retroviral Provirus,* Current Topics in Microbiology and Immunology, 103, 23-49, 1983.
Isenberg, I. *Histones,* Ann. Review of Biochemistry 48: 159, 1979.
Levan, G. Stahl, F., and Weltergren, Y. *Gene Amplification of the c-myc Oncogene in the Murine SEWA System,* Mutat. Res. 276(3): 285-290, May 1992.
Lewin, B. *Gene Expression,* Vol 2, Eukaryotic Chromosomes (Second Edition), Wiley, 1980.
Lewin, R. *Can Genes Jump Between Eukaryotic Species?* Science 217, 42-3, 1982.
Lycke,E., and Norrby, E. *Virus Induced Changes of Cell Structure and Function,* Textbook of Medical Virology, 83-104, 1983.
MacLean, N., Gregory, S.P., and Flavell, R.A. *Eukaryotic Genes – Their Structure, Activity and Regulation,* Butterworth, 1983.
Pintel, D., Dadachanji, D., Astell, C.R., and Ward, D.C. *The Genome of Minute Virus of Mice Encodes Two Overlapping Transcription Units,* Nucleic Acids Research 11, 1019-38, 1983.
Shapiro, J.A. *Mobile Genetic Elements,* Academic Press, 1983.
Simon, D., Munoz, S.J., Maddrey, W.C., and Knowles, B.B. *Chromosomal Rearrangements in a Human Primary Hepatocellular Carcinoma. Characteristic Abnormalities in Chromosomes 1, 5, 6, 9, 13, 16 and 22.* Cancer – Genet. – Cytogenet, 45(2), 255-60, 1990
Syvanen, M. *Cross-species Gene Transfer; Implications For a New Theory of Evolution,* Journal of Theoretical Biology 112, 333-43, 1985.
Various Authors, *Scientific American,* October, 1985.
Watson, J.D. *The Molecular Biology of The Gene,* W.A. Benjamin – Addison Wesley World Student Series, 1977.
Zimmern, D. *Do Viroids and Viruses Derive from a System That Exchanges Genetic Information Between Eukaryotic Cells?* Trends in Biomedical Sciences 7, 205-7, 1982.

CHAPTER 5

Ashburner, M., Carson, H.L., and Thompson, J.N., (Editors), *The Genetics and Biology of Drosophila,* Academic Press, 1982.
Haskell, G. *Practical Heredity with Drosophila,* Oliver & Boyd, 1961.

CHAPTER 6

Beale, G., and Knowles, J. *Extranuclear Genetics,* Arnold, 1978.

Dart, E.C., Pioli, D., and Atherton, K.T. *Genetic Engineering – Essays in Applied Biology,* John Wiley, 1981.
Day, M.J. *Plasmids,* Studies in Biology No. 142, Arnold, 1982.
Jinks, J.L. *Cytoplasmic Inheritance,* Carolina Biology Reader No. 72, Carolina Biological Supply Company, 1978.
Matsuura, E., Fukada, H., and Chigusa, S. *A Genetic Link between an mRNA-specific Translational Activator and the Translation System in Yeast Mitochondria,* Genetics 125,3, 495-503, July, 1990
Tribe, M.A., Whittaker, P.A., and Morgan, A.J. *The Evolution of Eukaryotic Cells,* Studies in Biology No. 131, Arnold, 1981.
Williamson, R. (Editor) *Genetic Engineering,* Vol. I, Academic Press, 1981.

CHAPTER 7

Jinks, J.L. *Extrachromosomal Inheritance,* Prentice Hall, 1964.
Kirk, J.T.O., and Tilney-Bassett, R.A.E. *The Plastids,* W.H. Freeman, 1967.
Preer, J.R. *Extrachromosomal Inheritance: Hereditary Symbionts, Mitochondria and Chloroplasts,* Annual Review of Genetics 5, 361-406, 1971.
Tilney-Bassett, R.A.E. *Genetics and Plastid Physiology in Pelargonium III,* Heredity 25, 89-103, 1970,

CHAPTER 8

Barnes, D.M. *Strategies for an AIDS Vaccine,* Science 223, 1149-53, 1986.
Butcher, D.N., and Ingram, D.S. *Plant Tissue Culture,* Studies in Biology No. 86, Arnold, 1976.
Cline, M.J. *Genetic Engineering of Mammalian Cells. Its Potential Application to the Genetic Diseases of Man,* Journal of Laboratory and Clinical Medicine 99, 299-308, 1982.
Dart, E.C., Pioli, D., and Atherton, K.T. *Genetic Engineering – Essays in Applied Biology,* John Wiley & Sons, 1981.
Day, M.J. *Plasmids,* Studies in Biology No. 142, Arnold, 1982.
Glover, D.M. *Gene Cloning – The Mechanism of DNA Manipulation,* Chapman & Hall, 1984.
Laurence, J. *The AIDS Virus,* Scientific American, 70-79, December 1985.
Letvin, N. *Human Protein CD4 and its Protection Against AIDS,* Proceedings of The National Academy of Sciences (USA), June 1991.
Newmark, P. *First Human AIDS Vaccine Trial Goes Ahead Without OK,* Nature, 325, 290, 1987.
Scott, A. *Pirates of the Cell,* Basil Blackwell, 1987.
Shay, J.W. (Editor) *Techniques in Somatic Cell Genetics,* Plenum Press, 1982.
Smith, J.E. *Biotechnology,* Studies in Biology No. 136, Arnold, 1981.
Sharp, J.A. *An Introduction to Animal Tissue Culture,* Studies in Biology No. 82, Arnold, 1977.
Various Authors, *Scientific American,* October, 1985.
Williamson, R. (Editor), *Genetic Engineering,* Volume I, Academic Press, 1981.
Winnacker, E.L. *From Genes to Clones,* Introduction to Gene Technology, VCH Verlagsgesellaschaft, 1987.

CHAPTER 9

Alberts, B., Bray, D., Lewis, J., Raff, M., Roberts, K., and Watson, J.D. *Molecular Biology of the Cell,* (2nd Edition), Garland, 1989.
Darlington, C.D., and Mather, K. *The Elements of Genetics,* Schocken, 1969.
Fincham, J.R.S. *Using Fungi to Study Genetic Recombination* Carolina Biology Reader No. 2, Carolina Biological Supply Company, 1983.
Fincham, J.R.S. *Genetics,* Butterworth Heinemann, 1983.
George, W. *Elementary Genetics,* Macmillan, 1965.
John, B., and Lewis, K.R. *Somatic Cell Division,* Carolina Biology Reader No. 26, Carolina Biological Supply Company, 1972.
John, B., and Lewis, K.R. *The Meiotic Mechanism,* Carolina Biology Reader No. 65, Carolina Biological Supply Company, 1976.
Kemp, R. *Cell Division and Heredity,* Studies in Biology No. 21, Arnold, 1970.
Lewis, K.R., and John, B. *The Matter of Mendelian Inheritance,* J. & A. Churchill Ltd., 1964.
Lloyd, D., Poole, R.K., and Edwards, S.W. *The Cell Division Cycle,* Academic Press, 1982.
McLeish, J., and Snoad, B. *Looking at Chromosomes,* Macmillan, 1959.
Menees, T.M., and Roeder, G.S. *ME14, a yeast gene required for Meiotic Recombination,* Genetics 123, 675-682, December, 1989.
Sybenga, J. *Meiotic Configurations,* Springer Verlag, 1975.

CHAPTER 10

Acton, A.B. *Crossing-over within the inverted regions in Chironomus,* American Naturalist 90, 63-65, 1956.
Acton, A.B. *Chromosome Inversions in Natural Populations of Chironomous tentans,* Journal of Genetics 55, 61-94, 1957.
Ayala, F.J. *Molecular Evolution,* Sinauer Associates, 1976.
Darlington, C.D. *The Evolution of Genetic Systems* (2nd Edition), Cambridge University Press 1958.
Darlington, C.D. (Editor: Creed, E.R.) *Ecological Genetics and Evolution (the evolution of polymorphic systems), 1-19,* Blackwell, 1971.
Darlington, C.D. *The Place of Chromosomes in the Genetic System,* Chromosomes Today 4, 1-13, 1973.
Erlich, P.R., Holm, R.W., and Parnell, D.R. *The Process of Evolution,* McGraw-Hill Kogakusha, 1974.
John, B. *Population Cytogenetics,* Studies in Biology No. 70, Arnold, 1976.
Kimura, M. *Evolutionary Rate at the Molecular Level,* Nature 217, 624-626, 1968.
Lewis, W.H., (Editor), *Polyploidy,* Plenum Press, 1980.
Lewontin, R.C. *The Genetic Basis of Evolutionary Change,* Columbia University Press, 1974.
Lewontin, R.C., and White, M.J.D. *Interactions Between Inversion Polymorphisms of Two Chromosome Pairs in the Grasshopper Moraba scurra,* Evolution 14, 116-129, 1960.
Okagaki, R.J., Neuffer, M.G., and Wessler, S.R. *A Deletion Common to Two Independently Derived Waxy Mutations of Maize,* Genetics 128, 425-431, July, 1991.
Ohno, S., Weiler, C., Poole, J., Christian, L., and Stenius, C. *Autosomal Polymorphism Due to Pericentric Inversions in the Deer Mouse (Peromyscus maniculatus),* Chromosoma 18, 177-178, 1966.

Pearson, P.L., (Editor: J. Huxley), *Evolution – The Modern Synthesis*, (3rd Edition). Man and Pongids: Evidence from Chromosomes Bearing on their Relationship, Allen & Unwin, 1975.
Scott, A. *Pirates of The Cell*, Basil Blackwell, 1987.
Spiess, E.B. *Chromosomal Adaptive Polymorphism in Drosophila persimilis II*, Evolution 12, 234-245, 1958.
Shorrocks, B. *The Genesis of Diversity*, Hodder & Stoughton, 1978.
Sturtevant, A.H., and Dobzhansky, T. *Inversions in the Third Chromosome of Wild Races of Drosophila pseudoobscura and their Use in the Study of the History of the Species*, Proceedings of the National Academy of Science, USA 22, 448-450, 1936.

CHAPTER 11

Berry, R.J. *Neo-Darwinism*, Studies in Biology No. 144, Arnold, 1982.
Bullini, L., and Colluzi, M. *Natural Selection and Genetic Drift in Protein Polymorphism*, Nature, 239, 160-161, 1972.
Bulmer, M. *Estimating the Variability of Substitution Rates*, Genetics 123, 615-619, 1989.
Clarke, B. *Darwinian Evolution of Proteins*, Science 168, 1009-1011, 1970.
Clarke, B. *Selective Constraints on Amino-acid Substitutions During the Evolution of Proteins*, Nature 228, 1970.
Clegg, J.B., (Editor **Creed, E.R.**), *Ecological Genetics – Gene Duplication and Haemoglobin Polymorphism*, Blackwell Scientific Publications, 298-307, 1971.
Dayhoff, M.O., Park, C.M., and McLaughlin, P.J. *Building a Phylogenic Tree: Cytochrome c.*, Atlas of Protein Sequence and Structure Vol. 5, National Biochemical Research Council, Washington D.C., 1972.
Fitch, M.W., and Margoliash, E. *A Method for Estimating the Number of Invariant Amino-acid Coding Positions in a Gene Using Cytochrome c as a Model Case*, Biochemical Genetics I, 65-71, 1967.
Hubby, J.L., and Throckmorton, L.H. *Protein Differences in Drosophila IV*, A Study of Sibling Species (18 gene loci were compared in 27 Drosophila species), American Naturalist 102, 193-205, 1968.
Milne, H., and Robertson, F.W. *Polymorphism in Egg Albumen Protein and Behaviour in the Eider Duck*, Nature 205, 367-369, 1965.
Parkin, D.T. *An Introduction to Evolutionary Genetics*, Arnold, 1979.
Various Authors, *Scientific American, October 1985.*
Yudkin, M., and Offord, R. *Comprehensible Biochemistry*, Longman, 1973.

CHAPTER 12

Antonoviks, J., and Bradshaw, A.D. *Clinal Patterns at a Mine Boundary*, Heredity 25, 349-362, 1970.
Ayala, F.J. *Genetic Differentiation During the Speciation Process*, Evolutionary Biology 8, 1-75, 1975.
Bates, H.W. *Contributions to an Insect Fauna of the Amazon Valley; Lepidoptera Heliconidae*, Transactions of the Linnaean Society, London 23, 495-566,1862.
Benson, W.W., Brown, K.S., and Gilbert, L.E. *Co-evolution of Plants and herbivores: passion flower butterflies*, Evolution 29, 659-680, 1976.
Berry, R.J. *Neo-Darwinism*, Studies in Biology No. 144, Arnold, 1974.

Bowman, J.C. *An Introduction to Animal Breeding,* Studies in Biology No. 46, Arnold, 1974.
Bradshaw, A.D., and McNeilly T.S. *Evolution and Pollution,* Studies in Biology No. 139, Arnold, 1981.
Buovine, J.R. *Evidence from Double Infestation for the Specific Status of Human Hair Lice and Body Lice (Anoplura),* Systematic Entomology 3, 1-8, 1978.
Bush, G.L. *Modes of Animal Speciation,* Annual Review of Ecology and Systematics, 6, 339-364, 1975.
Cavalier-Smith, T., (Editor), *The Evolution of Genome Size,* John Wiley & Sons, 1985.
Clarke, C., and Sheppard, P.M. *Super-genes and Mimicry,* Heredity 14, 175-185, 1960.
Cott, H.B. *Adaptive Coloration in Animals,* Methuen, 1940.
Darwin, C. *On the Origin of Species by Means of Natural Selection,* Murray, 1858.
Davidson, G., (Gear, J.H.S., Editor**),** *Anopheline Species Complexes,* in *Medicine in a Tropical Environment,* Balkema, 254-271, Cape Town, 1971.
de Beer, G. *Adaptation,* Carolina Biology Reader No. 22, Carolina Biological Supply Company, 1978.
Dickson, R.C. *Development of the Spotted Alfalfa Aphid Population in North America,* Internationaler Kongress für Entomologie (Wien) 2, 26-28, 1962.
Dillon, L.S. *The Genetic Mechanism and the Origin of Life,* Plenum Press, 1978.
Dobzhansky, T. *Genetics of the Evolutionary process,* Columbia University Press, 1970.
Dobzhansky, T., Ayala, F.J., Stebbins, G., and Valentine, J.W. *Evolution,* W.H. Freeman & Co., 1977.
Dobzhansky, T., and Powell, J.R., (King, R.C., Editor**),** *The willistoni group of sibling species of Drosophila.* Handbook of Genetics, Vol. 3, Invertebrates of Genetic Interest, 589-622, Plenum Press, 1975.
Dobzhansky, T., and Spassky, B. *Drosophila paulistorum, a cluster of species in statu nascendi,* Proceedings of the National Academy of Sciences of the United States of America, 45, 419-428, 1959.
Dowdeswell, W.H.* *The Mechanism of Evolution,* Heinemann, 1975.
Edwards, K.J.R. *Evolution in Modern Biology,* Studies in Biology No. 87, Arnold, 1979.
Ehrman, L., and Parsons, P.A. *The Genetics of Behaviour,* Sinauer Associates, 1977.
Endler, J.A. *Geographic Variation, Speciation and Clines 246,* Princeton University Press, 1977.
Ford, E.B. *Evolution Studied by Observation and Experiment,* Carolina Biology Reader No. 55, Carolina Biological Supply Company, 1974.
Ford, E.B.* *Ecological Genetics,* Chapman and Hall, 1979.
Ford, E.B. *Genetics of Adaptation,* Studies in Biology No. 69, Arnold, 1979.
Forey, P.L. (Editor)* *The Evolving Biosphere,* Cambridge University Press, 1981.
Fryer, G., and Iles, T.D. *The Cichlid Fishes of the Great Lakes of Africa. Their Biology and Evolution; 641,* Oliver & Boyd, 1972.
Futuyama, D.J. and Slatkin, M. *Coevolution,* Sinauer Associates, 1983.
Gale, J.S. *Population Genetics,* Blackie, 1980.
Haffer, J. *Avian Speciation in Tropical South America,* 390, Nuttall Ornithological Club, Cambridge, Mass., 1974.
Kettlewell, B. *The Evolution of Melanism,* Oxford University Press, 1973.
Levington, J. *Genetics, Palaeontology and Macroevolution,* Cambridge University Press, 1988.
Maynard-Smith, J. *Sympatric Speciation,* American Naturalist 100, 637-650, 1966.
Maynard-Smith, J.* *The Theory of Evolution,* Pelican, 1975.
Mayr, E. *Population, Species and Evolution,* 453, Belknap Press, Cambridge, Mass., 1970.

Papageorgis, C. *Mimicry in Neotropical Butterflies,* American Scientist 63: 522-532, 1975.
Parkin, D.T.* *An Introduction to Evolutionary Genetics,* Arnold, 1979.
Roughgarden J. *Theory of Population Genetics and Evolutionary Ecology: An Introduction,* Macmillan/Coller, 1979.
Scott, A. *Pirates of the Cell,* Basil Blackwell, 1987.
Sheppard, P.M. *Recent Genetical Work on Polymorphic Mimetic Papilios;* 20-29, Symposia of the Royal Entomological Society of London, 1961.
Shorrocks, B. *The Genetics of Diversity,* Hodder & Stoughton, 1973.
Tajuna, F. *The Effects of Population Size on DNA Polymorphism,* Genetics 123, 597-601, November, 1989.
Vane-Wright, R.I. *On the Definition of Mimicry,* Biological Journal of the Linnaean Society, London, 13, 1-6, 1980.
Vevers, G. *The Colours of Animals,* Studies in Biology No. 146, Arnold, 1982.
Von Frisch, O. *Animal Camouflage,* Collins, 1973.
Wallace, A.R. *The Geographical Distribution of Animals:* 1110, Macmillan, 1876.
White, G.B. *Anopheles gambiae Complex and Disease Transmission in Africa,* Transactions of the Royal Society of Tropical Medicine and Hygiene 68, 278-301, 1974.
White, M.J.D., (Editor), *Speciation in the Australian Morabine Grasshoppers – the Cytogenetic Evidence,* 57-68, (which appears in the book Genetic Mechanisms of Speciation in Insects), Australia and New Zealand Book Co., Sydney, 1974.
World Health Organisation, *Species Complexes in Insect Vectors of Disease (blackflies, mosquitoes, tsetse flies),* WHO Document/VBC/77, 656, pp56, WHO Geneva, 1977.
Zimmern, D. *Do Viroids and DNA Viruses Derive from a System that Exchanges Genetic Information Between Eukaryotic Cells?* Trends in Biomedical Science, 7, 205-7, 1982.

Chapter 12
EVOLUTION
SUPPLEMENTARY READING LIST

A few references are given in this supplementary reading list to some specialised areas of evolutionary studies. The headings are listed alphabetically.

i) **Absence of Competition**

Ford, E.B. *Ecological Genetics (the Arctic Hare in Iceland),* Chapman & Hall, 1979.

ii) **Camouflage**

de Ruiter, L. *Some Experiments on the Camouflage of Stick Insects,* Behaviour 4, 222-232, 1952.

iii) **Cepaea nemoralis**

   Arnold, R.W. *The Effects of Selection by Climate on the Land Snail Cepaea nemoralis,* Evolution 23, 370-378, 1969.
   Cain, A.J., King, J.M.B., and Sheppard, P.M. *New Data on the Genetics of Polymorphism in the Snail Cepaea nemoralis,* Genetics 45, 393-411, 1960.
   Day, J.C.L., and Dowdeswell, W.H. *Natural Selection of Cepaea on Portland Bill,* Heredity 23, 169-188, 1969.
   Ford, E.B. *Ecological Genetics (Fourth Edition),* pp 192-217, Chapman & Hall, 1979.

iv) **Clines**

   Bishop, J.A. *The Cline in Industrial Melanism in Biston betularia between urban Liverpool and rural North Wales,* Journal of Animal Ecology 41, 209-243, 1972.
   Haldane, J.B.S. *The Theory of a Cline,* Journal of Genetics 48, 277-284, 1948.
   Kettlewell, H.B.D., and Berry, R.J. *Gene Flow in a Cline,* Heredity 24, 1-14, 1969.

v) **Co-evolution**

   Breedlove, D.E., and Erlich, P.R. *Plant-herbivore Co-evolution: Lupins and Lycaenids,* Science 162, 671-672, 1968.

vi) **Divergency**

   Clarke, B. *Divergent Effects of Natural Selection on Two Closely Related Polymorphic Snails,* Heredity 14, 432-443, 1960.

vii) **Dominance**

   Clarke, C.A., and Sheppard, P.M. *The Evolution of Dominance under Disruptive Selection,* Heredity 14, 163-173, 1960.
   Sheppard, P.M., and Ford, E.B. *Natural Selection and the Evolution of Dominance,* Heredity 21, 139-147, 1966.

viii) **Drosophila pseudoobscura**

   Dobzhansky, T. *On the Dynamics of Chromosomal Polymorphism in Drosophila,* Symposia of the Royal Entomolgical Society of London No. 1, 30-42, 1961.
   Dobzhansky, T. *Genetics of Natural Populations XIX,* Genetics 35, 288-302, 1950.
   Dobzhansky, T. *Genetics of Natural Populations XXVII,* Evolution 12, 385-401, 1958.
   Sturtevant, A.H., and Dobzhansky, T. *Inversions in the Third Chromosome of the Wild Races of Drosophila pseudoobscura and their Use in the Study of the History of the Species,* Proceedings of the National Academy of Science, Washington 22, 448-450, 1936.

ix) **Early Stages of Speciation**

   Lewis, H., and Raven, P.H. *Rapid Evolution in Clarkia,* Evolution 12, 319-336, 1958.

x) **Genetic Drift**

Dobzhansky, T., and Pavlovsky, O. *An Experimental Study of Interaction between Genetic Drift and Natural Selection,* Evolution 11, 311-319, 1957.

Fisher, R.A., and Ford, E.B. *The Spread of a Gene in Natural Conditions in a Colony of the Moth Panaxia dominula,* Heredity 1, 143-174, 1947.

xi) **Isolating Mechanisms**

Hagen, D.W. *Isolating Mechanisms in Threespine Sticklebacks (Gasterosteus),* Journal of the Fisheries Reservation Board of Canada 24, 1637-1692, 1967.

xii) **Local Races and Species**

Camin, J.H., and Ehrlich, P.R. *Natural Selection in Water Snakes (Natrix sipedon) on Islands in Lake Erie,* Evolution 12, 504-511, 1958.

xiii) **Maniola jurtina**

Creed, E.R., Dowdeswell, W.H., Ford, E.B., and McWhirter, K.G. *Evolutionary Studies on Maniola jurtina: the English Mainland 1956 to 1957,* Heredity 13, 363-391, 1959.

Creed, E.R., Dowdeswell, W.H., Ford, E.B., and McWhirter, K.G. *Evolution Studies on Maniola jurtina: the English Mainland 1958 to 1960,* Heredity 17, 237-265, 1962.

Ford, E.B. *Ecological Genetics (Fourth Edition),* pp 61-77, Chapman & Hall, 1979.

xiv) **Melanism**

Clarke, C.A., and Sheppard, P.M. *Genetic Control of the Melanic Form 'insularia' of the moth Biston betularia,* Nature 202, 215-216, (1964).

Kettlewell, H.B.D. *Industrial Melanism in Moths and its Contribution to Our Knowledge of Evolution,* Proceedings of the Royal Institution of Great Britain 36, 1-14, 1957.

Lees, D.R., Creed, E.R., and Duckett, J.G *Atmospheric Pollution and Industrial Melanism,* Heredity 30, 227-232, 1973.

xv) **Mimicry**

Brower, J. van Z., *Experimental Studies of Mimicry in Some North American Butterflies,* Evolution 12 32-47, 123-136, 273-289, (1958).

Brower, L.P., Brower, J. van Z., and Collins, C.T. *Relative Palatability and Müllerian Mimicry among Neotropical Butterflies of the Sub-family Heliconiidae,* Zoologica 48, 65-84, 1963.

Brower, L.P., Brower, J. van Z., and Westcott, P.W. *The Reactions of Toads to Bumblebees and their Robberfly Mimics,* American Naturalist 94, 343-355, 1960.

Clarke, C.A., and Sheppard, P.M. *The Evolution of Mimicry in the Butterfly Papilio dardanus,* Heredity 14, 163-173, 1960.

Clarke, C.A., and Sheppard, P.M. *Supergenes and Mimicry,* Heredity 14, 175-185, 1960.

Southern, H.N. *Evolution as a Process. (Mimicry in Cuckoos' Eggs, 219-232),* Allen and Unwin, 1954.

xvi) **Mortality in Small Populations**

Dowdeswell, W.H., Fisher, R.A., and Ford, E.B. *The Quantitative Study of Populations in the Lepidoptera, Maniola jurtina on Tean, one of the Scilly Isles,* Heredity 3, 67-84, 1949.

xvii) **Numbers Fluctuations and Variation**

Ford, H.D., and Ford, E.B. *Fluctuations in Numbers and its Influence on Variation in Melilaea aurenia,* Transcripts of the Royal Entomogical Society of London 78, 345-351, 1930.

xviii) **Panaxia dominula**

Fisher, R.A., and Ford, E.B. *The Spread of a Gene in Natural Conditions in a Colony of the Moth Panaxia dominula,* Heredity 1, 143-174, 1947.
Sheppard, P.M., and Cook, L.M. *The Manifold Effects of the Medionigra Gene in the Moth Panaxia dominula and the Maintenance of a Polymorphism,* Heredity 17, 415-426, 1962.
Williamson, M.H. *On the Polymorphism of the Moth Panaxia dominula,* Heredity 15, 139-151, 1960.

xix) **Polymorphism**

Darlington, C.D., (Editor: Creed, E.R.), *Ecological Genetics and Evolution,* (pp1-19; The Evolution of Polymorphic Systems), Blackwell, 1971.
Ford, E.B. *The Theory of Genetic Polymorphism,* Symposium of the Royal Society of London 1, 11-19, 1961.
Ford, E.B. *Genetic Polymorphism,* All Souls Studies, Faber and Faber, 1965.
Harris, H. *Enzyme Polymorphism in Man,* Proceedings of the Royal Society 164, 298-310, 1966.
Semeonoff, R., and Robertson, F.W. *A Biochemical and Ecological Study of Plasma Esterase Polymorphism in Natural Populations of the Field Vole, Microtus agrestis,* Biochemical Genetics 1, 205-227, 1968.
Thoday, J.M., and Boam, T.B. *Effects of Disruptive Selection, 2. Polymorphism and Divergence without Isolation,* Heredity 13, 205-218, 1959.
Turner, J.R.G. *Evolution of Complex Polymorphism and Mimicry in Distasteful South American Butterflies,* International Congress of Entomolgy 12, 267, 1965.

xx) **Polyploidy in Marginal Populations**

Darlington, C.D. *Chromosome Botany (2nd Edition),* (The Cross between Spartina alterniflora from America and the native Spartina stricta in England to produce Spartina Townsendii), Allen and Unwin, 1963.
Skalinska, M. *Studies in Cyto-ecology; Geographic Distribution and Evolution of Valeriana,* Bulletin of The Academy of Poland, Series B, 1, 149-175, 1950.

xxi) **Primulas**

Dowrick, V.P.J. *Heterostyly and Homostyly in Primula obconica,* Heredity 10, 219-236, 1956.
Woodell, S.R.J. *What Pollinates Primulas?* New Scientist 8, 568-571, 1960.

xxii) Specialised Adaptations on the Extremities of Habitats

Carson, J.L. *The Genetic Characteristics of Marginal Populations of Drosopohila,* Cold Spring Harbor Symposium of Quantitative Biology 20, 276-287, 1955.
Ford, E.B. *Ecological Genetics (Fourth Edition),* pp 366-367, Chapman & Hall, 1979.
Ehrlich, P.R., and Mason, L.G. *The Population Biology of the Butterfly Euphydryas editha III,* Selection and the phenetics of the Jasper Ridge Colony, Evolution 20, 165-173, 1966.

xxiii) Super-genes

Turner, J.R.G. *The Evolution of Super-genes,* American Naturalist 101, 195-221, 1967.

xxiv) Sympatric Evolution

Antonovics, T. *Evolution in Closely Adjacent Plant Populations,* Heredity 25, 219-238, 1968.
Aston, J.L., and Bradshaw, A.D. *Evolution in Closely Adjacent Plant Populations II. Agrostis stolonifera in Marine Habitats,* Heredity 21, 649-664, 1966.
Bradshaw, A.D., McNeilly, T.S., and Gregory, R.P.G. *Industrialisation, Evolution and the Development of Heavy Metal Tolerance in Plants,* 5th Symposium of the British Ecological Society, 327-343, Blackwell, 1965.
McNeilly, T.S., and Antonovics, J. *Evolution in Closely Adjacent Plant Populations,* Heredity 23, 205-218, 1968.
Walley, K.A., Khan, M.S., and Bradshaw, A.D. *Potential for Evolution of Heavy Metal Tolerance for Plants,* Heredity 32, 309-319, 1974.

CHAPTER 13

Fraenkel-Conrat, H., and Williams, R.C. *Reconstruction of Active Tobacco Mosaic Virus from its Inactive Protein and Nucleic Acid Components,* Proceedings of the National Academy of Sciences of the USA 41, 690-8, 1955.
Howard, B.H. *Vectors for Introducing Genes into the Cells of Higher Eukaryotes,* Trends in Biomedical Sciences 8, 209-12, 1983.
Hughes, S.S. *The Virus: A History of the Concept,* Heinemann, 1977.
Hughes, S.S. *Synthesis, Integration and Transcription of the Retroviral Provirus,* Current Topics in Microbiology and Immunobiology 103, 23-49, 1983.
Lavorgna, G., Malva, C., Manzi, A., Gigliotti, S., and Graziani, F. *The Abnormal Oöcyte Phenotype is Correlated with the Presence of Blood Transposons in D. melanogaster,* Genetics 123, 485-494, November, 1989.
Lerner, R.A. *Synthetic Vaccines,* Scientific American, 48-56, February, 1983.
Lewin, R. *Can Genes Jump between Eukaryotic Species?* Science 217, 42-3, 1982.
Lycke, E., and Norrby, E., (Editors), *Virus Induced Changes of Cell Structures and Functions,* (in Textbook of Medical Virology, 304-322), Butterworth, 1983.
Matthews, R. *The Origin of Viruses from Cells,* International Review of Cytology, Supplement No. 15, 245-280, 1983.
Nash, H. A. *Integration and Excision of Bacteriophage λ; the Mechanism and Conservative Site Specific Recombination,* Annual Review of Genetics 15, 143-167, 1981.
Newmark, P. *Will Peptides make Vaccines?* Nature 305, 9, 1983.

O'Brien, S.H., Bonner, T.I., Cohen, M., O'Connell, C., and Nash, W.G. *Mapping of an Endogenous Retroviral Sequence to Human Chromosome 18*, Nature 303, 74-77, 1983.
Pintel, D., Dadachanji, D., Astell, C.R., and Ward, D.C. *The Genome of Minute Virus of Mice Encodes Two Overlapping Transcription Units*, Nucleic Acids Research 11, 1019-38, 1983.
Scott, A. *Pirates of the Cell*, Basil Blackwell, 1987.
Scott, A. *Viruses and Cells: A History of Give and Take?* New Scientist, 16 January, 1986.
Simons, K., Garoff, H., and Helenuis, A. *How an Animal Virus Gets Into and Out of its Host Cell*, Scientific American, February, 1982.
Syvanen, M. *Cross-species Gene Transfer; Implications for a New Theory of Evolution*, Journal of Theoretical Biology 112, 333-343, 1985.
Weinberg, R.A. *Integrated Genomes of Animal Viruses*, Annual Review of Biochemistry 49, 197-226, 1980.
Zimmern, D. *Do Viroids and Viruses Derive from a System that Exchanges Genetic Information between Eukaryotic Cells?* Trends in Biochemical Sciences 7, 205-207, 1982.

CHAPTER 14

Alberts, B., Bray, D., Lewis, J., Roff, M., Roberts, K., and Watson, J.D. *Molecular Biology of the Cell*, (2nd Edition), Garland, 1989.
Barrington, E.J.W. (Editor), *Hormones and Evolution*, Academic Press, 1979.
Behnke, J.A. (Editor), *The Biology of Ageing – 'Altered Biochemical Responsiveness and Hormone Receptors during Ageing'*, by Roth, G.S. Plenum Press 1978.
Behnke, J.A. (Editor), *The Biology of Ageing – 'The Origin, Evolution, Nature and Cause of Ageing'*, by Sonneborn, T.M., Plenum Press, 1978.
Curtis, H.J. *Genetic Factors in Ageing*, Advances in Genetics 16, 305-324, 1971.
Davidson, E.H. *Gene Activity in Early Development*, Academic Press, 1977.
Davies, I., and Sigee, D.C. (Editors), *Cell Ageing and Cell Death*, Cambridge University Press, 1984.
Davies, I., and Fotheringham, A.P. *Lipofuscin – Does it Affect Cellular Performance?* Experimental Gerontology 16, 119-125, 1981.
Friedberg, E.C. *DNA Repair*, W.H. Freeman, 1985.
Gurdon, J.B. *The Control of Gene Expression in Animal Development*, Harvard University Press, Oxford University Press, 1977.
Gurdon, J.B. *Gene Expression During Cell Differentiation*, Carolina Biology Reader No. 25, Carolina Biological Supply Company, 1978.
Hardin, J.W., Cherry, J.H., Morné, D.J., and Lembi, C.A. *Enhancement of RNA polymerase Activity by a Factor Released by Auxin from Plasma Membrane*, Proceedings of the National Academy of Science, USA 69, 3146 to 3150, 1972.
Hart, R.W., and Setlow, R.B. *Correlation Between Deoxyribonucleic Acid Excision Repair and Lifespan in a Number of Mammalian Species*, Proceedings of the National Academy of Science, USA 71, 2169-2173, 1974.
Hozumi, N., and Tonegawa, S. *Evidence for Somatic Rearrangement of Immunoglobulin Genes Coding for Variable and Constant Regions*, Proc. Nat. Acad. Sci. USA 73, No. 10, 3628-3632, 1976.
Hualt, C., Lange, H., Lohmann, L., Rissand, I., and Weidner, M. *Lag Phases in Phytochrome Mediated Enzyme Synthesis (PAL)*, Planta 83, 267-275, 1968.
Key, J.L., Barnett, N.M., and Lin, C.Y. *RNA and Protein Biosynthesis and the Regulation of Cell Elongation by Auxin*, Annals of the New York Academy of Science 144, 49-62, 1967.

Key, J.L. *Hormones and Nucleic Acid Metabolism,* Ann. Review of Plant Physiology 20, 445-474, 1969.
Lamb, M.J. *The Biology of Ageing,* Blackie, 1977.
Lane, C.D. *Rabbit Haemoglobin from Frog Eggs,* Scientific American 235(7), 60-77, 1976.
Letham, D.S., Goodwin, P.B., and Higgins, T.J.V., (Editors), *Phytochrome and Related Compounds – A Comprehensive Treatise,* Elsevier/North Holland Biomedical Press, 1978.
Lloyd, C.W., and Rees, D.A., (Editors), *Cellular Controls in Differentiation,* Academic Press, 1981.
Murray, V., and Holliday, R. *Increased Error Frequency of DNA Polymerases from Senescent Human Fibroblasts,* Journal of Molecular Biology 146, 55-76, 1981.
Nichols, W., and Murphy, D.G., (Editors), *DNA Repair Processes,* Symposia Specialists Inc., 1977.
O'Malley, B.W., and Means, A.R. *Female Steroid Hormones and Target Cell Nuclei,* Science 183, 610-620, 1974.
O'Malley, B.W., and Schrader, W.J. *The Receptors of Steroid Hormones,* Scientific American 234, 32-43, 1976.
Orgel, L.E. *The Maintenance of the Accuracy of Protein Synthesis and its Relevance to Ageing: a correction,* Proceedings of the National Academy of Science, USA 67, 476-479, 1970.
Ray, P.M. *Auxin Binding Sites of Maize Coleoptiles are Localised on Membranes of the Endoplasmic Reticulum,* Plant Physiology 59, 594-601, 1977.
Rayle, D.L., and Cleland, R. *Enhancement of Wall Loosening and Elongation of Acid Solutions,* Plant Physiology 46, 250-253, 1970.
Romanchikov, L. *Trigger Approach to Proliferation, Ageing and Immortalisation of Animal Cells,* Genetika (Russia): 1744-1753, October 1991.
Smith, H. (Editor), *Regulation of Enzyme Synthesis and Activity in Higher Plants,* Academic Press, 1977.
Thimann, K.V. *Hormone Action in the Whole Life of Plants* – Chapter 14, University of Massachusetts Press, 1977.
Tonegawa, S. *The Molecules of The Immune System,* Scientific American, 104-110, October, 1985.

CHAPTER 15

Breitenbach, M., Achatz, G., Hegber, S., Wallner, J., Eichler, M., Sperak, W., Schweiger, C., and Rumpold, H. *The Contribution of Yeast Genetics to Tumour Biology and Tumour Diagnostics,* Wien-Klin-Wochenschr., 101 (15): 495-504, 1989.
Croce, C.M. *Integration of Oncogenic Viruses in Mammalian Cells,* International Review of Cytology 71, 1-17, 1981.
Czerniak, B., Herz, G., Wersto, R.R., and Koss, L.G. *Asymmetric Distribution of Oncongene Products at Meiosis,* National Academy of Science, USA: 89 (11): 4860-4863, 1st June 1992.
Deuel, T.F., Huang, J.S., Stroobant, P., and Waterfield, M.D. *Expression of a Platelet-derived Growth Factor-like Protein in Simian Sarcoma Virus Transformed Cells,* Science 221, 1848-50, 1983.
Dohi, Y., Sunada, S., Aoki, M., Morigushi, A., Okabayashi, M., Miyati, M., and Matsuda, H. *Eradication of Metastatic Tumour Cells from Lymph Nodes by Local Administration of Anti-CD3 Antibody,* Cancer Immunology, Immunotherapy, Vol. 36 (6), 357-63, June 1993.

Doolittle, R.F., Hungapillar, M.W., Hood, L.E., Devare, S.G., Robbins, K.C., Aaronson, S.A., and Antionades, H.N. *Simian Sarcoma Virus onc gene v-sis, is derived from the gene (or genes) encoding a platelet-derived growth factor.* Science 222, 275-7, 1983.

Downward, J., Yarden, Y., Mayes, E., Scrace, G., Totty, N., Stockwell, P., Ullrich, A., Schlessinger, J., and Waterfield, M.D. *Close Similarities of Epidermal Growth Factor Receptor and v-erb-B oncogene protein sequence,* Nature 307, 521-7, 1984.

Ferran, C., Bach, J.F., Chatenoud, L *Monoclonal Antibodies: Diagnostic and Therapeutic Use and Prospects for the Future,* Revue Française de Transfusion et d'Hémobiologie, Vol. 36 (2), 149-77, April 1993.

Freeman, R.S., and Donoghue, D.J. *Protein Kinases and Protooncogenes:* Biochemical Regulators of the Eukaryotic Cell Cycle, Biochemistry 30 (9), 2293-302, 1991.

Kemshead, J.T., Hopkins, K. *Uses and Limitations of Monoclonal Antibodies (moAbs) in the Treatment of Malignant Disease: a Review,* Journal of the Royal Society of Medicine, Vol. 86 (4), 219-24, April 1993.

Kipreos, E.T., and Wang, J.Y. *Cell Cycle Regulated Binding of the Proto-oncogene Product cAbl Tirosine Kinase to DNA in Chromatin,* Science: 256(5055): 382-385, 17 April, 1992.

Novikov, L.B., Tievlesnov, N., Timoshenko, M.G., Kalinovskii, V.P., and Federov, S.N. *Protooncogene Expression in the Organs of Intact Rats (src, sis, fos, myc, Ha-ras, Ki-ras, N-ras, mos and abl protooncogene expression in different mammalian tissues),* Eksp.–Onkol. 11 (3): 14-7, 1989.

Pagano, M., Pepperkok, R., Verde, F., Ansorge, W., and Draelta, G. *Cyclin A Is Required at Two Points in the Cell Cycle,* Embo. J. 11(3): 961-971, March, 1992.

Prescott, D.M., and Flexer, A.S. *Cancer – The Misguided Cell,* Sinauer Associates Inc., 1982.

Rechsteiner, M. *PEST Regions, Proteolysis and Cell Cycle Progression (kinases, selective proteolysis, phosphorylation and their regulation of mitosis),* Revis.-Biol.-Cellular 20: 235-53, 1989.

Rozengurt, E. *Bombesin (a neuropolypeptide) Stimulation of Mitogenesis: Specific Receptors, Signal Transduction and Early Events,* Amer. Rev. Respir. Dis. 142 (6 pt 2): 511-5,1990.

Rozengurt, E. *Signal Transduction Pathways in Mitogenesis (bombesin-like peptides stimulate a number of cellular activities that encourage mitotic cell division,* Brit. Med. Bulletin 45 (2): 515-28, 1989.

Rozengurt, E., and Ober, S.S. *Mitogenic Signalling in Murine 3T3 Cells: The Cyclic AMP Pathway,* Biochem. Soc. Trans. 17 (4): 629-32, 1989.

Scott, A. *Pirates of The Cell,* Basil Blackwell Ltd., 1987.

Shapiro, J.A. *Mobile Genetic Elements,* Academic Press, 1983.

Simon, D., Munoz, S.J., Maddrey, W.C., and Knowles, B.B. *Chromosomal Rearrangements in a (Human) Primary Hepatocellular Carcinoma (characteristic abnormalities in chromosomes 1, 5, 6, 9, 13, 16, and 22 – chromosome number = 46),* Cancer Genet. Cytogenet. 45 (2): 255-60, 1990.

Sinkovics, J.G. *Oncogenes and Growth Factors,* CRC-Crit.-Rev.-Immunol. 8 (4): 217-98, 1988.

Skinner, M.A., and Inglehart, J.D. *The Emerging Genetics of Cancer,* Surg. Gynecol. Obstet. 168 (4): 371-9,1989.

Storms, R.W., and Bose, H.R. (Jr.) *Oncogenes, Protooncogenes and Signal Transduction: towards a Unified Theory?* (Ca^{++}/phospholipid activation of ser/thr Kinase C by way of membrane receptors and point mutations), Adv. Virus Res. 37:1-34, 1989.

Strauss, M., Hereng, S., Lieber, A., Herrman, G., Griffin, B.E., and Arnold, W. *Ti Stimulation of Cell Division and Fibroblast Focus Formation by Antisense Repression of Retinoblastoma Protein Synthesis,* Oncogene: 7(4)- 769-773, April, 1992.

Szmuness, W. *Hepatocellular carcinoma and the Hepatitis B Virus – Evidence for a Causal Association,* Progress in Medical Virology 24, 40-69, 1978.

Tsuda, H., Hirohashi, S., Shimosoto, Y., Hirota, T., Tsugane, S., Watanabe, S., Terada, M., and Yamamoto, H. *Correlation between Histologic Grade of Malignancy and Copy Number of c-erb B-2 gene in Breast Carcinoma. A retrospective analysis of 176 Cases,* Cancer 65 (8): 1794-800, 1990.

Wang, J.Y., Kipneos, E.T., and Lin, B.T. *Cell Cycle Specific Phosphorylation of Oncogene and anti-Oncogene Products,* (ser/thr polypeptide kinase-cell division regulation of eukaryotic cells – cdc 2 gene), FASEB–J (Meeting Abstract), 4 (7) A 2334, 1990.

Waterfield, M.D., Scrace, G.T., Whittle, N., Stroobant, P., Johnsson, A., Wasteson, A., Westermark, B., Heldin, C-H., Huang, J.S. and Deuel, T.F. *Platelet-derived Growth Factor is Structurally Related to the Putative Transforming $p28^{SIS}$ of Simian Sarcoma Virus,* Nature, Vol. 304, No. 5921, 35-39, 7th July 1983.

Waterson, A.P. *Human Cancers and Human Viruses,* British Medical Journal, 284, 446-8, 1982.

Weinberg, R.A. *Cancer,* Scientific American, 102-116, November, 1983.

Whitfield, J.F. *Ti Calcium Signals and Cancer,* Crit. Rev. Oncol. 3(1-2), 55-90, 1992.

Wyke, J.A. *Oncogenic Viruses,* Journal of Pathology 135, 39-85; 51-55, 1981.

CHAPTER 16

Aird, I., Bentall, H.H., and Fraser Roberts, J.A. *A Relationship Between Cancer of the Stomach and the ABO Blood Groups,* British Medical Journal (1), 799-801, 1953.

Carlson, E.A. *Human Genetics,* D.C. Heath, 1984.

Chiarelli, A.B., (Editor), *Comparative Genetics in Monkeys, Apes, and Man,* Academic Press, 1971.

Chung, D.S., Matsunaga, E., and Morton, N.E. *The MN polymorphism in Japan,* Japanese Journal of Human Genetics 6, 1-11, 1961.

Chung, C.S., and Morton, N.E. *Selection at the ABO Locus,* American Journal of Human Genetics 13, 9-27, 1961.

Ciochon, R.L., and Corniccini, R.S., (Editors), *New Interpretations of Ape and Human Ancestry,* Plenum Press, 1983.

Clarke, C.A. *Human Genetics and Medicine,* Studies in Biology No. 20, Arnold 1980.

Cline, M.J. *Genetic Engineering of Mammalian Cells. Its Potential Application to Genetic Diseases in Man,* Journal of Laboratory and Clinical Medicine 99, 299-308, 1982.

Darwin, C.R. *The Descent of Man,* John Murray, 1871.

Dobzhansky, T. *Mankind Evolving,* Yale University Press, 1962.

Erlich, P.R., Holm, R.W., and Parnell, D.R. *The Process of Evolution,* McGraw Hill, 1974.

Ford, E.B. *A Uniform Notation for the Human Blood Groups,* Heredity 9, 135-142, 1955.

Goodfellow, P.N., Ellis, N.A., Goodfellow, P.J., Pym, B., Smith, M., Palmer, M., and Frischauf, A-M, *The Pseudoautosomal Boundary in Man is Defined by an Alu Repeat Sequence Inserted on the Y Chromosome,* Nature, Volume 337, No. 6202: 81-84, 5th January, 1989.

John Hopkins, University Press, *Mendelian Inheritance in Man. A Catalogue of Autosomal and X-linked Defects in Man, Including a List of Enzyme Defects in The Human Population,* John Hopkins University Press, Sixth Edition, 1983.

Koopman, P., Gubbay, J., Vivian, N., Goodfellow, P., and Lovell-Badge, R. *Male Development of Chromosomally Female Mice Transgenic for Sry,* Nature, Volume 351, 9 May 1991.

Mourant, A.E. *Associations Between Hereditary Blood Factors and Diseases*, Bulletin of the World Health Organisation 49, 93-101, 1973.
Mourant, A.E., Kopec, A.C., and Domaniewska-Sobezak, P. *The ABO Blood Groups: Comprehensive Tables and Maps of World Distribution*, Blackwell, 1958.
Noitski, E. *Human Genetics*, Macmillan, Second Edition, 1982.
Page, D.C. *Is ZFY the Sex-determining Gene on the Human Y Chromosome?* Phil. Trans. of the Royal Society, London, 1988.
Page, D.C., Mosher, R., Simpson, E.M., Fisher, E.M.C., Mardon, G., Pollack, J., McGillivray, B., de la Chapelle, A., and Brown, L.G. *The Sex Determining Region of the Human Y Chromosome Encodes a Finger Protein*, Cell, Vol. 51, 1091-1104, 24 Dec. 1987.
Palmer, M.S., Sinclair, A.H., Berta, P., Ellis, N.A., Goodfellow, P.N., Abbas, N.E., and Fellous, M. *Genetic Evidence that ZFY is Not the Testis-determining Factor*, Nature, Volume 342, No. 6252: 937-939, 21st December 1989.
Sinclair, A.H., Berta, P., Palmer, M.S., Ross-Hawkins, J., Griffiths, B.L., Smith, M.J., Foster, J.W., Frischauf, A-M, Lovell-Badge, R., and Goodfellow, P.N. *A Gene from the Human Sex-determining Region Encodes a Protein with Homology to a Conservative DNA-binding motif*, Nature, Volume 346: No. 6281, 240-4, 19th July, 1990.
Therman, E. *Human Chromosomes – Structure, Behaviour, Effects*, Springer-Verlag, 1980.
Woolf, B. *On Estimating The Relation Between Blood Groups and Disease*, Ann. of Human Genetics 19, 251-253, 1954.

APPENDIX 1

Alberts, B., Bray, D., Lewis, J., Roff, M., Roberts, K., and Watson, J.D. *Molecular Biology of The Cell (2nd Edition)*, Garland, 1989.
Fawcett, D.W. *The Cell*, W.B. Saunders Company, 1981.
Star, M.A. (Editor), *The Prokaryotes*, Springer-Verlag, 1981.
Thorpe, N.O. *Cell Biology*, John Wiley & Sons, 1984.
Tribe, M.A., Morgan, A.J., and Whitlaker, P.A. *The Evolution of Eukaryotic Cells*, Studies in Biology No. 131, Arnold, 1981.

APPENDIX 2

Bulletin of Mathematical Biology (Many Contributors), Pergamon Press, A Regular Publication.
Donnelly, P. *Theoretical Population Biology*, 30, No. 2, 2 October, 1986, Academic Press, 1986.
Donnelly, P., and Tavaré, S. *The Population Genealogy of the Infinitely-many Neutral Allele Model*, Journal of Mathematical Biology, 25: 381-391, Springer Verlag, 1987.
Ewens, W.J. *Mathematical Population Genetics*, Biomathematics, Volume 9, Springer Verlag, 1982.
Jacquard, A. *The Genetic Structure of Populations*, Biomathematics, Volume 5, Springer Verlag, 1970.
Marly, B.F.J. *The Statistics of Natural Selection*, Chapman Hall, 1985.
Schefler, W.C. *Statistics for the Biological Sciences (Second Edition)*, Addison Wesley, 1979.
Wardlow, A.C. *Practical Statistics for Biology*, John Wiley & Sons, 1985.
Zar, Z.H. *Biostatistical Analysis (Second Edition)*, Prentice Hall, 1984.

LIST OF PHOTOGRAPHS

Chapter 1
Scanning electron micrograph of Escherichia coli   77
Transmission electron micrograph of Escherichia coli   77

Chapter 3
Transmission electron micrograph of an animal white blood cell   140
Pancreatic acinar cell fixed with osmium and stained with lead
   hydroxide   140
Electron micrograph of a portion of dissociated nucleolar core
   isolated from an oöcyte of *Notophthalimos viridescens*   141

Chapter 4
Transmission electron micrograph showing part of a
   mammalian gene   164
Part of a protein framework and surrounding DNA   164

Chapter 6
Mitochondrial DNA molecule from an oöcyte   216

Chapter 8
Several PBR322 plasmids taken from E. coli   262
Scanning electron micrograph of human T-lymphocyte   262

Chapter 9
Transmission electron micrograph showing the fork in
   DNA where replication is taking place   289

MITOSIS

Interphase   290
Prophase   290
Prophase   291
Metaphase   292
Anaphase   293
Anaphase   293
Telophase   294
Interphase   294

MEISOSIS

FIRST MEIOTIC DIVISION

    a) Leptotene   296
    b) Early Zygotene   297
    c) Pachytene   297
    d) Diplotene   297
    e) Diakinesis   298
    f) Metaphase I   298
    g) Anaphase I   298
    h) Telophase I   298
    i) Interkinesis   299

SECOND MEIOTIC DIVISION

    j) Metaphase II   299
    k) Early Anaphase II   299
    l) Late Anaphase II   299
    m) Telophase II   300

A particularly clearly analysable meiosis diplotene from locust   300

Chapter 12

COLOUR PLATE
    Agaura spp., a katydid camouflaged as a leaf.
    Camouflaged light form Biston betularia
    Camouflaged Hymenopus coronatus, a *flower* mantis
      from Borneo
    Bematistes model and their Pseudacraea eurytus mimic
                                  between pages 416-417

Chapter 16

Trypsin-Giemsa Staining in human chromosomes   519
The karyotype as shown by trypsin giemsa staining – Cri du
    Chat syndrome   523
Karyotype – Trisomy in Human Chromosome 16   524
DNA fingerprinting   526

Appendix 1

Transmission electron micrograph of Escherichia coli   562

INDEX

A

A blood groups 151, 154, 197-8, 550
A_1 chimpanzee blood sub-group 513
A_1 human blood sub-group 513
A_2 chimpanzee blood sub-group 513
A_2 human blood sub-group 513
α^+ and α^- streptomycin resistance alleles in Chlamydomonas 226-7
1A1 region of the TDF 538
1A2 exon region of the TDF 537
1A2 Y chromosome locus of the TDF 536-8
♂ – amylase 478
α – haemoglobin 364, 367, 511-2
α – helix of polypeptides 100
α – particles 498
A – site of polypeptide synthesis 117-20
AB blood groups 151, 550
ABA (abscisic acid) 479
aberrations in human chromosomes 519-24
abl proto-oncogene 493
abnormalities in sex chromosomes 205, 521-3, 533-4, 538-9, 546-8
abnormalities of sexuality in humans 521-3, 533-4, 538-9, 546-8
abnormalities in laboratory bred hybrids 399
ABO blood group interactions with the Rhesus factor 553-4
aborigines' blood groups 513
abortion of human foetuses 320, 524
abscisic acid (ABA) 479
absence of:-
  chemical punctuation between codons 112, 116
  chloroplasts 223
  chromosome pairing 333
  crossings-over 310
  crossings-over in male Drosophila melanogaster 174
  crossings-over in Sordaria 301
  DNA replication during the second meiotic division 285, 299
  gene exchange 374, 376, 400
  genetic wastage in male gamete production 285
  mutation 194
  rejunction 377
  sex chromosomes in some species 205, 331
  virus particles in cancer cells created by viruses 493
Acarina spp. 386
acidic groups in amino-acids 93, 96-7, 101
Acnida spp. 331
acquired characters 31

acridines (for causing mutations) 81-85
Actinomycetes 232
actinomycin D 448
activation energy 98
activation of genes 444
active repressors 158-60
active site of polypeptides 99, 343, 346
activity levels in cells – responses by the nucleolus 139
acute transforming retroviruses (ATRs) 491-2
adaptability 32, 318, 390-2
adaptation 49
added variability 375
adding an amino-acid during polypeptide synthesis 119
additions of amino-acids in polypeptides after mutations 345
additions of bases to DNA 82-3, 313
additions of DNA to chromosomes 313, 318, 322, 345, 370, 533-4
additions of DNA to human females 533-4
additions of DNA to human males 521-2, 533-4
additions to genomes 375
adenine 40-6, 48-55, 87, 103, 112, 126, 238, 448-9, 478
Adenovirus 494
adrenal cancer 487
adjacent controlling genes 376
Aegilops spp. and the breeding of modern wheats 331-3
aerobic respiration 211-2, 339, 344, 364
African horned toad 132, 216, 450, 453, 456-7, 460
agar jelly 33, 61, 63-6, 70-1, 242, 504
ageing 311, 444, 452, 480-4
agents of mutation 320
agglutination of polypeptides 92, 101
agglutination reactions in blood 512
agglutinogens 514
agitation in a blender 61
Agrobacterium tumifaciens 257
Agrostis tenuis 397, 418-22
AIDS 258-61, 434, 436, 495
alanine 94, 342
alanine repeated codes in human DNA 503, 506
albinism 527
aliphatic amino-acids 94-6
alleles:-
  analysis of base sequences in 423
  best sets 370, 395
  changes in 319
  codominance 151
  controlling enzyme synthesis 193

creation 371
crossings-over in Sordaria brevicollis 302, 305
definition of 155-6
dominant 148, 167, 178, 203
effects of mutations on 154
environmental control of the expression of 315
establishment in a population forming a genotype 167
frequencies 371, 381, 397-8, 512, 515-6
frequency changes in a population 371, 396, 398, 406-12
heterozygous 169
homozygous 318, 395
incomplete dominance of 151
in genetic crosses. Chapters 5, 6, 7 and 16
in meiotic gene scrambling 288, 370, 395
in Rhesus factor control in humans 550-1
in successful mutations 314-5
natural selection of 414
new combinations 334-5
orientation during meiosis 169
predominance of, in a community of a species 354
ratios of 396
recessive 148, 167, 203, 482, 550
recombination of, in a new line of plant cells 257
replacement of 371
stability 319
allelic genes 288, 390
allopatric speciation 383, 396
allopatry 380-1, 383-4, 421
alteration to:-
 centromeres 326
 chromosome structures 323-30
 gene transcription 472
 gene translation 472
 number of chromosomes 330-4, 519-21, 523-4
alternative alleles 154, 310, 312, 315, 371, 375
alu repeated DNA sequences in humans 539, 541-2
ambiguity of the genetic code 126
American National Cancer Institute 506
amide– group 95
amino–acid:-
 additions 69
 analysis in polypeptides 337-8
 base sequence control 83
 coding 112
 codons 86, 89, 211
 data banks 339
 dipeptide bond formation 97
 /DNA association for gene control 536
 invariable sequence in cytochrome c 344
 positions in zinc fingers 535-6

residues 129
sequence comparisons in primates 511-2
sequence differences 357, 359
sequences 50, 85, 87-8, 91-2, 99, 108-12, 245, 337-40, 357-8, 363, 375, 423, 455, 512, 543
sequences in data banks 339
side chains 101
sites 343
sites – functionally conservative 346
sites in cytochrome c 344
sites – invariable 343-5
sites – standard numbers 345
sites – variable 346
solubility 93
structures 94-6
substitutions 339-68
substitutions – mathematical probability of 346
substitutions – rates of 356-7
amino-acids:-
 biosynthesis of 101
 charged polar 96
 containing sulphur 94-5
 established in sites 342
 fairly soluble 95
 hydrophobic 93-4
 in animal feed synthesis 233
 in chloroplasts 213
 in modifications of gene expression 479
 in mutational consequences 313-4
 in zinc finger polypeptides 535
 naturally occurring 93
 new/predominant 346
 soluble 96
 uncharged polar 95
 used for genetic mapping 60-6, 76
 used for identifying mutations 56-7
amino- end 113, 117, 119
amino- group 93, 96, 113, 117
aminopterin 249-50
amphibia 132, 216, 331, 445, 452-4, 456-7, 460
amplification of genes 153, 446, 489, 490-6
amplication of genes to create oncogenes 489, 559
AMV reverse transcriptidase 238
anaerobic respiration 211, 222
analysis of polypeptides 339
anaphase in mitosis 139, 274-5, 293
anaphase in meiosis 298
ancestors of modern wheats 332
ancestral trees in evolution 358, 363, 512
aneuploidy 330-1
animal antibody production 254-5
animal behaviour 385
animal cell genetic engineering 253-5
animal embryos 253
animal female gamete formation 280

633

animal hormones 473-6
animal male gamete formation 279
animal pole of fertilised eggs 454
Anopheles gambiae 386
Anopheles maculipennis 386
antennapedia genes 455-6
anthers 23, 256, 278-9, 411
anthocyanins 479-80
antibiotics:-
  biosynthesis 234
  genetic engineering of 232
  resistance by protozoa 221-2, 234
antibodies 153, 197, 259-61, 313, 440, 467-8
antibody/cancer cell recognition 255
antibody genes 254
antibody heavy chain polypeptide 465
antibody light chain polypeptide 465
antibody splicing 153, 468
anticodon/codon recognition 115
anticodon modification by cytokinin 478
anticodons 110-123, 125, 478
antigens 197, 255, 258-9, 440, 467, 528, 551
anti-haemophilia globins 532
anti-Müllerian hormone 547
anti-Rhesus 512
anti-Rhesus antigen 552
Antirrhinum majus 223
apes 508
aphids 391
apomixis 331
aposematic (warning) patterns 415
apparatus for mitosis and meiotis 310
apparent reversion to the wild-type in sweet peas 192
appearances of chromosomes 265
arabinose fermentation gene in genetic mapping 70-3
arginine 78, 96, 455, 512, 536
arising of new viral particles 436
'arms' of chromosomes 272, 534
aromatic amino-acids 94
artificial synthesis of a gene 229
Ascarii 22
ascomycete fungi 301, 309
ascospores 301-9
ascus 301, 304, 307
asexual breeding systems 390-2
asexual genetic drift 353
asexual reproduction 224, 226
asexual reproduction-reversion to 391
asexual species 315, 391-2
asexual tautomeric mutations 353
asexuality 92
asparagine 95, 117, 118, 128
aspartic acid 96
Aspergillus oryzae 239
asymmetric carbon atom 93

A = T repeated base pairings in genetic engineering 235-41
ATP 119, 211
ATPase 211
attachment of chromosomes to spindle fibres 272
AUG initiation codon for polypeptide synthesis 113-4
autosomal genes 173, 181-5, 188, 322, 331, 525-8, 550-1
autosomes 181, 188, 322, 331, 551
auxin 476-9
Avery, O. 38, 41
avian myeoblastosis virus (AMV) transcriptidase 238
avidin 474
avirulent bacteria 34-37

B

β haemoglobin 364, 367, 511
β polypeptide chains of haemoglobin 101
B blood group 151, 197-8, 550, 553-4
B lymphocyte white blood cell nucleoli 140
1B region of the TDF gene 538
B-cells in blood 259, 440
B ring of anthocyanins 480
baboon blood groups 514
back-cross 178, 182-4, 195
bacteria:-
  as haploid organisms 167
  crossings-over in 57-80
  for genetic investigations 19
  genetic mapping in 39, 56-78
  in transformation experiments 38, 237, 241
  insertion of genes into 234, 237
bacterial:-
  cell structure 209, 211, Appendix 1
  mutation rate 316
  plasmid genes 216-7, 234
  production of mutant genes 241
  properties 40
  reproduction time span 315
  ribosomes 135
  rRNA 105
  triplicate codes 92
Balbiani, E.G. 22
baldness in human males 532
banding patterns on the shells of the snail Cepaea nemoralis 201-3, 410
bands to identify chromosomes after using stains 518-9, 523-4
Barnett, L. 80
Barr bodies in human female cells 525
barriers to cross-breeding 32, 377-80, 382-3, 385, 396, 415
base:-
  additions to DNA 82-5, 313, 357

634

deletions from DNA 85, 87, 313
insertions during DNA replication 268-70
pair homology 451
pairing 39, 45-55, 150, 267-8, 348, 350-2
sequence analysis in DNA 80-9, 534-5
sequence in a cancerous human DNA fragment 506
sequences in DNA 40, 155, 375
sequences in the homeobox 456
sequences in mRNA 87
sequences in sex determination 533, 535
sequences in viruses 492
substitutions 82-3, 85-6, 314, 357
bases 148, 229, 258
bases added in a gene 83-5
bases deleted from a gene 85
basic group inter-actions in amino-acids 101
Bates, H.W. 413
Batesian mimicry 413-5
Bateson, W. 22
Beadle, G. 56
behaviour 378, 380, 385-7, 397, 401
benign growths 486
best sets of alleles 316, 370-1
Bethesda National Institute of Health 80
bimodal maxima of butterfly wing spots 409
binding of HIV 260
binding of polypeptides to DNA 158-9
binding sites for antibodies 261
biochemical pathways 192, 249
biological environment 379
Biomedical Research Council, Washington D.C. 339
biosynthetic pathway 154, 158, 160
bird sex determination 206
Biston betularia 154, 157, 233, 370, 412
bithorax genes 456
Bittner (B-type) virus 494
black coat colour in mice 191-2
black rats – chromosome numbers 330
black rats – pericentric inversions 327
black spores in Sordaria 303-5, 307-9
blackflies 331, 386
bladder cancers 498
blastoderm 461-2, 464
blastula 253
blends of parental phenotypes 33, 317
'blocked' bases 83-4, 86-8, 105, 108, 113
'blocked' RNA strand 432
blocking factor 106-7, 122
blocking of methionine amino-group 113
blood 486-8, 496
blood and cancer 486
blood B-cells and HIV 259
blood groups 151, 154, 197, 373, 512, 514, 549-50, 576
blood in old world monkeys 514
blood sub-groups in humans and apes 512-4

blood vessels and cancer 487, 497
blue/green pigments 566
blue/greens 19, 534
bluebells 334
body temperature change and DNA base losses 481
bond lengths in pairing bases 50
bone marrow cancer 499
bone marrow genetic engineering in humans 556-7
bony fish haemoglobin 366-7
boundaries between introns and exons 150
boundary regions in sex chromosomes 539-40, 542
brain cells in mice 450-1
bread wheat 332
breakage in DNA to cause ageing 481
breakage points in donor bacteria's chromosome 65
breakdown of a dominant super-gene 157
breakdown of mRNA 447
breakdown of nuclear membranes during cell division 265
breaking and rejoining of chromosomes during meiosis 282
breaking and rejoining of chromosomes during mutation 323-8
breaking the rules of Mendelian inheritance 560
breaks within centromeres 329
breast cancer 487
breeders 21, 32, 229-30
breeding barriers 32, 377-80, 382-3, 385, 396, 415
breeding mechanisms 373
breeding mechanisms and evolution 390
breeding of wheats 332-4
breeding systems and mutations 316-7
breeds 229
Brenner, S. 80, 88, 108
'brights' in Paramecium aurelia 215
broken chromosome fragments 322
broken ends of plasmid DNA – use in genetic engineering 241
Brown, R. 22
brown coat colour in mice 191-2
brown snail shells 201
Burkitt's Lymphoma 495
Bush, G.L. 381
butterfly ecological genetic studies 409, 414-5
butterfly mimic rings 415

C

C gene set for antibody production in mice 465-8
C human Rhesus allele 551
c human Rhesus allele 551

635

C type virus 494
^{14}C 448
1C region of the TDF 538
Caenorhabditis elegans 464
calcium phosphate and genetic engineering 246
California Institute of Technology 108, 501
calls – mating 378
calluses 257
Calvin CO_2 cycle 213
Cambridge University 41, 88
camouflage 398, 410-1, 417-8
cancer:-
  Chapter 15, pp 486-507
  cell antigens 255
  cell locations 255
  cell recognition by antibodies 255
  cell/spleen fusions 254-5
  cells 254-5
  combating drugs 255-6
  research classic experiments 499-506
cancerous:-
  gene amplification 153
  gene transfers 245
  genes 78, 152, 245, 252, 376
  mutations – experimental proofs of 488-9
  point mutations 488, 506
  polypeptide 506
  retroviruses 434
  transformation 502-6
  translocations 245
capture of mimetic species 414, 417
3 carbon atom (3') 42
5 carbon atom (5') 42
carboxyl- end of of a polypeptide 117, 119, 544
carboxyl group 96, 113
carcinogens and ageing 481
carcinogens and cancer 487-8, 491, 494, 498
cardiovascular cancers 498
carmine tobacco flowers 194-5
carriers of harmful recessive alleles in humans 517
carrion crows 388
carrot cell totipotency 445
cat leukemia virus 495
catalases 481
catalysis 98, 172
catarrhine primates 509-10
categories of ecological studies 398
categories of mobile genetic elements 426-34
categories of mutations 313-4, 320
cdc 2 gene 493
CD4/antibody/HIV complex 260-1
cDNA 237-43, 489-91
cell:-
  cluster differentiation 472
  cultures 254

differentiation 206, 311
division 132, 138-9, 142, 257, 310-1 and Chapter 9
division in response to damage 311
division – nucleoli at 139
fusion 227, 244, 248-51, 254-5, 302, 555
lines 350-2, 355
mortality 353
organelles 215 and Appendix 1
streaming 462
transformation to cause cancer 502-6
walls 38, 257, 569
walls (removal for genetic engineering) 256
cellular inheritance studies 22, 369
cellular requirements for establishing a new mutation 390
cellular stress 138
cellulases 257, 477
cellulose cell walls 256, 477
'central dogma' of transcription and translation 126-7
centric fission of chromosomes 328-9
centric fusion of chromosomes 329-30
centric shift in chromosomes 328
centrifugation 111
centrioles 209
centromere:-
  breakage 303-305
  /gene/cross-over points in Sordaria 303-6
  position changes 327
  positions 326
centromeres:-
  at anaphase 275
  at metaphase 274
  at mitosis and meiosis 310
  at the first meiotic division 283-4
  crossing-over positions relative to 303, 305
  defined 272-3
  frequencies of occurrence 246
  in a dividing cell 321
  in Y chromosomal analysis 534
  positional alterations by mutations 325-7, 329-30
centrosomes:-
  at anaphase 275-6
  at metaphase 274
  at mitosis and meiosis 310
  at prophase 271-3
  in a dividing cell 284, 286, 321
Cepaea nemoralis 154, 200-1, 370, 409-11
cereal breeding 333
chance and evolution 169, 312, 372, 393
change in the number of chromosomes 314
change in ploidy 370
characters 144, 167, 180, 189, 218, 223, 257, 317
Chase, M. 39
chemical bonds 45-6

chemical concentrations in female gametes 311, 353-4, 359-60, 463
chemical environment 320
chemical H in blood 512
chemical mutagens 498
chemical product analysis in genetic engineering 243
chemical stains 22, 265
chi–squared statistics 576-85
Chiarelli, A.B. 508
chiasma(ta) 282-3, 300
chimpanzee blood groups 513
chimpanzee blood sub-groups 512-3
chimpanzee Down's Syndrome 511
chimpanzee/human chromosome comparisons 510-1
chimpanzees 509-10, 512, 540
Chironomus 22
chlorophyll 172
chlorophyll allele 315
chloroplast:-
 control of inheritance 213-4
chloroplasts 209, 213, 218, 223, 566
Chorthippus parallelus 334
'Christmas tree' branch 135
'Christmas tree' fibrils in the nucleolus 141
'Christmas tree' shapes on the nucleolus organiser 133-4
chondrodystrophic dwarfism 526
chromatids 270, 300
chromatin 133, 137, 139, 163, 475, 477
chromatin of the nucleolus 132
chromatin potency 472
chromoplasts 209, 566
chromosomal:-
 aberrations in human chromosomes 518
 behaviour at cell division 265-77, 285, 290-4, 296-300, 310
 DNA 213-4, 216
 genes 162, 217
 genetic crosses 166-206
 mutations 314, 320-34
 replication 36, 39, 256
chromosome:-
 alleles 155-6
 appearance 22, 519, 523-4
 circular in bacteria 19, 106, 122, 144, 146
 coiling 271
 compatibility 392-3
 definition 20, 47
 DNA double helix 48
 donation of transforming DNA 35-6
 exchange of parts 324-6, 329-30
 genes 142-5, 151, 153, 162
 groups in hybrid wheats 332
 fragments 504, 523
 fusion 314
 in E. coli 76

in interrupted bacterial mating 59, 61, 64, 75-7
length related recombination frequency 67
loss 330-1
mobile genes 426
movement in cells 274, 285, 287-8
number 285
number doubling 333-4
order of bases 47
rearrangements 380-1, 390
separation by centrifugation 245
shedding 248-55
shortening 271-82
size and shape in primates 510
size at cell division 265, 271, 290-300
structural changes 314
structure 58
super-alleles/super-genes 155
tails, attachment of 273-4
thickening 282
viability after structural changes 320
viability of equality 321-2
chromosomes:-
 homologous 167
 lying flush 281
 pair V in modern wheats 333
 pairing 296
 /plasmid relationship 209, 216, 435
 scrambled 282
cichlid fish 388-90
cigarette smoke and cancer 498
ciliated protozoa 209
cinnamic acid 480
circadian rhythms 452
circular mitochondrial DNA 212
circular plasmids 211, 235-6
circles of DNA 19-20, 209, 213, 216, 235, 252
Clark, B. 88
Clarke, B. 347
Clarkia ameona 330
classic experimental work in genetics 21-89
classic experiments of cancer research 499-506
classification 21, 372, 385, 387, 402
Claytonia virginica 331
Clean Air Act and selection 412
cleavage losses from polypeptides 112
cleavage in zygotes and embryos 446
climatic changes 397
clines 383-5, 397, 399, 400-1
cloned bacterial genes 245
clones 237, 242
closely linked genes 67-8
clusters of genes 57
Cnemidophorus spp. 331
CO_2 fixation 213-4
co-adaption 390, 422
co-adaptive fitness 378-9

coat colour in mice 191
codes (genetic) 105, 314
co-dominance in human blood groups 197-8
co-dominance of alleles 151
codon/amino-acid table – the genetic code table 89
codon/anticodon pairing 119
codon/anticodon recognition 115
codon/anticodon recognition ambiguities 125-6
codon – initiation, for polypeptide synthesis 121
codon sequence 501
codons:-
  definition 112
  for amino-acids 86
  for terminating the synthesis of a polypeptide 88-9, 120-1, 123, 211
  in extra-chromosomal DNA 211
  in gene expression 479
  in human genetics 535
  in mitochondria 211
  in mRNA 87-9, 105
  in relation to anticodons 110-1
  in single-base mutations 340-2
co-evolution 381, 385, 418, 426, 428, 430
co-evolution of hosts and parasites 431
cofactors 117
coiling of chromosomes 272
coiling of Limnea pereger shells 469-71
Col E1 plasmid 235
Col factor in killer bacteria 216, 219
Col factor in E. coli 221
colchicine 245, 332-3
Cold Harbour Laboratory, New York State 503
cold resistance 333
colicin poison 221
colinearity of amino-acids with codons in mRNA 112, 123
colinearity of the blocked DNA strand and the amino-acids in a polypeptide 113
colinearity of the genetic code in DNA with the amino-acid sequences in polypeptides 97, 101, 112-3
collagen 98
collections of polypeptides in cells 91
colonies of bacteria 71-2
colour-blindness in humans 528-31
coloured chickens 198-200
coloured feathers in chickens 198
coloured sweet-pea flowers 193
comb shape in poultry 189
combinations of switched on genes 472
common ancestors 343, 358-9, 372, 382, 387-8, 398, 421, 540
common ancestral species 399

comparative amino-acid sequences in widespread polypeptides 364-8
comparative chromosome studies in primates 508-11
comparative globin evolutionary studies 364-7
comparative polypeptide studies 337-368
comparative primate polypeptide amino-acid sequences 511-2
comparison of allopatric and sympatric speciation 379-80
comparison of base sequences in viruses, viroids and host cell genomes 435-6
comparison of DNA and RNA 102-4
comparison of eukaryotic and prokaryotic polypeptides 245
comparison of human genes with a viral oncogene 496-7
comparison of mitosis and meiosis 310
compatible Rhesus inheritance in human mothers and babies 554
compatibility of chromosomes at fertilisation 392
compatibility of gametes 392
'competence' in bacteria 36, 239, 242
'competence' in eukaryotic cells 244
competition 380, 422
competition avoidance 334
competition reduction 382
competitive instincts in man 515
complementary DNA (cDNA) 237-43
complementary RNA 435
complementary strand of DNA 432-3
complete cDNA 243
complex polypeptides 80
computers in genetic analysis 76, 497
concentration of end products 459
concentration of polypeptides 157
confidence levels in statistics 574-5, 582-5
conjunction 220
conserved base sequences 543-4
conserved exons 544
contemporary species relationships 357
contiguous populations 383
continuous variation 172, 576-7, 585
contraction of spindle fibres 274, 284
control genes 258, 365, 376, 490, 532, 546, 555
control of differentiation 446-452
control of embryonic development 458-64
control of gene expression 543
control of genes 446-52
control of transcription 446-7
control of translation 448
cooling to cause mutations 320
Cooper, G.M. 505
copper in soils as an agent of selection 418

copper tolerance in Agrostis tenuis (a grass) 418-21
co-repressors 157-8
Correns, C. 22
correlation of data in amino-acids 363
corroboration of amino-acid sequencing and the fossil record 363
cortisol 475
cortisone 473
Corvus cornix 388
Corvus corone 388
cotransformation 78, 248
cotransformation frequencies in genetic engineering 78, 251
cotransformation frequencies in humans 555
cotransformation frequencies of linked genes 556
cotransformation/independent transformation ratios 556
cotyledons 209
courtship rituals 378
CpG 'islands' containing initiation and termination codons 128, 537-8
creation of genetic variation 152
Crepis capillaris 296-300
cri-du-chat syndrome 521, 523
Crick, F.H.C. 41, 47, 80, 83, 85-6, 125-6
critical folding in polypeptides 345
crop disease-resistance 233, 257
cross-breeding 374, 422
cross-breeding – prevention 335
cross-fertilisation 386, 393-4, 399-400, 404, 421
cross-fertilisation barriers 380, 415
cross-fertilisation potential 387
cross-linkages and DNA repair 483
cross-linkages and transcription 482-3
cross-linkages of DNA to DNA 482-3
cross-linkages of DNA to chromatin 482-3
cross-over values 185
cross-overs and genetic mapping 67-76
crosses – genetic:-
    Chapters 5, 6, 7, 14 and 16
crossings-over:-
    absence of, in male D. melanogaster 288, 310
    comparison with mutations in creating variation 154-6
    comparison with translocations 325
    definition 282
    during meiosis 51, 265, 312, 323
    after interrupted bacterial mating 62-3
    even numbers of 67-8
    frequencies 67-8
    in D. melanogaster 173, 181-5, 316
    in genetic mapping 71, 73, 75-6, 246
    in genetic transformation 36-7, 39
    in humans 502
    in inversions 323-4
    in Locusta migratoria 300
    in Sordaria brevicollis 301, 303-5, 309
    in the first meiotic division 282-3
    odd numbers of 67-8
    of single-stranded DNA 61
    of viral DNA with the host cell genome 489
    probability of 68
    to cause recombination 58
    within genes 149
cross-pollination 378, 387, 412, 419
crows 388
crustaceae 386
crystallography 41
cultures of antibiotics 260-1
cures for AIDS 260
cures for cancer 499
cures for human genetic defects using genetically engineered products 231-2
curled-wing in D. melanogaster 182
cysteine 95, 343-4, 346, 535
cytochrome b 212
cytochrome c 339, 343-4, 346, 357, 480, 511
cytochrome oxidation enzymes 212
cytochromes 211, 363
cytogenic incompatibility 415
cytokinin 478-9
cytological changes 384
cytological isolation 384
cytology 372, 380, 386
cytoplasm:-
    containing circular DNA 209
    definition 564
    /gene interactions 445
    in eukaryotic cells 124, 128, 564
    in genetic crosses 168, 453-4, 459-63, 469-71
    /nucleus interactions 449
    in oöcytes 459
    in ribosome synthesis 131, 135-7, 139
    in sex determination 206
cytoplasmic chemicals 166, 311
cytoplasmic control of differentiation 446, 449-50
cytoplasmic endosymbionts 215
cytoplasmic filaments 567
cytoplasmic inheritance Chapters 6, 7 and 14
cytoplasmic polypeptides 137
cytoplasmic ribosomes 137
cytoplasmic tRNA 211
cytosine 40-1, 44-6, 48-55, 112
cytosol 128, 140

D

D gene sets for antibody synthesis in mice 465, 469
δ haemoglobin 364, 367
D human Rhesus allele 551
d human Rhesus allele 551

damage to cells 157, 311
damage to cellular components caused by free radicals 105, 480
damaged DNA 271, 318, 480
damaging alleles 229, 313
damaging alleles – replacement of 556
damaging environmental factors 315
damaging mutations 244, 312, 314, 371, 375
dandelions 331
dangers of human genetic engineering 230, 244
dark super-allele in Biston betularia 157
Darwin, C. 21-3, 31-3, 288, 369, 381, 514
Darwin's finches 32, 381
data banks for amino-acid sequences in polypeptides 339
data on computer tapes for classification 372
dating the era when mutations took place 357, 363
daughter nuclei 293
de Vriess, H. 22
defective alleles 232
defective genes 230
defective human alleles 253, 525-54
defective respiration 223-4
defining numbers for amino-acids 339-40
definition of a gene – an attempt 148-50
definition of races 373
definition of species 374
degenerate genetic code 88, 112, 314, 340
degeneration of gametes 311
degeneration of small chromosomes 327
degrees of freedom in statistics 580-83, 585, 593-6
degradation gene for toxins 233
degradation of carbon compounds 234
dehydrogenase deficiency 532
deletions from chromosomes 212, 318, 322, 325-6, 345, 370, 521-3, 534
deletions of amino-acid sites in polypeptides 345
deletions of bases 85, 313
deletions of DNA from human males 533
deletions of genes 153
deoxyribose 41-6, 50, 103, 229
deoxyribonucleic acid 102
deregulation of genes 245, 252, 370, 429, 434, 491, 555
derepression of genes 153, 472, 474, 480
derepressors 158, 482, 490
desirable gene insertion 262
desirable genes 231, 248, 251, 257-8
desired cDNA 242
desmosomes 568
detoxification using genetic engineering 233
Deuel, T.F. 496
developing embryo 278
development 277-8, 354, 452

development control genes 454
Dhar, R. 488, 506
diakinesis of cell division 298
diet and cancer 496
diet of Drosophila melanogaster 174
differences between sample means 601
differences of amino-acid sequences 346
different amino-acid substitutions 341-2
different genera of monkeys mating 510
different species 387, 510
different species cell fusions 247, 250-1, 257
different species gene exchange 391
differential centrifugation 245
differentiation 138, 153, 160, 206, 311, 384, 414, 444-7, 458, 464-9, 472, 533
difficulties of classification 372, 402
difficulties of equal migration of chromosomes at cell division 326
dihybrid crosses 27-31, 179-81, 189, 202
dimethyl sulphoxide and genetic engineering 246
dimorphism 412
dipeptide formation 97
Diplococcus pneumoniae 33-9
diploid:-
  gene location 145
  genetic map 147
  grasses 331
  meiosis 277-88, 296-300
  mitosis 265-77, 289-95
  nuclei 301
  number of chromosomes 172, 310
  /octoploid cross 334
  set of chromosomes 331
diploids 22, 40, 42, 76, 143-4, 148, 155, 167-8, 226-7, 246, 265, 310, 326, 331, 334, 348-9
Diptera eggs 454
directional change of environment 397
directional selection 407
discontinuous expression of characters 31, 154
discontinuous phenotypes 409
discontinuous variation 172, 474-5, 576-7
disease 174, 232-3, 253-4, 257-8, 260, 373, 426-7, 429, 437, 442
disease:-
  antigens and human blood group distribution 555
  combating 439, 499
  control by antibodies 464-9
  resistance 233, 257, 333
  resistance improvements by genetic engineering 233
  transmission 436
  vectors 386
  vectors that are sibling species 386
disruptive selection 408-9
dissociation of chromosomes 273

distal chromosome parts 539-40
distances between genes 66, 556
distribution geographically of humans with different blood groups 555
distribution of cytoplasmic chemicals in cells 311
divergence date of origin for homologous X and Y chromosome regions 541
diverging allele frequencies 371, 397
diverging phenotypes 319
diversity 385
dividing cells Chapter 9
dividing membranes 275-6, 285, 321
DNA:-
   as the material of inheritance 33
   ase 38
   bacterial 19
   base analogues 436
   base pairing with ribose nucleotides 107
   base sequences 337, 442
   bases in nucleic acids 40-55, 92, 148, 229
   binding polypeptide 546
   cancerous 502-3
   cancer viruses 493-6, 499
   chemical stability 104
   – chloroplast 213
   /chromatin cross-linkages 481
   circular 213
   codes 48, 91, 340-2
   detailed structure 49-55
   determining a species 423
   dissociation from histones 162
   /DNA cross-linkages 481
   elongation 449
   eukaryotic 150-163
   extra-chromosomal 209-10
   fingerprinting 526
   fragments 153, 251, 375, 426
   genes 155
   genetic code 80-3, 86, 111
   genome 20
   glycolases 318
   helicase 268-70
   highly repetitive genes 163
   /histone structure 161
   /histones 563
   in ageing 480
   in embryos 254
   in eukaryotic genetic engineering 246
   in genetic mapping 77
   in mutations 313-4, 322
   in ribosome synthesis 131-3
   in the genetic engineering of fungi 256
   in the genetic engineering of plants 256-7
   insertion into a bacterial plasmid 238-41
   ligase 318
   limitations as the material of inheritance 49
   loop to accommodate inversions 323
   mitochondrial 209, 211-3
   mobile 151-3, 425-30

   mutations Chapter 10
   mutations in super-genes 156
   particles 436
   /plasmid interactions 216, 236
   polymerase 240, 266-7, 318, 483
   properties 40, 334
   rate of transfer from donor to recipient bacteria 64
   redundant 150, 312, 322
   repair 40, 483
   replication 214, 268, 270, 310, 348
   /repressor interaction 157
   reverse transcription 127, 433-4, 490
   /RNA comparisons 103
   shape 48
   shedding 247
   single copy genes 163
   single-stranded 76, 105, 108, 237
   slightly repetitive genes 163
   strand separation 106, 116
   structure 41, 265
   synthesis 105, 126, 237-9, 250
   synthesis from viral DNA 431-2
   synthesis pathways 247, 249
   tautomeric shift in the bases of 59, 61, 349
   transformation 79
   viral 20
   viruses 431-2
   without histones 563
   zinc-finger control of the genes in 160
Doolittle, R.F. 497
dominance 23-28, 148, 157, 167, 171, 175, 185, 188, 202-3, 257, 409, 412
dominant alleles 154, 167, 170-1, 178, 198, 251
dominant human damaging alleles 525-6
dominant inhibitor allele in chickens 198-200
dominant model species 417
dominant super-alleles 154
donation of DNA 35, 38-39, 59-63, 66-69, 75
double helix structure of DNA 41, 47-8, 50, 61, 86, 267
double recessive 178-9, 182, 315
double stranded cDNA 239-40
double stranded DNA 39, 77, 149, 431
double stranded DNA analogue 258
double stranded DNA viruses 431
double stranded RNA 110
doubling of chromosome number 334
Down's syndrome in chimpanzees and man 511, 519-21
downstream gene regulation 535-6, 548
downstream genes 452, 492, 533
downwind Agrostis tenuis plants 419
drift (genetic) 154, 156, 212, 214, 220, 312, 335, 343, 353, 357, 371-3, 379, 390, 393-7, 402-3, 406, 423, 426, 437
Drosophila melanogaster:-
   absence of crossings-over in males 173

autosomal linked genes 181-5
cells of fertilised eggs 562
cross-linkages in the DNA of 482
chromosome pair 1 147,
curled wing allele 182-4
diet 174
differences in development 462-4
discontinuous variation in 577
ebony body colour 169-70
embryo development 458
eye colour 56, 185-8
for genetic crosses 174-88
genetic mapping 147, 246
grey body 169-70
homeobox 455-8
homeotic genes 461
life cycle 173-4
meiosis 310
meiosis in males 173
mitosis 138, 265
monohybrid cross 176
mother effect in 458-60
mutation 318
puffs in salivary gland chromosomes 446
red eye 185-8
segmentation genes 461
steroid hormones in 475
sex determination 205-6
sex linked genes 185-8
  TDF regions 534
  virgin females 174
  white eye 185-8
drug resistance 211-2, 244
drug sensitivity 212
drugs for slowing the processes of ageing 484
Drws-y-Coed copper mine 418-22
drying and wetting of DNA 320
Dulbecco, R. 501
dumpy wing in D. melanogaster 179, 181
duodenal ulcers 549-50
dwarfism in humans 526
dysentery 439

E

E human Rhesus factor allele 551
e human Rhesus factor allele 551
e^- 481
early embryo cells for genetic engineering 251
early embryo cells that control differentiation 453, 460
early stages of sympatric evolution 421
East Cornish type of Maniola jurtina 405
'ebony' body colour allele in D. melanogaster 169-70, 175-6, 177, 179, 181-4
ecdysone 473, 475
ecological genetics 397-422

ecological niche 380-1, 388
ecological population studies 369, 398
edge of a habitat range – speciation at 334, 377
egg mimicking swellings on plants 415
egg nuclei 463
egg organelles 223
egg production 168, 183-4, 277, 287, 311
elaioplasts 566
electric charge 346
electron microscope 265-6
electron transport 339, 344
electrostatic attraction in amino-acids 100
elephantiasis 386
elongated chromosomes 266
'elongation factors' in polypeptide synthesis 119
elongation of a polypeptide 116-9, 123
elongation of DNA 265
embryo cells in genetic engineering 251
embryo control genes 454-5
embryo development 206, 251, 458
embryo genetic engineering 517-8
embryo segmentation 458, 463
embryo spatial orientation 458
embryonic cleavage 446
embryonic development genes 206, 458
embryonic development patterns 464
embryonic differentiation 452-5
embryonic mouse antibody gene sets 467
embryos 253, 278, 287, 441, 455
emergence of new races and species 371
Emrich, J. 86
3' end of nucleotide chains 42, 110, 267-8, 270, 323
5' end of nucleotide chains 42, 110, 267-8, 270, 323
endemic species 388, 390
end-product gene control 157, 447, 449
ending a polypeptide chain 119
endocrine glands 472, 474
endonuclease 240-1, 266-7
endoplasmic reticulum (E.R.) 124, 129, 137, 139, 477-8, 483, 489, 564-5
endosperm 140, 209
endosymbiosis – mitochondria and eukaryotic cells 212
endosymbiosis – Paramecium aurelia 215
enucleated eggs 453
environment 31-2, 172, 312, 315, 319-20, 397
environmental:-
  changes 31
  control of gene expression 172
  factors 157
  pressures 319
  selection 369, 371, 393
  stability 319
  stress 311

enzyme:-
  activation 259
  components 79-80
  control 319
  production 231
  synthesis 78-9, 83
enzymes:-
  /antibody interactions 259
  control of synthesis in chloroplasts 214
  determining eye colour 56
  determining rates of biochemical reactions 91
  DNA polymerase 266, 318
  for inserting DNA into plasmids 236
  in base pairing 267
  in biochemical pathways 56, 78, 172, 192
  in DNA synthesis 126, 238-9
  in reversing DNA transcription 229, 237
  in repairing damaged DNA 40
  in ribosome synthesis 135
  manufacturing chemical compounds 172
  metabolic 244
  'one gene – one enzyme' theory 78-9
  polynucleotide ligate 266
  polypeptides 91, 98-9, 128
  separating DNA strands 106
  tryptophan synthesis 101
  uses in genetic engineering 231-2
Ephrussi, B. 56
episomes 221
epistasis 189, 192, 200, 202
epistatic genes 189, 191, 194
epistatic interaction of two genes in poultry 189-91
epistatic modification to gene expression in mice 191-2
epistatic reversion to the wild-type 191-2
epithelial cells 459
Epstein Barr virus 440, 496
equal reciprocal translocations 325
equality of chromosomes at cell division 310, 320
equator at cell division 271-4, 283-4, 292-3, 310
equilibrium of allele ratios 396
erythroblastosis foetalis 552
Escherichia coli:
  F^+ sex factor 220-1
  genes 145, 149, 151-2, 154, 159
  genetic map 146
  genetic mapping 57-80
  introduction of synthesised genes into the plasmids of 230, 233, 235, 238-9, 262
  K12 strain 56
  mating 74
  mobile genes 151-2
  mutant strains 57, 79
  polypeptide synthesis 105-6, 109, 111, 122, 126

rRNA 109
ribosomes 111
used for reconstituting cancerous viruses 503
established allele 343, 356
established amino-acids in a site 342
established cell cultures 244-5
established cell lines 253
establishment of a mutation in asexual species 392-5
establishment of a mutation in sexual species 392-5
establishment of an allele in a population 393
esterification of amino-acids 110
estimated mean 594-7
estimated standard deviation 592-6, 598, 600, 603
ethological barriers to reproduction 378
ethology 372, 378
eukaryotes 19, 149, 214
eukaryotic cell ffusions 247
eukaryotic cells:
  Appendix 1 561-69
  cell fusions 247
  chromosomes 161
  cytoplasmic ribosomes 212
  DNA 150, 163
  DNA codes 89, 92
  extra-chromosomal DNA 209
  extra-chromosomal inheritance 218, 222-7
  genes 160, 231, 236-7, 241
  genetic engineering 244-61
  mobile genetic elements 427, 430
  ribosome synthesis 131, 135-6, 139
  transposons 427
eukaryotic:-
  chromosomes 161
  DNA 150, 163
  extra-chromosomal DNA 209
  extra-chromosomal inheritance 218
  extra-chromosomal inheritance – examples 222-7
  genes 160, 231, 236-7, 241
  genetic engineering 244-61
  genetic mapping 78
  genome 76
  polypeptide synthesis 124-5, 128
  ribosomes 135-7
  species 40, 127, 154, 163
  transcription of viral genes 434
Euphydrias aurinia 409
Euplotes spp. 289
evolution:-
  Chapter 12
  and mobile genetic elements 436-9, 442
  base sequence comparisons 435
  comparison studies of polypeptides 561

during a known period of time 398, 418-21
of differences in offspring 287
of rational species 515
of primates 514
of sexual differentiation 533
of X and Y chromosomes 540-1
evolutionary:-
  changes 334-5, 369, 418
  consequences of genetic change 335
  distances between species measured by amino-acid sequences differences 359-64
  divergence 408
  genetics 33, 143, Chapter 12
  implications of mutations 156, 312-3, 318-20, 341, 348
  limitations of DNA 49
  processes 395-6
  relationship between chloroplasts and eukaryotic cells 214
  relationship between mitochondria and eukaryotic cells 212
  significance of tautomerism 55, 348-357
  time span of species 417-8
  trees 339-41, 357-64
exceptional spore patterns in Sordaria brevicollis caused by mutation 308-9
exceptional viable hybrids 377
exchange of chromosome parts 324-6, 329-30
exchange of genes – absence of 374
exon/intron boundary base sequences 435
exons:-
  conserved 544
  description of 472, 492
  effect on by mobile genes 438
  electron micrograph 164
  in gene transcription 127
  in genetic mapping 76
  in human genetics 509, 535
  in human Y chromosomes 150
  in 'species genes' 374-5
  in the genetic causes of cancer 500
  in the synthesis of cDNA 237
  mutations in 312, 316
  problems of discovery for gene expression 472
  pY 53·3 exon 542-4
experiments that identified the genetic causes of cancer 495-506
explosive evolution 388-90
exposed ends of zinc fingers 434-57
expression of damaging homozygous alleles 515-8
expression of genes 446
extending range of a species 384
extinctions:-
  being the rule rather than the exception 319, 383-5
  Darwin and 32

determined by mobile genetic elements 438-9
hermaphrodites and 391-2
limitations of DNA and 49
extra autosomes 521
extra-chromosomal alleles 166
extra-chromosomal DNA 20, 166, 209, 234
extra-chromosomal inheritance 209, 227
extra genes in nucleoli 138
extra set of chromosomes 331
extra X chromosome in males 521
extra Y chromosome in males 522
extractions and transplants of cell nuclei 444-5, 449, 453

F

F absent bacteria 219
F^+ bacteria 219
F^- bacteria 219
F factor 34, 65
F in Hfr bacteria 65, 216
F plasmid 216
failure of differentiation 486-7
failure of gametes to fuse 378
failure to regulate negative hormone feedback in animals 474
failure to transform XX mice with human SRY 548
fairly soluble amino-acids 95
falling numbers in populations 394
far-red light 480
fats 91
fatty-acids 233
Fawcett, D.W. 140, 164
Fe^{2+} and Fe^{3+} 480
feather pigmentation in chickens 198
feedback controls of polypeptide synthesis 449
feeding habits 387, 421
female gamete chemicals 206, 224, 453-4, 459-60, 463
female gametes 175, 177, 203, 213, 223-4, 278, 280, 311
female mutation rate 316
female polar body decay 287-8
female to male transformation in mice 545
females 186-8
fermentation 70
fertile hybrids 386, 510
fertilisation 167-8, 277-8, 280, 287, 317, 354, 355, 454
fertilised eggs 270, 277, 463
fertility in primate hybrids 510
fibre producers 209
fibres of the spindle 272
fibrils in nucleoli 132-4

fibrinopeptide A 357, 364, 511
fibrinopeptide B 357, 511
fibroblasts 497, 502
fibrocystic disease 515, 527
fibrous polypeptides 98
filamentous polypeptides 99
filaments in primroses 411
finches 32, 381
first generation hybrids 393-4
first meiotic division 279-85, 296-303
first polar bodies 280
fish 331, 388-9
fish-cichlids 388-9
fishu tarazu homeodomain 458
fishu tarazu mutant 457
fission of chromosomes 370
fitness – lack of in humans 517
five bands allele in Cepaea nemoralis 202, 204
flower pigments in sweet peas 151, 192-4
flowering stages of Agrosis tenuis 420
fluorescence of chromosomes 509
'flush' homologous chromosomes at meiosis 310
foetal cells in humans 553
foetal mouse liver cells 450
foetal haemoglobin 364
folded haemoglobin 99
folding of cytochrome c 343
folding of polypeptides 99, 343, 535-6
follicle for egg production 459
foot-web (amphibia) nuclei 450
Ford, E.B. 409
Ford, H.D. 409
'foreign' desirable genes 252
'foreign' DNA 236, 254
'foreign' genes 230, 234-5
forest refuges during dry eras 414-5
forward mutations 341, 346-7
fos gene 493
fossil/bone morphology 514
fossil/contemporary species relationships 357
fossil evidence for evolution 387
fossil fuels 233
fossils 92, 339, 387, 514
founder member of a new allele in a population 394
founding few in small populations 396
four cell embryos 441, 556
four cell embryos – cell separation in genetic engineering 251
fragmentation of DNA 237, 246
fragments of cDNA 242-3
frameshift mutations 86, 313
Franklin, R.E. 41, 47
free energy 435
free radicals 480-1
freezing of cells 556
frogs 387, 399, 400-01, 455-6

fruiting bodies 301
full tautomeric mutations 350-2, 354, 356
functional regions of polypeptides 98, 343, 374-5
functionally conservative amino-acid sites 346
fundamental genes 484
fungal cell genetic engineering 256
fungal cell fusions 256
fungal gametes 224
fungal plasmids 235
fungi 19, 224, 232, 234, 253, 255-6, 277, 391, 562
further reading for evolutionary genetic studies 369
fused cells 248, 255-6
fused nuclei 251
fusion 254-5
fusion of:-
 animal cells 254
 of cells 226-7, 244, 247, 249-51, 302
 of centromeres 330
 of chromosomes 314, 326, 370, 509, 511
 of eukaryotic cells in genetic engineering 254-8, 555
 of haploid fungal cells 301
 of human and mouse cells 247, 555
 of hyphae 256
 of mitochondria 211
 of nuclei in Sordaria brevicollis 301-2
 of nucleoli 132
 of plant cells 257

G

γ haemoglobin 367
γ radiation 481, 488, 498
GA (gibberellic acid) 478-9
Galapagos Islands 31-2, 381
galls 258
gamete compatibility 372, 392, 397, 399
gamete forming tissues 348
gamete mother cells 354
gametes:-
 and mitochondrial inheritance 213
 in Chlamydomas reinhardii 226
 in Cepaea nemoralis 202-4
 in D. melanogaster 168-70, 176-84, 186-88
 in meiosis 277-8, 287, 310-1
 in Neurospora crassa 224
 in poultry 191-2, 199
 in snapdragons 197
 in the Mendelian Law of Independent Segregation 28-9
 in the Mendelian Law of Random Assortment 30
 in tobacco 194-6
 incompatibility 378

inviable 326-7
mutations 317, 353-6
produced by mother cells 169
receiving alleles 168-9
sex determination by 205-6
Weissman's observations of 22
Garen, A. 88
gastrulation 454, 462
gene:-
  activation 444
  amplification 153, 376, 446, 452, 489
  amplification – cancerous consequences 153, 559
  associated disorders in humans 549
  banks 236-7
  base sequences 155
  /centromere/crossing-over points – relative positions in Sordaria brevicollis 303, 305
  clusters 57, 78
  control 157, 159-60, 426, 473-6, 545
  control chemicals 450
  controlling polypeptides 160
  /cytoplasm interactions 445
  definition – an attempt 112, 148-9, 156
  deletion 153
  deregulation 370
  derepression 472
  DNA/probe pairing in eukaryotic genetic mapping 556
  exchange 373, 386-7, 391, 396-7, 403
  exchange – absence of 376
  expression 155, 171-2, 311, 316 and Chapters 5, 7, 14 and 16
  flow 383
  insertion into eukaryotic cells 247
  insertion into plasmids 237
  interactions 172
  length 79
  libraries 231
  linear sets of mutable sites 148
  location 144, 156
  loci 201, 246, 376
  mobility 467-9
  positions 149, 175, 314, 316
  regulating polypeptides 533
  regulation 245, 475, 533, 535
  repression 157, 444
  sets 310, 390
  shuffling 152
  storage 236-7
  synthesis by genetic engineering 236-43
genera of monkeys 377
general requirements for mutation viability 320
genes:-
  antibiotic 232
  autosomal linkage of 181-4
  coding for polypeptides 340

col factor 221
controlling inheritance 20
defective human 253, 377, 525-54
desirable, in genetic engineering 258
different mutation rates in 315
duplication of 322, 366-7
epistatic interactions of 189-94
eukaryotic 160
homeobox 456
human sex determination 533-48
in Cepaea nemoralis 409-10
in chloroplast DNA 217
in Drosophila spp. 138
in genetic crosses, Chapters 5, 6, 7, 14, 16
in hybrids 378
in mitochondria 212
in plasmids 221-3
in plastids 209
in population studies 423
in tephritid fruit flies 381
in translocated DNA 489
multiple 162
nuclear 225-9
positions of, in E. coli 65
prokaryotic 56-80, 105-23
regulating polypeptide synthesis 102, 124, 126
regulating polypeptides 533
regulating ribosome synthesis 132-5, 137, 139
regulating transcription 106-8
scrambling at meiosis 312
substitutions 356
widely separated on bacterial chromosomes 61
genetic:-
  analysis 80-9
  basis of cancer 487-97
  change 335, 395-6
  code 23, 83-4, 86, 89, 97, 111, 313-5, 337
  code ambiguity 125-6
  code in mitochondria 92
  control of a biochemical pathway 192
  crosses 23, 155, 166-206, Chapters 5, 6, 7, 14 and 16
  differences created by the first meiotic division 288
  drift 154, 156, 212, 214, 220, 312, 335, 343, 353, 357, 371-3, 379, 390, 393-7, 402-3, 406, 423, 426, 437
  engineering 126-7, 210, 212, 217, 229-61, 347, 439-42, 499, 517, 533, 555-6
  homozygosity 229
  isolation 32, 379, 380, 401
  map of E. coli 145-6
  map of D. melanogaster 147
  mapping 39, 56-78, 146-7, 184, 212, 246-7, 251, 254, 534, 556

646

markers 255
polymorphism 154
probes 534
recombination in Sordaria brevicollis 301-9
screening of embryos 517
stability 333, 385, 395
transformation 35, 38, 425
triplicate code 80-89
uniqueness in gametes 288
genetically isolated groups 396
genome 20, 76, 105, 132, 151-3, 244, 246-7, 252, 256, 258, 312, 322, 375, 379, 426-7, 429-30, 434-5, 437-8, 442, 489
genotype 30, 49, 66, 167, 170, 171, 175, 181, 183-4, 188, 193, 199, 204, 219, 224, 239, 250, 253, 287, 383, 394, 407
genotype ratios 171
genotype symbols – convention in genetic crosses 180
geographical distribution of human blood groups 555
geographical distribution of larid gulls 403
geographical isolation 32, 373, 377, 384
germ cell:-
  fertilisation 278
  genetic engineering 230, 253, 441
  mutation rates 315-7
  transformation 254
germ cells 22-3, 27, 169, 254, 277, 316, 438
germ plasm 454
gibberellic acid (GA) 478-9
gibbon blood groups 513
giemsa stain 518-9, 524
giraffes 31
glandular cancers 494
globin evolution 364-8
globin genes 367
globin mRNA 450-1
globin – rabbit 448
globins 339, 447-8, 532
globular polypeptides 99-100
globulins 553
glucose 128
glucose agar 71, 74
glucose – 6 – phosphate control genes 532
glucose – 6 – phosphate dehydrogenasee deficiency 532
glutamate 233
glutamic acid 96
glutamine 95, 341
glutathione 481
glycerate – 3 – phosphate dehydrogenase 79
glyceric acid – 3 – phosphate 79
glycine 95, 343, 345-6, 506
glycine and polypeptide structure 343
glycopolypeptides 128-9
GM factor in blood 511
golgi apparatus 129, 477, 565

gonad development chemicals 454
gonad mother cells 355, 441
gonad production 461
gonads 253-4
Goodfellow, P.N. 544
gorilla blood groups 513
gorillas 509, 512, 541
grana 213
granules in nucleoli 132-3
grass/wheat crosses 332-3
grasses 331-3
grebes 422
Greek terminology for the stages of meiosis 265, 295-300
green algae 19
greenhouse studies of Agrostis tenuis, 420, 422
grey body in D. melanogaster 169-71, 176-81, 183-4
grey crescent in fertilised eggs 454
Griffiths, F. 33-35, 37, 39, 61
grooves in DNA structure 48
Gros, F. 108
groups of chromosomes in hybrid wheat 332
growth 277, 452
growth hormone 232, 473
growth hormone gene in mice 252
growths – benign 486
growths – malignant 486
GTP 118-9
guanine 40, 43-46, 48-55, 112, 126, 249, 483
guard cells 479
gulls as 'ring species' 402-4
gut cells – dividing 488
gut cells – reversion to totipotency 453
gut lining adhesion by bacteria 234
Gymnopais spp. 331
gypsy blood groups 513

H

H^+ 477
H chain in antibodies 465
H chemical in blood 512
H polypeptide 489
habitat 379-81, 388, 396
haem- ring 344-5, 364
haemoglobin 91, 99, 101, 163, 344, 357, 364-7, 511
haemolytic disease of the new born 552
haemophilia 532
hairpin loops in genetic engineering 238-9
half-mutations 349-52, 354-5
hamster embryo cells 502
haploid cells 22, 143-4, 226-7, 334
haploid gene location 145
haploid genome 132, 142
haploid nucleus 225

haploid number of chromosomes 277, 310
haploid species genetic map 146
haploid vegetative parents 226-7, 285
haploids 92, 148, 167-8, 256
haploidy 143, 277, 285-6, 301, 310
Hardy-Weinberg equations 515
Harvard Medical School 260, 505
Harvard University 108
HAT medium 250
Hayes, F. 56
HC allele 525
head size in Rana spp. 399
heart failure in incompatible Rhesus babies 553
heat killed bacteria 34
heating (and cooling) DNA 320
heavy chain antibody synthesis 469
heavy metal mines and evolution 418
Heldin, C.A. 496
helicase for DNA synthesis 268-70
heliconid butterfly eggs 415
helminths 386
hen oviduct chemicals produced in response to hormones 474
hepatitis B virus 496
herbicide resistance 257
herbicides 233
hermaphroditic Limnea pereger crosses – 'mother effects' 470-1
hermaphroditic species 391-2
hermaphrodism 471
herpes 495
herpes II virus 499
herring gulls 402
Hershey, A. 39
heterochromatin 140, 317
heteropycnotic bodies 266
heterostyle polymorphism in primroses 411-2
heterozygosity 155, 203
heterozygote – selection in humans 532, 548
heterozygotes 29, 178-9, 182, 199, 315, 356, 516, 517, 532
heterozygous carriers of damaging alleles 515
heterozygous foetus 549
heterozygous parents 169
H fr (and F^+) sex factor in E. coli 216, 220-1
HGPRT 250
'hi' allele 317
high frequency recombination 216, 221
high mobility polypeptides 543
highly regulated development in Caenorhabditis elegans 464
highly repetitive DNA 163
Hind III restriction site for genetic engineering 236, 240-1
Hippocratic Oath 230
histidine 57-66, 78, 341, 344, 535
histidine synthesis 154

histone/DNA structure 161-2
histone H_4 357
histones 161-3, 211, 448
HIV 258-60, 262, 425, 436-7, 450
HIV viral strains 261
HMG 1 543
HMG 2 543
HMS *Beagle* 31
H_2O_2 480
homeobox 453-8
homeobox genes 462
homeodomain 455-7
homostylistic primroses 412
homeotic genes 458, 462, 464
Homo sapiens 508, 514
homologous chromosomes 133, 142-4, 147-8, 155, 167-9, 176, 265, 271-2, 281, 284, 310, 324-8, 349-52, 355
homology in bases 456
homozygosity 155, 315, 319, 356, 375
homozygosity in genetically engineered plants 256
homozygous alleles 395
homozygous dominants 315, 516-7
homozygous expression of damaging alleles 515
homozygous parents 169
homozygous recessives in humans 515-7
homozygous wild-type failures 549
horizontal transmission of viruses 494
hormonal control 449
hormonal modification of gene expression 471-80, 482, 491
hormone production and sexuality 547
hormones 91, 138, 244, 257, 311, 444, 447, 449, 471-80, 482, 484, 491, 494
hormones produced by genetic engineering 232
horses 378
host cell genome 435
host cells 492
host plant coevolution 381
hosts of parasites 381
Huang, J.S. 496
human:-
 alanine codons repetition 504-5
 autosomal super-genes 528
 autosomes 518
 baby abortions 320
 blood group geographical distributions 555
 blood groups 151, 154, 197-8, 512-3, 549-55
 blood sub-groups 512
 bone marrow cell genetic engineering 556
 cancer 497-9
 cells and HIV 258
 /chimpanzee chromosome comparisons 510
 chromosome aberrations 518
 chromosome indentification 518

cotransformation frequencies 555
defective genes 313
DNA alanine codon repetition 504-5
DNA/bacteriophage recombination 504-5
DNA that transforms mouse cells into cancer cells 502-6
dominant defective alleles 525-6
embryo genetic defects 232
embryos 253
ethnic groups 373
exon similarities in the X and Y chromosomes 533
fragments of DNA 153, 251, 375, 426, 506
gene expression 251
genes 230, 253
genetic engineering 253, 555-7
genetic engineering dangers 230, 243-4, 556
genetic mapping 61, 246, 251, 254
genetic studies 205
genetics 508-57
genetics system 21
genome 20, 78, 492, 502, 545
genome mapping 556
genome project 506-7
germ cell mutation rate 316-7
globins 364
haemoglobin 364, 366-7
heterozygotes – selection 549
/HIV infection 258
homeobox genes 455
immune system 259
independent transformation frequency 556
inherited allelic defects 231, 525-32
inherited diseases 231, 525-32
linked genes- mapping of 556
male determination 538
mitochondrial analysis 514, 561
/mouse cell fusions 247, 251
/mouse DNA recombination 502-3
mutant defective allele 231
oncogenes 489
oncogenic viruses 495
platelet genes 496-7
point mutations in a proto-oncogene 506
/primate chromosome comparisons 508-12
races 373
recessive defective alleles 527-8
Rhesus factor inheritance 550-4
sex chromosomes 512, 523-4, 528-48
sex determination 150, 205-6, 533-48
sex-linked genes 528-32
sexual abnormalities 521-2, 539
testicular determining factor (TDF) 534, 537-9, 542-5
T-lymphocytes 262
transplants 528
trisomy in chromosomes 519, 521
wound cells 496

X chromosome 205-6, 518, 521-2, 528-34, 536-42, 544, 548
X chromosome short arm 534
Y chromosome 205-6, 521-2, 529-34, 536-44
Y chromosome short arm 150, 534
zygotes 253
humans 142
Huntingdon's chorea 525
hybrid:-
  abnormalities in Rana spp. 399
  crosses 399
  DNA 503
  fitness in primates 510
  grasses 331
  infertility 378
  inviability 378, 398, 404-6, 408
  /non-hybrid gamete compatibility 392
  phage particles 504-5
  populations 387
  viability 372, 400
  zones 385, 387-8, 403, 415
hybrids 23-4, 28, 333-4, 374, 377-8, 381, 384, 386, 393, 398, 400, 415, 422, 510
hybrids – first generation viability 392
hybrids – selection against 404-5
hydrogen atoms and tautomeric shift 51-5
hydrogen bonds 50-5, 103
hydrogen peroxide 480
hydrolytic enzymes 479
hydrophobic associations in amino-acids 93-4
hydrophobic associations in polypeptides 100
hydrophobic groups in polypeptides 536
hydrophobic polypeptides 93-4, 213
hydroxyl radical 480
hydroxylysine 128
hyphae fusion in fungi 256
Hyponomeuta padella 421
hypothesis for scientific investigations 575
hypoxanthine 250

I

i (blood group allele) 550
I^A (blood group allele) 151, 197, 550
I^B (blood group allele) 151, 197, 550
I^O (blood group allele) 197
IAA (auxin) 476-7
ice refuges for seagulls 403
identical alleles 203
identical base sequences 203
identical chromosomes in somatic cells 277
identification of human chromosomes 518, 523-4
immune response to specific antigens 464
immune system 152-3, 254
immunisation 259
immunity 260
immunodeficiency 260

Imperial Cancer Research Fund Laboratories 205, 538, 544-5
improved antibiotic properties of fungi 256
improved wheats 333
improvements to animal stocks 253
improvements to plant stocks 256-8, 331-3
inability to use G or T in DNA synthesis 249-50
inactive repressors 158
inborn errors of metabolism 231, 556
inborn errors of metabolism – corrections of 441
inbreeding 317, 390-2, 396
incompatible Rhesus inheritance – human mothers and babies 552-4
incompatible gametes 378
incompatibility – cytogenetic 415
incomplete dominance 151, 196-7
incorporation of tautomeric point mutations 348-53
incorporation of viral DNA into a bacterial chromosome 152
incorrect gene repair related to ageing 483
incorrect migration of a chromosome 331
incorrect translation of a gene 252
independent segregation of chromosomes 392
Independent Segregation – Mendelian Law of 28
independent transformation frequencies in humans 556
indirect evidence that DNA is the material of inheritance 40
individuals in populations 394
indole acetic acid (auxin) 476-7
Indonesian archipelago 33
inducers 138, 158-60, 447, 449, 452, 455, 533
induction 454, 458, 493
industrial melanism 412
infectious diarrhoea 232
infectious lymphoma 495
infectious polypeptides 428
infertile hybrids 378
ingested DNA 246
ingestion of cancer cells 487
inheritance particles 23, 166
inherited human diseases 253
inherited number of nucleoli and rRNA genes 139
inhibition of DNA synthesis 249
inhibition of translation 448
inhibition saliva test 512, 514
inhibitor allele 198-9
initiation codon – AUG 114-6
initiation codons 88-9, 115, 128
initiation codons contained in 'islands' 128, 537
initiation of rRNA synthesis 138

inner mitochondrial membrane 211, 481
inoperative enzymes 57
inoperative genes 57
inoperative mutations 57
inosine 126
Inouye, M. 86
insect/plant co-evolution 381
insecticides 233, 412
insertion of synthesised genes into micro-organisms 230-1, 236-44
insertion of whole genes – screening in genetic engineering 243
insertion of Y derived chromosome segments into X chromosomes 540
Institut Pasteur, Paris 56, 108
insulin 232, 337-8, 364
interacting genes 317, 404, 423
interactions of alleles 189
interactions of genes and cytoplasm 445
interactions of nucleus and chloroplasts 225
interactions of oncogenes and other gene products 493
interactions of super-genes in Cepaea nemoralis 200-4
interbreeding 396, 405
interbreeding barriers 335
intercellular mobile genetic elements 427
interfaces for hybrid production 405-6
interkinesis 299
interferon 259
interphase of cell division 290
interrupted mating in bacteria – genetic mapping 56-66
inter-relationship of inducers and repressors 493
inter-sexes 331, 542
inter-specific cancer infection 502
inter-specific genetic translocation 437
interspecies 379
intestine differentiated cells 450
intestine epithelial nuclear transplants 450
intra cellular mobile DNA 427
intron/exon boundary 501
introns 76, 127, 150, 164, 231, 236-7, 312-3, 316, 375, 472, 492, 500-1, 509, 535
invaginations in eggs after fertilisation 454
invariable amino-acid sites 343-5
invariable genes (new) 371, 376
invariable (so called) genes 315, 370-1, 374-6
invariable species gene (discussion) 438
inversion loops at meiosis 324
inversion viability 323-5
inversions in chromosomes 370
inversions – paracentric 323-5
inversions – pericentric 327
inversions – rotation to make viable 323
inviable alleles 244
inviable gametes 327

650

inviable genes – increasing proportion of in humans 244
inviable hybrids 396
inviable mutations 320, 396
inviable offspring 378
iodine deficiency 474
ion transport in membranes 483
iron 480
iron atom in haem- ring 344
'islands' (CpG for initiation and termination of polypeptide synthesis) 128, 537
isolated carrot cells – return to totipotency 445
isolated groups of individuals (populations) 396, 407
isolating barriers – removal of 335
isolation 384, 388, 406, 421-2
isolation:-
  allopatric 319, 380
  degree of 394
  genetic 401
  geographical 373, 377, 379
  parapatric 382
  of prokaryotic genes 231
  of races 371, 373
  origins of 377-8
  sympatric 319
isoleucine 89, 94, 342
isomers 93

J

J gene set for mouse antibody synthesis 465-8
Jacob, F. 56, 108
Johnsson, A. 496
Jewish Hospital, St. Louis, U.S.A. 496
joining of amino-acids to make polypeptides 97, 119
joining of centromeres 330
joining of nucleotides 43-4, 138

K

K (cytokinin) 478
K strain of E coli 56
K^+ ion movement in plants 479
κ antibody chain 465, 468-9
kappa particles in Paramecium aurelia 214-5
karyotypes of chromosomes 382, 386-7, 523-4
keto forms of DNA bases 52-5, 349-51, 355
Kettlewell, K.B.D. 412
Keyacris scurra 327
Ki-ras gene (cancer associated) 493
kidney cancers 498
killer paramecia 215
kinases (cancer associated) 493
kinetoplasts (cell organelles) 209, 566

Klinefelter's syndrome 521
known time period for evolutionary change 418-420

L

λ antibody gene sets 465
λ chain 465-6
λ polypeptides 465-6
L- amino-acids 93
laboratory cross-fertilisation 400
laboratory population studies 369, 373, 397
Laboratory of Eukaryotic Molecular Genetics, London 545
lac region of E. coli's chromosome 159
lack of genetic fitness 517
lack of hormones 474
lactate dehydrogenase 511
lagging chain of replicating DNA 268-70
Lamarck, J-B. 21-2, 31
lampbrushes 452
lamprey globins 366-7
large alterations to chromosomes 51, 315, 319
large fragments of chromosomes 322, 326, 329-30
large mutations 370
large populations – evolution studies 357-9, 402, 408
large samples in statistical analysis 587
large vacuoles 567
larid gulls 402-4
Larus argenatus 402-3
Larus fuscus 402
Larus glaucoides 404
larynx cancers 48
late stages of sympatric evolution 421
late anaphase in meiosis 299
lateral binding of polypeptide filaments 98
lateral DNA loops 452
latitude and frog populations 397, 399-400
Laws of Mendelian Inheritance 23-31
leading chain of replicating DNA 267-70
leaf insects 418
Lederberg, J. 56
left-handed coiling of Limnea pereger shells 470
left-handed helix spiral 98
left-handed polypeptide twist 100
legal aspects of genetic engineering 230
leghorn chickens 198
legume nodulation 234
lemurs 512
leopards 418
leptotene of meiosis 296
lesser black-backed gulls 402
lethal alleles 232
lethal genetic changes 312

leucine 57-66, 70, 78, 88, 94, 340, 536
leucine gene position by genetic mapping 71, 73-4
leukemia 491, 494-5, 521
leukoplasts 209, 566
Leydig cells 547
libraries for genes 231
lice 381
life – what makes it 91
life cycle of D. melanogaster 173-4
life cycles 157, 316
life span of a species 438
ligase 240-1
ligate 267
light 315, 480
light chain antibody polypeptide 466-9
light microscope 265
light super-allele in Biston betularia 157
Limnea pereger 'mother effect' genetic inheritance 469-71
Limnea pereger – shell coiling directions 470
linkage of genes 154-5, 158, 180-5, 248, 404, 532, 550
linked genes – cotransformation frequencies 556
Linnaean classification 372
Linnaean Society of London 413
lipid/protein membranes 132
lipid/protein ratios in cell membranes 483
little grebe – rejunction 422
liver cells 450-1
location of a gene in a polyploid 145
location of a gene in homologous chromosomes 144
location of a human oncogene 502-6
location of cancer cells 255
location of genes 143
location of transforming cancerous DNA 502
location of tumours 255
loci of human sex-determining genes 533
locus of a gene 200, 316
long wing in D. melanogaster 179-81, 183-4
loops in DNA that allow viable inversions 323-4
loosening of plant cell walls in response to auxin (IAA) 447
loosening the specificity of symbiosis by genetic engineering 234
loss of chromosomes after cell fusions 251
loss of T4 cell membranes 260
lowest free energy 435
lowland gorillas 512
lumen of the endoplasmic reticulum 128-9
lung cancer 487, 498
lymph 486-8
lymph system 464
lymphocytes 262, 495
lymphomas 494-5

lysine 87, 96, 455, 512, 536
lysis of bacteria 149, 252, 490, 504
lysosomal enzyme 482
lysosomes 88, 566
lysozyme gene 86
lysozyme polypeptide 87

M

M blood group 512
macaque monkeys 512
Macleod, C. 38
main populations 396
maintenance of chromosome number 277
maize breeding barriers 380
malaria 386, 497
malaria resistance 532, 549
malate dehydrogenase 511
Malay archipelago 33
male determination 206, 538-52
male Drosophila spp. – absence of crossings-over in 173, 310
male gamete production in plants and animals 279
male gametes 175, 177, 223, 279, 311
male hormone production 205
male mutation rate 316
male sterility in flowers 222
males 186-8
malignant growths 486-7
mammal genetic engineering 253
mammalian antibodies 153
mammalian gene – photo 164
mammalian haemoglobin 366-7
mammals 234
man 315-6, 455, 508, 514, chapter 16
Maniola jurtina 404-6, 409
mannose 128-9
map (genetic) distances 69, 74-5, 251, 254
mapping eukaryotic genes 246, 555
mapping gene clusters 78
mapping human genes 246, 555
mapping linked genes on autosomes 182-5
mapping mitochondrial genes 212
mapping prokaryotic genes 56-80
Marker, K.A. 88
marker genes 244
markers – genetic 255
marrow killing by anti-cancer drugs – reduction of 556-7
marsh fritillary 409
Massachussetts Institute of Technology 502
master genes in the homeobox 455-8
matching nucleotide chains 45-9, 267-71
mate selection 381, 386, 415
mathematical analysis of quantified data Appendix 2, 571-606

mathematical probability of amino-acid
   substitutions in polypeptides 346
mating calls 378, 387, 401
mating of bacterial strains 57
Matthaei, H. 80
mature mouse antibody gene sets 466-9
maturity 377
McCarty, M. 38
McTurk, P.A. 164
meadow brown butterflies 404
meadow plants 419-20
mean mean difference 602
mean values 373, 589
mechanical reproductive barriers 378
mechanism of recombination 58-61
Research Council Laboratory of Molecular
   Biology, Cambridge 80
medical uses of genetic engineering 231-4,
   255
medicinal antibody production 439
megaspores 280
meiosis:-
   chromosome behaviour during 265, 277-88,
      296-311
   crossings-over in 51, 76, 149, 156, 246
   evolutionary advantages of 390
   in cancer cells 488
   in Crepis capillaris 296-300
   in D. melanogaster 172
   in gene modification 452-4
   mutations in 312, 315, 317, 323-4, 354
   semi-conservative DNA replication in 40, 50,
      127, 138
   set of photos 296-300
   tautomeric shifts in 354
   type of cell division 168
meiosis/mitosis comparison 310
meiotic apparatus 310
meiotic division 277-88, 295-300, 354
meiotic mutations 308-9
meiotic recombination 395
Meladrium spp. 331
melanism (in moths) 412
membrane-bound organelles 561-2
membrane-bound viral particles 260
membrane destruction 259
membranes 19, 59, 91, 124-5, 128, 131-2, 137,
   168, 211, 233, 239, 245, 260-1, 265, 272,
   275-6, 284-5, 291, 321, 472, 483
Mendel, G. 22-31, 369
Mendel's hybrid experiments 23-31
Mendelian Law of Independent Segregation
   28-9
Mendelian Law of Random Assortment 30-1
Meselson, M. 108
mesosomes 568
messenger molecules 91

messenger RNA (mRNA):-
   and the AIDS virus (HIV) 258
   bacterial 160
   base sequences in 556
   codons 340-2
   codons table 89
   colinearity of amino-acids with codons in 123
   controlled synthesis by chromosomal DNA
      214
   exon/intron relationship 150
   in cell differentiation 446-50
   in genetic mapping 76, 81-3, 86-8
   in mice 467-8
   in polypeptide synthesis 102, 105-6, 112-3,
      116, 120-3, 125-8, 479
   in Xenopus laevis 132
   increased synthesis of, caused by auxin 476
   inducers and 449
   /mobile genetic elements interactions
      428-34
   modification of 153
   nucleotides in 116-7
   repressors and 157, 449
   /ribosome association 139
   splicing of 153
   synthesis 108
   /tRNA association 110
   wild-type determination 138
metabolism in cells 91
metal ions 483
metamorphosis 462, 473
metaphase in Euplotes 292
metaphase of mitotic cell division 272-3
metaphase I of meiosis 298
metaphase II of meiosis 299
metastases 255, 286-7, 499
metastasis 490
methanol 233
methionine 57-66, 70-1, 73-4, 78, 88-9, 94,
   113-6, 115-6
methionine/AUG association for starting
   polypeptide chains 113
7 – methyl- guanine 448-9
methylcholanthrene 502
Methylophilus methylophilus 233
mice 34, 191, 252, 254, 316, 387, 450-1, 455,
   457, 464-9, 494, 502-3, 545-8, 555
microbodies 567
microfossils 92
microinjection of DNA into cells 251
microscopes 22, 265-6, 518
microscopy limitations 22, 265
microscopy using u/v radiation 518
microtubules 567
microvilli 262, 568
migrant arrival 372
migration of chromosomes 331, 521
migration of populations 396-7

migration of replicated chromosomes 271
migratory herd 396
Miller, O. 141
millipore filters 242
mimetic species 413
mimetic systems 413
mimicry 398, 412-7
mine tips and evolution 397-8, 420, 422
minimal medium (MM) 56-8, 62, 69, 74
Ministry of Health's Pathological Laboratory, England 33
mispairing of G and T related to ageing 483
mitochondria 89, 209, 211, 212, 218, 223, 225, 430, 480-1, 483, 514, 561, 566
mitochondrial extensions – kinetoplasts 209
mitochondrial genetic code 92
mitochondrial ribosomes 211-2
mitosis:-
  chromosome behaviour during 278-81
  colchicine effects on 245
  crossings-over during 149, 265, 310
  in aphids 391
  in cancer cells 488, 491-2, 497
  in cell differentiation 445, 452-4, 456
  in D. melanogaster 459-6
  in diploids 265-77
  in genetic engineering 248-9, 251, 441
  in humans 521-2
  in mice 467
  in Neurospora crassa 224
  in Nicotiana tabacum 194
  in plants 256-7
  in Sordaria brevicollis 301-2
  /meiosis comparison 310-1
  mutations during 315, 317, 325
  photos 289-94
  producing identical chromosome sets 206
  semi-conservative replication during 40, 49-51, 127, 138
  tautomeric shift during 350, 352-3, 355
mitotic apparatus 272-5, 310
mitotic crossings-over 265, 310
mitotic dance 271-2
mitotic mutations in Nicotiana tabacum 194
mitotic mutations in Sordaria brevicollis 308-9
mixed marriages 373
MM3 gene in Xenopus laevis 456
MM10 gene in mice 457
MM10 homeodomain 457
MMG1 543
mobile:-
  DNA 78, 151-3, 218
  extra-chromosomal DNA 210, 219
  genetic elements 127, 151, 218, 261, 313, 335, 375-6, 379, 425-42, 509
  groups of individuals 396
  molecules 104

mobility of cancer cells 487
mobility of DNA and RNA – a comparison 104
mocker swallowtail butterflies 414
model species in mimicry 412-5, 417
modern bread wheats 331-2
modern understanding of genetic transformation in bacteria 39
modification of allele expression 116
modification of the 'one gene – one enzyme' theory 80
modifications of mRNA 153, 468
modifications to multiple genes 163
modifications to successful genes 314
modifications to super-gene expression 156
modifying dominant third locus allele 203
modifying effects of other genes on a super gene 156
modifying genes 410
molecular mass 93, 111
molecular stability of different types of RNA 104
mongolism 511, 519-21
monkey permissive cells 252
monkeys 252, 344, 377, 496, 508-15
monocotyledons 209, 258
monocytes 497
Monod, J. 108
monohybrid back-cross 178-9
monohybrid crosses 25-6, 176-7
mononuclear white blood cells 495
monosomy 330
Moore, J. 399
morabine grasshoppers 382
moral aspects of genetic engineering 230
morph 381
morphological differences 387
morphological change caused by genetic drift 382
morphology 372, 376, 386, 401, 406-7, 421, 514
mortality 205
mortality of cells 353
mos gene 493
mosquitoes 386
mother cells 169, 279, 354, 438
mother effect 206, 210, 222, 458-60, 469-71
moths 412, 418
mountain gorillas 513
mouse:-
  antibodies 254
  antibody genes 465
  brain cells 450
  cancer cells 254
  coat colour 191-2, 545
  fibroblast cells 502
  /human cell fusions 251, 555
  /human recombination 502-3

immune system antibodies 153
  mammary virus 494
  species 387
  spleen 254
  Sry gene 546
  transformation from female to male 545
mouth cancers 498
movement of chromosomes at anaphase 275, 298-9
mRNA *(see messenger RNA)*
mt^+ and mt^- 226-7
mules 378
Müllerian mimicry 413-5
multicellular species 316
multifactorial variation 404
multiple:-
  copy genes 312, 482
  effects of alleles 156
  effects of genes 315
  effects of genes altered by genetic engineering 243
  effects of mutations 317
  effects of mutations in a super-gene 156
  effects of super-genes 154
  finger polypeptides 535
  mutations 392
Mus domesticus 387
Mus musculus 387
muscle 97, 483, 486
Musée d'Histoire Naturelle, Paris 31
mutable sites 48
mutagenic properties of the human platelet repair gene 496-7
mutant:-
  allele 375
  allele – natural selection of 548
  bacterial strains 56-9
  cancer cells 255
  cells 249-50, 317
  chromosomes 321
  embryonic development 463
  enzyme genes 231, 244
  genes 86, 211, 395
  Sordaria brevicollis 302, 308-9
  strains 250
  strains of bacteria 56-9
  super-genes 156
mutations:-
  312-368 (Chapters 10 and 11)
  absence of 194
  and genetic drift 372
  and the breeding system 316-7
  and the rate of ageing 481-2
  base change 82, 84-5, 87
  cellular requirements for viable 390, 392
  creating differences between individuals 154
  creating new alleles 156-7

  damaging 371, 375
  during gamete formation 354
  during the scrambling of chromosomes 287
  identified by amino-acids added 52
  in aphids 391
  in Batesian mimics 414
  in creating cancerous cells 492, 498, 500-2
  in E. coli 57
  in gamete mother cells 353
  in gametes 312
  in Kappa particles 215
  in karyotype visible changes 382
  in multiple copy genes 162, 312
  in Nicotiana tabacum 194
  in sex determination 533
  in Sordaria brevicollis 368-9
  in transformation experiments 34
  large 370
  mapping the positions of 79
  mitochondrial 212
  neutral 371, 375
  point 50-55, 82-3, 87, 315, 318, 340-1, 343-5, 348-55, 357, 370, 437, 488
  rates 315-6, 319
  repair 318
  single base (point) (see point)
  storage of genes to prevent 237
  two base 84, 342
  three base 84-5, 342
  viable 370-1
  viability – general requirements for 320
  viral 430
myc gene 493
myoglobin 163, 364
myoglobin evolution 366-7
myosin 98

N

N-acetylglucosamine 128
N blood groups 512
n chromosome number (haploid) 144, 146, 168, 277-8, 286-7, 299, 310
2n chromosome number (diploid) 143-4, 147-8, 167-70, 332-4
3n chromosome number (triploid) 145, 331
4n chromosome number (tetraploid) 274, 331
5n chromosome number 331
8n chromosome number (octoploid) 334
NAD 79
$NADH_2$ 79
nail patella syndrome/blood group linkage 550
nasal cavity cancers 496
National Institute of Health, Bethesda, U.S.A. 80
natural gas 233

Natural History Museum, Kensington, London 514
natural selection:-
  as collective term for environmental factors 312
  by the environment 319
  Darwin's understanding of 32
  causing variation among offspring 287-8
  determining successful characters 314
  following gene creation 342-3, 348, 356
  in development of mitochondria 212
  in evolutionary change 335
  in genetic evolution of races and species 371, 373, 379, 393-7, 402-7, 414, 417, 423
  interactions with mutation, meiosis and genetic drift 154
  modifying X and Y chromosomes in humans 533
  of amino-acids in polypeptides 93, 342-3, 348
  of bases in nucleotide chains 47
  of mutant phenotypes 318-9
  of human globin genes 367
  of mobile genetic elements 347, 442
  stabilising allele frequencies 156
naturally occurring amino-acids 89, 93
naturally occurring mutations 50-5, 312-331, 334-5, Chapter 11
negative feedback – failure of 474
negative Rhesus allele combinations 551
negatively charged amino-acids 96
nematodes 386
nerve impulses across synapses 477
Neurospora crassa 213, 223, 229
Neurospora spp. 56
neutral mutations 371, 375,
neutrophils 497
new:-
  alleles 156, 356
  DNA 322
  gene creation 40
  genes 40, 357
  invariable genes 371
  mutations 315, 373, 392-4
  mutations – cellular requirements for establishing 390, 393-5
  predominant animo-acid 346
  races 371
  recombinants 407
  set of chromosomes 392
  species 384-5, 396
  species evolution 383, 396
  viable alleles 375, 396
  viable genes 322
  wild-type allele 356
new world monkey 514
New York University School of Medicine 80
NEWAT protein data computer 497
Newton J. 86

-NH_2 group 128
niche – ecological 381
Nicotiana tabacum 194
Nirenberg, M. 80
nitrogen fixation 234
nitrogenous bases 41
N-linked oligosaccharides 128
nodulation in legumes 234
non-
  alignment of centromeres 327-8
  allelic genes 374-6
  'brights' 215
  cichlid fish families 388-9
  complex DNA 160
  dividing cell chromosomes 266
  dividing cells 498, 561-9
  flush chromosomes at mitosis 310
  functional exon regions 316
  functional regions of polypeptides 374-5
  glycosylated polypeptides 128-9
  histones 161, 450, 543
  homologous chromosomes 329-30
  mendelian results of reciprocal crosses 210
  mobile plasmid genes 222
  mobile prokaryotic gene 222
  permissive host cells 252
  polar amino-acids 93-4
  specific recognition of mRNA by ribosomes 139
  suppressor allele 201
  viral DNA 216
nonsense codons 313
normal pathway for DNA synthesis 249
normal embryonic development 463
North American example of a cline in Rana pipiens 399-402
not very soluble amino-acids 95
Notophthalmos viridescens 141
N-ras gene 493
nuclear:-
  fusions 226, 251, 256, 302
  genes 225
  inheritance 226
  membranes 131, 137, 265, 271-2, 276, 284, 477, 564
  mitotic divisions 459-60
  pores 125, 131, 139
  proteins 543
  substitutions 449
nuclei 522
nuclei – haploid 301
nuclei of embryo cells 252
nuclei produced in fertilised D. melanogaster eggs 459-60
nucleolar fusions 132
nucleolar gene replacement 138
nucleoli 453
nucleoli at cell division 139, 276

656

nucleoli – fibrils 132
nucleoli – fusions 132
nucleoli – human 142
nucleoli – number per nucleus 132
nucleoli – number related to activity level 139
nucleolus 124-5, 131-42, 271, 276, 563
nucleolus – activity and differentiation 138
nucleolus – DNA 137-8
nucleolus – gene replacement 138
nucleolus – organiser 125, 133, 135, 137, 142
nucleolus – ultrastructure 133
nucleoplasm 133, 135, 518, 563
nucleoprotein 133
nucleosomes 161-3
nucleotide chains 43-4, 47-8, 106, 266, 268, 270
nucleotide units 42-3, 108
nucleotides 106, 110-1, 116-7, 123, 238
nucleotides pairing 45-6, 87, 107, 267-70
nucleus:-
  containing chromosomes that carry inheritance factors 22
  containing nucleoli 131
  in C. reinhardii 227
  in polypeptide synthesis 124-5, 128
  in ribosome synthesis 131, 136-7
  in S. brevicollis 301-2
  in Xenopus laevis 139
  in zygotes 168
  injected with DNA 251-2
  organelle 563
null hypothesis 575-6, 581-2, 597-8, 600
number of base substitutions 346
number of base pairs per human genome 502, 509
number of chromosomes in primates 509-11
number of chromosomes – changes to the 314, 330-1, 511, 519, 521-2
number of cytoplasmic mitochondria 213
number of new species 385
number of nucleoli per nucleus 132
number of shell bands in Cepaea nemoralis 200-4, 410
numbers in populations – their influence on evolution 394-6
nurse cell chemicals 459-60, 463
nurse cells 459-60

O

O blood groups 197-8, 549-50
oak trees and evolution 387
objective decisions 572
obligate apomicts 331
obligate parasites 430
occupational cancers 498
Ochea, S. 80
octoploid / diploid cross 334

octoploidy 334, 377
Oenothera hookerii 225
Oenothera muricata 225
oesophagus cancer 550
oestradiol 473
oestrogen 474
oestrous cycle - absence of 522
offspring inviability 378
offspring sterility 378
–OH group 95
OH^- 480
Okada, Y. 86
old world monkey blood groups 514
old world primate chromosome numbers 509-10
oligosaccharides in glycopolypeptides 128-9
oncogenes 487-9, 491-3, 496-7, 499
ocongenic viral hybrid DNA 505
one centromere per chromosome 320
one gene – one enzyme 78, 112
one gene – one polypeptide 56, 80, 112
Onogracaeae spp. 326
oöcyte chemicals for differentiation 460
oöcyte gene amplification 452
oöcyte nucleus 459-60, 462
oöcytes 138, 252, 280, 452, 459
open reading frame of a gene 545
operator 159-60, 162, 413
orang-utang / human comparisons 541
order of bases 47
organelle replication 275
organelles 19, 209, 215
organelles of non-dividing cells Appendix 1
orientation in space of amino-acid chemical groups 93
orientation of cells in animal embryos 463
orientation of chromosomes at cell division 285-6, 303-4
orientation of chromosomes – random at meiosis 285-8
orientation of spore colour alleles in Sordaria 303-8
origins of chromosome complements in wheats 331-3
origins of genetic isolation 379
origins of viral oncogenes 491-2
osmosis 472
outbreeding alternative 392
outbreeding species 317, 331, 333-4, 390
outer cell membranes 259-61, 272, 567
ovalbumin 153, 474
ovarian cancer 487
ovaries 278
ovary development 546
overlapping genes 80, 149, 243, 375
overlapping ranges 388
ovotestes 542
ovules 278

oxidase 480
oxidation 79, 91, 345
oxygen:-
　as an ageing factor 481
ozone:-
　atmospheric, effect on DNA 319-20
　layer 319

P

^{32}P 39
^{35}P 39
(P) – cleavage of 79
P site of polypeptide synthesis 117-20
pachytene 297
Page, D.C. 534
pairing nucleotide chains 266, 270
pairing of homologous chromosomes at meiosis 333
pairing of initiation codon with tRNA anticodon 115
pairing of mRNA and DNA 76
pairing of RNA or DNA probes with the unblocked cellular DNA strand 541
pairs of homologous chromosomes 144
PAL 480
pale spores in Sordaria brevicollis 301-9
pancreatic acinar cell 140
Papilio dardanus 414-5
Papilio phorcas 414
papilloma viruses 496
papoviruses 494
paracentric inversions in chromosomes 323-5
paramecia 215
Paramecium aurelia 214-5
parapatric distribution 380
parapatric isolation 379-85
parapatric processes of speciation 383
parapatric ranges 385
parapatric species 380, 387
parapatry 380-1, 383-4, 422
parasites 380-1, 430
parthenogenesis 331, 391
particles – inheritance 23-31, 166
particulate nature of inheritance 22-3, 27
Passifloraceae spp. 415
passion flower vines 415
pathogenic properties of viruses 247
pathogenic properties – removal of 247
pathogens 232
PBR 322 plasmid 262
PCR plasmid 235
'pea' comb shape in poultry 190
pea plants (sweet peas) 23
Pediculus capitis 381
Pediculus humanus 381
pentose sugars 41, 103
peppered moth 154, 157, 412

peptide bonds 92, 97, 119
peptides 117
pericentric inversions in chromosomes 327-8
permissive host cells 252
peroxidase enzymes 481
persistent damaging alleles 313
pesticides and their effects on primroses 412
Pfr 480
pH 472
phage particle hybrids 504-5
phage particles 439
phage particles that can induce cancer 503-6
phages 20, 86
phages that can insert desirable genes into bacterial genomes 440
phagocytes 259-60, 483
phagocytosis 245, 261
pharyngitis 439
pharynx cancers 496
phenotype:-
　affected by changes in allele frequencies 397
　changes 398
　changes caused by amino-acid substitutions 343
　changes caused by genetic engineering 230
　changes caused by transformation 239
　characters 154
　definition 167
　determined by a dominant allele 178
　divergence 319, 335
　in Cepaea nemoralis 202, 204, 411
　in D. melanogaster 167, 181, 183-8
　in E. coli 219
　in Mendelian experiments 25
　in prokaryotes 216, 219
　in sex determination 205-7, 521, 532-48
　in sweet peas 192-3
　mutations to create new 314-5, 319, 393
　ratios 26
　stabilisation 319, 407-8
　variation 92
　written description of 171
phenotypic expression of genes 253, 256, 312, 317
phenotypic variation 318
phenotypic viability 314
phenylalanine 34, 480, 536
phenylalanine ammonia lyase (PAL) 479
phenylketonurea 528
Philadelphia chromosome leukemia 521
phosphate 42-6, 229
phosphate residues 41
phosphoglucomutase 511
phosphohexose isomerase 511
phosphoserine phosphatase 79
phosphoserine transaminase 79
photoperiod 380
photoperiodic modification of gene expression 479

photosynthesis 172, 212-3, 319
Phthiris pubis 381
phyla 364
physical environment 379
physiology 372
phytochrome 480
pig insulin 338
pigment synthesis 192-4
pigmentation of chicken feathers 198-200
pigments in sweet pea flowers 192-3
pigs 338, 344
pin type of primrose 411
pink expression of genes 194-7, 200-202
pinocytotic vesicles 568
Pipra aurelia 382
Pipra fasciicauda 382
Pipra filicauda 382
Pisum sativum 26
plague 438
plant:-
 cell fusions 257
 cell genetic engineering 256
 cells 256-8, 486, 563-69
 disease resistance 257
 female gamete formation 280
 hormones 257
 male gamete formation 279
 tumours 486
 viruses 33
plants 19, 253, 257, 278, 391, 562-9
plasma 168
plasma membranes 258
plasmadesmata 569
plasmalemma 477
plasmid:-
 acceptance of 'foreign' DNA 234
 alleles 314
 circular DNA 209
 DNA 234, 241, 257
 equivalents from viruses 252
 gene insertion 230
 gene positions 314
 genes 143, 217, 233
 in cells 564
 in E. coli 219-20
 insertion of genes into 229
 mobile genetic elements 426, 430
 mutation 334
 R factor 221
 recombination of antibiotic genes 232
 replication 217
 sticky ends base pairing 240
 synthesis of single stranded DNA 239
 uses in genetic engineering 234-243
plastid DNA 143
plastid inheritance in eukaryotes 214
plastids 209, 223
platelet derived growth factor PDGF 497

platelet gene in humans 496-7
platelets in blood 497, 498
ploidy change 370
PMB 9 plasmid in E. coli 235
point mutation to cause cancer in humans 506
point mutations 50-55, 82-3, 87, 315, 318,
 340-1, 343-5, 348-55, 357, 370, 437, 488
pointer to extra-chromosomal inheritance
 210
pointer to mitochondrial inheritance 213
poison 221, 414
poky Neurospora crassa 224
polar amino-acids with overall electrostatic
 charge 95
polar bodies from female meiosis 280, 285,
 287, 354
polar nucleus 280
polarised light 93
pole cells that form gonads 461, 463
pole plasm 454
poles in cells 271-2, 274
pollen 23, 195, 277, 311, 378
pollen precursor cells 256
pollen tube nucleus 279
pollen nucleus 280
pollination 387, 412
pollution 498
polyethylene glycol and genetic engineering
 247-8
polymerase (DNA) 238, 240, 267
polymorphic snails 410
polymorphism 154, 157, 381, 409, 414, 549
polynucleotide ligate 266-7
Polyoma SV-40 494
Polyoma virus 235
polypeptide:-
 α chain of haemoglobin 100
 amino-acid sequence/fossil comparisons
  339
 amino-acid sequences 131
 analysis in cell organelles 514, 561
 β chain of haemoglobin 100
 binding to DNA 158-9
 chain elongation 116-9, 121, 123
 chains 337
 coats around viruses 425
 comparisons for systematics 337-68
 differences 387
 fibres 272
 for immunity against HIV 261
 functional properties after mutations 85
 functional regions 98, 343
 hydrophobic groups 94, 100, 536
 invariable sites 343
 loss of function 84
 left-handed twist 100
 non-functional region 98
 one from each gene 148

production rate 158
release factor 120
shape 98
structure 96-101, 343
structure – primary 99
    – quaternary 100-1
    – secondary 100
    – tertiary 100
sugar residues 128, 337, 555
sulphur bridges 337-8, 343, 346
synthesis 31, 48, 83-4, 102-129, 131-2, 137, 139
synthesis – central dogma 126
synthesis – controls 157-60, 449
synthesis direct from DNA 127
synthesis in eukaryotes 124-5
synthesis – less usual pathways 127
synthesis – summary for prokaryotes 122-3
synthesis – usual pathway 127
that binds to DNA 546
polypeptides:–
  analysis of amino-acid sequences in 423
  as mobile 'genetic' elements 436
  binding to DNA 158-9, 536
  cellular sets of 91
  conferring character and life 91-3
  critical folding for functional properties 343, 345-6
  distinguished by electrophoresis 243
  effects of water on their shape 98
  encoded by chloroplast DNA 213
  filamentous 99
  histones 161
  in defining a gene 148
  in E. coli 122-3, 126-8
  in gene control 157
  in genetic engineering 244-5, 252, 258-60
  in genetic mapping 76, 79, 81
  in human genetic engineering 556
  in ribosomes 111, 131, 135-7, 139
  infectious 428-9
  multiple finger 535
  non-histones 161
  preference for the descriptive word 'polypeptides' instead of 'proteins' 92
  structure of haemoglobin A 364
  substitutions of amino-acids in 339-40, 345
  sulphur bridges in 100
  synthesis of 50, 84-5, 88, 102, 119, 122-3, 153, 155
  synthesis by mobile genes 151, 153, 155
  with sulphur bridges 100
  zinc-finger 160, 452, 534-5
polyploids 143, 145, 148
polyploidy 143, 167, 331-4
polysomy 521-2
population 354
population genetic studies 369

population genetics 33, 373, 404, 422
population isolation 407
population numbers 418
population size 372-3, 394
population size and mimicry 417
populations – large 397, 399
populations – rate of incorporation of a viable new mutation into 394
populations – small 397, 407
position change of a centromere 327
positions of genes on chromosomes 148
positive Rhesus allele combinations 551
positively charged amino-acids 96
posterior parts of insect embryos 456
Postgraduate Medical School, London 56
potential uses of human genetic engineering 231-2, 253-5, 260-1, 556-7
poultry comb shape genetic crosses 189-91
practical uses of genetic engineering 231-4, 254-61
precise order of amino-acids in polypeptides 97
precursor molecules 249-50
predators 413-4, 417-8
predicted phenotype ratios 26
predominant allele 354
prevention of cancer 498-9
prevention of chromosome pairing at meiosis 333
prevention of cross-breeding 335
prevention of fertilisation 377
prevention of interbreeding 335
prey 417
primary barriers to reproduction 377-8
primary cell cultures 245, 248
primary oöcytes 280
primary polypeptide structure 99
primary reproductive barriers 377-8
primary screening in prokaryotic genetic engineering 242
primary spermatocytes 279
primate blood groups 512
primate chromosomes 509-11
primate evolution 508, 514-5, 541
primate genera hybrid fertility 377, 510
primate genetics 508-12
primate haemoglobin 366-7
primates 508-515
primitive species 91
primitive wheat 332
primordial germ cells 280
primroses 154, 411-2
Primula vulgaris 154, 411-12
prions 426, 428
probability – mathematical 581-3, 594-5, 605, 608-9
probability – of amino-acid substitutions 347
probes for genetic mapping 541, 543, 556

problems for eukaryotic genetic engineering 245-6
problems of cereal breeding 331-3
problems of having uneven numbers of chromosome sets 331
processes of population evolution Chapter 12
Prochloron 19
product feedback 449
product feedback controls 449
product of genetic engineering 237
production of gametes 278-9, 285-8, 299-300
production of medicinal antibodies 254-5
progesterone 474
prokaryotes 38, 124, 128, 217, 222
prokaryotic:-
  cells 19-20, 561, 563-9
  cells containing extra-chromosomal DNA 210
  cells containing plasmids 216
  cells containing ribosomes 120
  cells – insertion of genes into 236
  cells, modified by mobile RNA and DNA 425-7, 430
  extra-chromosomal DNA inheritance 209, 218
  genes 231, 241
  plasmid genes 219, 222
  polypeptide synthesis 113-23, 127
  ribosomes 111
  species 429
proline 94
promoters 160, 438, 440, 484
pronuclei 252
proofs:-
  chromosomal DNA mutations cause cancer 502
  DNA is the material of inheritance 33-40
  one gene produces one enzyme 78-80
  viruses can cause cancer 501
properties of DNA 40
prophase of meiosis 296
prophase of mitosis 271
proportion of changes in phenotypes 394
proportion of genes with alternative or several alleles 288
proportion of heterozygotes 517
proportion of homozygous recessives 517
proportion of wild-type homozygotes 517
protein-coat loss in viruses 429
proteins 33, 38, 92, 132-3, 475
proto-gonad cell transformation 546
proto-oncogenes 490-3
protoplasm 107, 216, 564
protozoa 209, 215, 222, 386
proximal sex-chromosome parts 539-41
pseudoautosomal boundary 545
pseudoautosomal region of sex chromosomes 339-42

$p28^{SIS}$ oncongene 497
puffs in D. melanogaster salivary gland chromosomes 446-7, 449, 473, 475
punctuation in the reading of the genetic code 112, 116
punctuation of codons 116
pupae 174
pure bred dominants 179
pure breeds 229
purification of prokaryotic genes 237
purified DNA 19, 38, 237
pY53.3 exon in the male determining gene 542-6
pygmies 373
Pysella mali 421

Q

qualitative control of gene expression 449
quantified experimental results 574-9
quantitative control of gene expression 449
quantity of DNA 509
quantity of DNA in primate chromosomes 511
quaternary polypeptide structure and amino-acid substitutions 348
quaternary structures of proteins 92, 100
Quercus douglasii 387
Quercus dumosa 387
Quercus durbinella 387

R

R (rough, avirulent) bacteria 34-7
R factor transfer genes 221
R- group in amino-acids 93
R plasmid 216
R (resistance) factor 216
R strain of Diplococcus pneumoniae 34-7
rIIA and rIIB gene in T4 virus 81
rabbit globin 448
rabbits 543
races 331, 371-4, 377-8, 385, 387, 397-8, 400-2, 404, 423
radiation 320
radiation and cancer 498
radioactive cDNA 242
radioactive elements 255
radioactive elements in DNA 39
radioactive mRNA 76
radioactive probes 555-6
radioactivity 320
Rana pipiens 383, 387, 399-402
Rana spp. – a large population study 400-2
Random Assortment – Mendelian Law of 30
random orientation of chromosomes at meiosis 287

range (of a species or population) 334, 379-80, 384-5, 388, 396
rapid polypeptide synthesis in oöcytes 138
rare recessive alleles 313
rare viable mutations 314
rate of:-
　ageing 484
　amino-acid substitutions 356
　base pair substitutions in primates 540-1
　biochemical reactions 91
　mutation 315-6
　polypeptide synthesis 449
　reverse mutations 318
　ribosome production 138
　transfer of donor bacterium's chromosome 64
ratio of full mutation cells 353-4
ratio of males to females 205
ratio of half-mutation cells 353-4
ratios of DNA bases 40
rats 252
Rattus rattus 327
reading frame 543
readjustment of range 385
rearrangements of chromosomes 314
receptor sites 259
receptors for HIV 258
recessive allele spreading through a population 315
recessive alleles 154, 167, 171, 229, 244, 251, 313, 482
recessive homozygotes in humans 515, 527-8
recessive super-alleles 154
recessiveness 23-8, 148, 167, 171, 188, 193, 199, 202, 251, 317, 409
recipients of mobile genetic material 35-9, 61-5, 67-9, 75, 220, 246
reciprocal genetic crosses 177, 210, 222-3
reciprocal translocations 325-6
recombinant bacteria 71
recombination 56, 58-76, 149, 173, 185, 221, 232, 257, 309, 370-1, 390, 414, 505
recombination frequency 67, 73-5, 502
recombination mechanism 58-61
recombination best sets of alleles 370
recombined gene sets 390
red blood cells 447
red eye in D. melanogaster 171, 185-8
red snapdragons 196
redistribution of range 380
reduction (and oxidation) 91, 344
reduction of chromosome number in a species 327, 330
reduction of chromosome number – diploid to haploid 278-80, 286-7, 299-300
reduction of competition 380
reduction of selection 407
reduction of symbiotic specificity 234
reduction of tRNA specificity 478-9

redundant loops in DNA 150, 163, 236, 312, 322
refractile body 215
regular position for breaking a donor bacterium's chromosome 63
regulation of embryonic development 464
regulation of genes 245, 535
regulatory effects 434
regulatory genes 153, 245
rejoining of chromosomes after meiotic crossings-over 283
rejunction 374, 377, 379-80, 383-5, 421-2
rejunction – absence of 377
release factor in polypeptide synthesis 120
removal of bases from DNA 82-3
removal of breeding barriers 335
removal of isolating barriers 335
removal of sterility in wheats 333
removal of viral pathogenic properties 252
repair of damaged tissues by platelets 496-7
repair of DNA 483
repair of mutations 318
repeating genes 134, 153, 162, 316, 509, 559
repeating genes and cancer 489, 559
repetition of alanine codes in human DNA 503
repetitive DNA 163
replacement of defective alleles by genetic engineering 230
replacement of nucleolar genes 138
replicated chromosomes 271-2, 274-5
replicates 284, 303
replicating chromosomes 266-71, 273, 304
replicating chromosomes and cancer 488
replicating DNA – photo 289
replication of DNA 36, 39-40, 215, 219, 252, 256, 267, 271, 277, 310
repressed gene 158-9
repression of genes 454, 458, 472, 480, 493
repressor gene 153
repressors 138, 157-60, 447, 452, 455, 473, 482
reproductive barriers 377-80, 383
reproductive isolation 369, 385
reproductive systems 316-7
requirements for mutation viability 320-2
residues of sugars in polypeptide 128-9, 337, 555
resistance to disease 333
resistance to streptomycin 226
respiration 211, 223-4, 339
respiratory defects in Neuropora crassa 223-4
response to damage 488
'resting' phase of mitosis 266
restriction enzyme analysis 212
restriction enzymes 212, 236, 240-1, 252
restriction sites 236-41, 243
retroviral DNA 437

retroviruses 127, 258, 427, 433-6, 440-2, 492, 496
reverse cline 405
reverse mutation rates 318
reverse mutations 341, 347-8, 357
reverse transcriptidase 229, 231, 258, 433, 489-90, 499
reverse transcriptidase (AMV) 434-5
reverse transcription 126-7, 489-90
Rhesus factor 512, 549-554
Rhesus factor allele combinations 552
Rhesus factor antibodies 552-3
Rhesus factor – table 554
Rhesus monkeys 514
ribonuclease 364
ribonucleic acid (RNA) structure 33, 102-4
ribonucleoprotein 150
ribose 103, 110
ribose nucleotides 105, 107
ribosomal:-
  large sub-unit 111
  non-specific recognition of mRNA 139
  polypeptides 131, 136-7, 211, 213
  RNA (rRNA) 109, 111, 474, 476
  30 S fragment 109, 111, 113-4, 116-8, 120-2, 135-7
  40 S fragment 135-7
  50 S fragment 109, 111, 116-8, 120-2
  60 S fragment 137
  80 S fragment 136
  5 S rRNA 136-7
  5·8 S rRNA 109, 135-7
  18 S rRNA 109, 135-7
  28 S rRNA 109, 135-7
ribosome:-
  attachment to mRNA 114-23
  fragments 135-6
  fragments moving to the endoplasmic reticulum 137
  functions 111, 131
  movement along mRNA 117-8, 120-3
  /mRNA complex 109
  polypeptides 212, 214
  production rate 138
  recognition of mRNA 139
  recycling 121
  transport 137
ribosomes:-
  bacterial 212
  eukaryotic 135
  formed in response to gibberellic acid 478
  in chloroplasts 213-4
  in E. coli 109, 111
  in eukaryotic cells 140
  in human cells 142
  in mitochondria 211
  in polypeptide synthesis 114-25
  in translation of mRNA into a polypeptide 446, 449
  non-specific recognition of mRNA molecules 139
  transport of 139
rift valleys and evolution 388, 390
right-handed amino-acid spiral in polypeptides 98, 100
right-handed coiling of Limnea pereger shells 470
rigorous natural selection 407
ring species 402-3
rituals of courtship 378
river blindness 386
RNA 20, 33, 40, 102, 105, 111, 124-5, 127, 132
RNA:-
  as a material of inheritance 38-40, 425, 427-30, 432-6, 442
  base pairing 104
  base sequences 337, 435, 442
  cancer viruses 432-4, 489-90, 494, 499
  chemical stability 104
  controlled DNA synthesis 127
  controlled RNA synthesis 127
  /DNA comparison 103
  in HIV 260
  genome fragments 426
  in mitochondria 211
  molecular mobility 104
  polymerase 107-8, 451, 473, 476-7, 482
  size 104
  structure 104
  synthesis 105, 126-7
  template 126
  viruses 430, 432-4, 489
Rockefeller Institute, New York 38
rose comb shape in poulty 189-90
rotation of DNA 267
rotation to accommodate inversions 323
rough endoplasmic reticulum 125, 140
Roux, W. 22
Royal Society of London 22
rRNA (ribosomal RNA) 102-5, 109, 111, 113, 122, 125, 132, 134-6, 139, 142, 430, 451
rRNA:-
  genes 453
  genes related to activity levels in cells 138
  genes related to the number of nucleoli per nucleus 139
  mitochondrial 211
  synthesis initiation 138
Rules of Mendelian Inheritance 27

S

S strain of Pleurococcus pneumoniae 34-7
S_1 nuclease 239

Saccharomyces cerevisiae mitochondrial
   genes 211-2
salivary gland chromosomal puffs in D.
   melanogaster 446-7
same species cell fusions 247, 250-1, 257
same species gene exchange 391
sample means 586-91, 594-606
San Diego University, California 339
sandflies 386
Sanger, F. 337, 339
sarcomas 494, 496
scatter diagram of amino-acid substitutions
   347
scavenger pathway for DNA synthesis 249-50
Schizosaccharomyces pombe 543-4
scientific method 571-3
Scilla sibirica 334
scorpions 326
Scp 1 plasmid 235
Scrace, G.T. 496
'scrambled' chromosomes 283-8, 310
'scrambling' of genes 281-3, 312
scrapie 428-9
screening after genetic engineering 242,
   244-5, 254, 257
sea squirt 19
seasonal barriers to reproduction 377
seasonal differences 386
second DNA strand 237
second meiotic division 277, 279-80, 285,
   301-2, 304
second polar body 280
secondary constrictions 139
secondary male characters 547
secondary oöcytes 280
secondary polypeptide structure 100
secondary reproductive barriers 377-8
secondary screening in prokaryotic genetic
   engineering 243
secondary spermatocytes 279
sedimentation characteristics 111, 135-6, 211,
   213
seed copper tolerance in Agrostis tenuis 419
seemingly invariable genes 375
segment differentiation in embryos 458
segment number in embryos 461
segmentation 463
segmentation genes in D. melanogaster 461-2
segregation (Mendelian) 33
segregation of chromosomes 392
selection 32, 343, 371, 393
selection:-
   against a hybrid 380, 398, 404
   and the Clean Air Act 412
   artificial 229
   by predation 412
   by the environment 369
   directional 406, 409
   disruptive 408

   of mates 386, 415
   stabilising 407-8
selective advantage 49, 314, 394
selective advantage of the heterozygote 532,
   548
selective breeding 21
selective disadvantage 49, 314, 346, 394,
   548-9
selective neutrality 314, 394
selective pressures 393
self-fertilisation 25, 412
self-pollination 26, 194
semi-conservative DNA replication 40, 127,
   267-70
semi-species 372, 378, 388
seminal vesicles 547
seminiferous tubules 279, 547
Sendai virus and genetic engineering 247
senescence 452, 479
sensitivity to drugs 211-2
sequence of bases 82-87
sequences of amino-acids 337, 357-64, 423
serine biosynthesis – genetic mapping of the
   necessary genes 79
serotonin 477
Sertoli cells 547
sets of chromosomes 276, 284-5, 286-7
sets of chromosomes – uneven numbers of
   331
several possible alleles 154, 315
sex chromosome:-
   base homology 540
   boundary region in man's ancestors 540
   genes that control autosomal genes 532, 548
   inheritance 205
sex:-
   chromosomes 181, 185, 205
   chromosomes – absence of 331
   determination 205-6
   determination in birds 206
   determination in humans 150, 160, 533-48
   determining base sequences 533
   determining loci 533-48
   determining polypeptides 533
   (F) factor in E. coli 216, 219-20
   linkage 173, 185-8, 210, 222
   of disease vectors 386
sexual breeding systems 312, 315-6
sexual maturity 377
sexual reproduction 224, 226
sexual reproduction and evolution 390
sexual reproduction – mutations 317
shallow clines 385
sharp segregation 172
shedding of chromosomes 248-50, 256-7, 555
shedding of DNA at cell division 247
shedding of DNA at cell fusion 502
shedding of plasmids 256

shell bands in Cepaea nemoralis 410
shell colours in Cepaea nemoralis 201, 409-10
Shih, C. 502-3
short arm of human Y chromosome 534
shortening of chromosomes at cell division 271
shuffling of genes 152, 426, 467
Shulman, L. 80
sibling species 372, 378, 385-6
siblings (brothers and sisters) 32
sickle-cell allele 532
sickle-cell anaemia 532, 549
side chains in polypeptides 99
side effects of drugs 255
side groups in polypeptides 100
significance (statistical) 575-6
Simian sarcoma virus 40, 252, 497
'similar' amino-acid substitutions in polypeptide sites 346
'similar'/'different' amino-acids 342-3, 346
simple goitre 474
simultaneous equations for deducing evolutionary trees 359-62
Simulum damnosum 386
single base (point) cancerous mutations 488
single base (point) mutations 82, 156, 340-1
'single' comb shape in poultry 190
single-copy genes 163, 317
single gene Mendelian inheritance 23-7, 166-71, 176-7
single-stranded:-
  cDNA synthesis 238
  DNA 39, 61, 63, 318, 434
  DNA radioactive probes 546-7
  DNA viruses 432
  retroviral RNA 434
  RNA 258
  RNA synthesis 122, 149
  RNA viruses types 1, 2 and 3 433
sinistral coiling of Limnea pereger shells 471-2
sis gene 493
sister replicates 273-4
sisters 170
site numbers of amino-acids in polypeptides 345
size of populations 394
skin cancers 374
slightly repetitive genes 163
slow transforming retroviruses (STRs) 491
slowing down the processes of ageing using drugs 484
small:-
  chromosome fragments 322
  chromosomes degenerating 327
  genetic changes 380
  population studies 385, 395-8, 404, 406-12

ribonuclear particles (snRNP) 150
  samples 592-606
  supernumary chromosome segments 334
  vacuoles 567-8
smoking and cancer 498
smooth muscle 497
snRNA 150, 436, 482, 500-1
snRNP 150
snails 154
snapdragons 197
solubility 346
solubility of amino-acids 93
somatic:-
  cell division 348
  cell transformation 254
  cells 40, 168, 254, 270, 277, 311
  cells – crossings-over 149, 265, 310
  cells – genetic engineering of 441
  cells – human base pairs 509
  mother cell mutations in gonads 353-4
  mutations 312, 444, 469
  recombination 469
Sordaria brevicollis 301-9
Sorex araneus 329
Southern English type of Maniola jurtina 405
spacer regions 163, 375, 509
spacer segment 465
spacers 134, 162, 468
Spartina alterniflora 334
Spartina stricta 334
Spartina townsendii 334
spatial control by homeobox mastergenes 455
spatial organisation of cells in embryos 458
spatial orientation of animal embryos cells 455, 458, 460
specialisation of cells 277
specialised cells 277
speciation 314, 374-5, 381, 383, 396, 423, 438
species:-
  amino-acid sequencing in cytochrome c 343, 346
  and natural selection 314, 318-9
  classification 372
  co-adaptive peaks of fitness 378
  definition 376-9
  development 91-2
  extinctions 383-5
  flocks 390
  'gene' concept 315, 374-5
  genome 370
  mutations 354
  new 371
  range – edge of 334
  rate of extinction 319
  stability 314-5
  time span 371

variation 314-5
viable mutations in 391-7
specific antibodies 467
specific antigens 467
specific character 91
specific locus – binding of a polypeptide to 534
spermatids 279
spermatocytes 279, 300
spermatogenesis 538, 545, 548
spermatogonia 279, 542, 544
spermatozoa 279
sperms 168, 174, 277, 311
spindle fibres 98, 272-5, 284, 321, 327, 333
spindle of dividing cells 271, 276, 292-3, 310, 321, 327-8
spindle – problems 327-8
spindle site for chromosome attachment 321
spleen cells/cancer cells fusions 254-5
splice site 544
splicer regions 545
splicing 127, 153, 426, 430, 468
spoil heaps and selection in the grass Agrostis tenuis 418-20
spontaneous abortion of human babies 524
spore colour alleles in Sordaria brevicollis 301-9
spread of an allele 372
spread of HIV 260
src gene 493
SRY-derived polypeptide 544, 546
SRY gene in humans 545
Sry gene in mice 545-6
SRY/Sry exon codon differences between mouse and man 547-8
SSV 40 viral genome 252
stabilising selection 407-8
stability – genetic 49, 319
stability of basic form 319
stability of DNA structure 40, 49-55
stability of genetic inheritance in wheat 333
stains 132, 265-6, 271, 314, 386, 518-9, 524
standard amino-acid site numbers 345
standard deviation 587-92
standard error 587, 596-7, 600
standard error of the difference 576-7, 587, 594, 597-605
start codon 121, 123, 537, 538, 545
start codon within a CpG island 537
starting point of transfer of a donor bacterium's chromosome 64
starting polypeptide synthesis 113
statistical significance 575-6, 578-9, 581-2, 584, 598-9, 605
steatopygia 374
steep clines 385
stem cell genetic engineering 444
stem colour – an example of 223

stepped clines 383, 385
sterile offspring 378
sterility 331
sterility in male flower parts 222
sterility in males 334, 521, 534, 547-8
sterility – removal of from wheats 333
steroid hormone cell receptor complexes 474-5
steroid hormones 138, 474
stick insects 418
'sticky ends' of cDNA 240-1, 427, 440
stigmata 23, 411
stillborn babies 553
stomach cancer 487
stop codons 544-5
streaming of embryo cells 461-2
streaming of blastoderm cells 461
Streisinger, G. 86
streptomycin resistance in Chlamydomonas reinhardii 225-6
streptomycin susceptibility in Chlamydomonas reinhardii 226-7
Stroobant, P. 496
structural alterations to chromosomes 314
structural polypeptides 91
structure of DNA 41-51
structures of DNA and RNA compared 103
Styella spp. 453
styles 411
subjective decisions 572
sub-populations 381
sub-species 372, 378, 385-8, 398, 402-4, 423
substitution of bases in an allele 340-2
substitutions of amino-acids in polypeptides 339-368
substitutions of bases in an allele to elucidate the genetic code 83-9
substitutions of wild-type alleles to replace damaging alleles (using genetic engineering) 556
success of a species 390
successful mutations 314-5
sugar residues in DNA 41-3, 103
sugar residues in polypeptides 128, 245, 337, 555
sugar residues in RNA 103
sugars 41, 103, 128, 245, 337, 555
sulphur in amino-acids and polypeptides 94-5, 100, 337, 343, 346
summary of prokaryotic polypeptide synthesis 122-3
super-allele frequency 381
super-alleles 143, 156-7, 166-7, 200, 202, 317, 370, 381, 409, 411-12, 423, 577
super-coiling of DNA 163
super-genes 143, 154-7, 166, 172, 200-1, 317, 319, 371, 409-12, 423
supernumary chromosome segments 334
super-oxide radical 481

super-species 378, 385, 387-8
super-super-genes 370-1, 376
suppressors for banding in snail shells 200-4
surprising, rare cross-fertilisation 377, 510
survival of the fittest 32
susceptibility to cancer caused by genetic inheritance 488
susceptibility to streptomycin in Chlamydomonas reinhardii 226-7
SV 40 virus 494
Svedberg units of sedimentation 111
swallowtail butterflies 414
sweet pea flower colour 192-4
sweet pea genetic cross table 193
sweet pea pigments 192-4
sweet peas 192-4
sweet peas apparently reverting to the wild-type 192-4
'switched on' gene combinations 472, 480
'switched on' genes 453
switching genes on 158
switching genes on and off 157-60, 449-52
symbiosis 212
symbol convention used to describe alleles 171, 180
symbol for the wild-type allele 175
symbols for dominant alleles 161, 171
symbols for genetic crosses 167, 169-71
symbols for recessive alleles 167, 171
sympatric evolution 418
sympatric evolution in Agrostis tenuis 418-21
sympatric isolation 319, 379-85
sympatric speciation – early stage 421
sympatric speciation – late stage 421
sympatry 381-4, 422
synapses 477
synergistic actions 478
synthesis 473, 482
synthesis:-
  of cellular chemicals 171
  of DNA and RNA – a comparison 105
  of extra genes in animal oöcytes 138
  of genes Chapter 8
  of serine 78-9
  of single-stranded cDNA 238
  of the second DNA strand 238-9, 267-70
  of tryptophan 101

T

T cells 259
T 4 cells 258-60, 496
T 4 DNA ligase 240-1
T lymphocytes 262
T 4 virus 80, 83, 86-8, 149
tables for genetic crosses – convention for use 175

Tachybaptus ruficollis 422
Tachybaptus rufolavatus 422
tadpoles 453
'tails' of chromosomes 273, 275
Taraxacum officinale 331
target cells for hormones 475
Tatum, E. 56
tautomeric point mutations 348-355
tautomeric shift 50-5, 126, 348-9, 353
tautomerised adenine 52
tautomerised cytosine 55
tautomerised guanine 54
tautomerised thymine 53
tautomerism 50-5, 126, 348-9, 353, 355
tautomerism in diploid somatic cells 348-53
taxonomists 372
Taylor, A.L. 146
TDF 541, 545
TDF – 1A1, 1A2 regions 538
TDF base sequence 545
TDF fragment 535
TDF gene in humans 539, 542, 544
TDF male determining loci 543-4
TDF region of the Y chromosome 537-8
telomere 296
telophase 293, 298
temperature and DNA breakage 481
temperature and frogs 400
temperature control in making 'puffs' 446-7
template 126, 238, 267-9, 423
tephritid fruit flies 381
termination codons 88-9, 112, 116, 120-1, 123, 128, 537
tertiary polypeptide structure 99-100, 343
Terzaghi, E. 86
testes 277
testes derived pY53.3 human RNA analogues 544
testes test 543
testicular determining factor (TDF) 534-5, 537
testicular development 542-3
testis chords 546
testis development 546
testis mass 547
testis mother cells 138
testosterone 473, 476, 547
tetramers 364-7
thickening of chromosomes 271
Thoman, M.S. 146
three base mutations 83-5, 342
threonine 67-70, 95, 128, 536
thrum primroses 411
thylakoid 481
thymidine 250
thymine 40-1, 44-6, 48-55, 81, 86, 103-4, 238, 249-50, 483
thyroxine 473-4

667

Ti plasmid 257-8
tigers 418
time intervals for interrupted mating 62-6
tobacco flowers 194
Tobin, C.J. 503
tolerance to copper in Agrostis tenuis seeds and vegetative parts 418-21
tortoises 32
totipotency 258, 444-5, 472
totipotent cells 257, 391
toxic free radicals 480-1
trans cinnamic acid 480
transamination 79
transcribed strand of DNA 87
transcriptidase (AMV) 238
transcription:-
  affected by cross-linkages 483
  control during 446-7
  controlled by non-histone proteins 450
  controlled by the TDF base sequence 534
  controlled by ZFY genes 534, 545
  /DNA relationship 83-8
  essential for gene expression 446, 449, 472
  from retroviruses 433-4
  from viral DNA 431
  from viral RNA 431-2
  homeobox 455
  in Crick's Central Dogma 126-7
  in differentiated gene expression 467-8
  in E. coli 122, 160
  in polypeptide synthesis 50, 91, 102, 105, 122
  in ribosome synthesis 135
  in transposons 427
  in upstream and downstream genes 452
  in virinos 429
  in X and Y chromosomes 537
  increased by plant hormones 479
  problems in genetic engineering 245
  /redundant DNA relationship 150, 162
  reverse, in the synthesis of DNA 105, 126, 229, 237
  splicing 153
transfer of genetic material from one species to another 437
transfer RNA (tRNA):-
  /amino-acid attachment in chloroplasts 214
  /amino-acid complex 115-7, 119-20
  associations 111, 214
  chloroplast 213-4
  control by polymerase 451
  in eukaryotic polypeptide synthesis 125-6
  in glycopolypeptide synthesis 128
  in prokaryotic polypeptide synthesis 122-3
  in Saccharomyces cerevisiae 134
  in Xenopus laevis 132
  mitochondrial 211
  origins of RNA viruses 430

shape 110
stimulated synthesis of 474, 476
transformation 33-9, 245, 248, 251-2, 254, 257, 440-1, 489-96, 502, 505
transformed bacteria 219
transformed cells 486
transforming:-
  cDNA 241, 246
  chromosome 248
  DNA 244
  DNA fragment 36-9
  gene 241
  human oncogene 506
  SRY gene 546
translation:-
  control 448-9, 455, 543
  for gene expression 472
  in bacteria 241
  in Crick's Central Dogma 126-7
  in eukaryotic cells 124, 244
  in genetic engineering 234, 236-7, 241
  in polypeptide synthesis 50, 88, 91, 102, 105, 111, 113, 126, 131, 139, 162
  in prokaryotic cells 113-9, 121, 123
  in relation to an inducer, a repressor, an operator and a gene 159
  in the synthesis of light chain antibodies in mice 467, 469
  in upstream and downstream genes 452
  of DNA viral genes 431-2
  of DNA virino genes 429
  of retroviral genes 433-4
  of RNA viral genes 432-3
  of transforming genes 242-3
  of undamaged copies of genes 482
  problems of, in genetic engineering 246, 252
  screening of products 244
  sequence determination of amino-acids in polypeptides 340
  splicing before 468-9
translocation 152, 256, 313, 325, 326, 370, 376, 426, 465
translocations of genes for antibody synthesis 464-9
translocations of genes that cause cancer 489
transplants in humans 528
transplants of nuclei 453
transport across membranes 472
transport of ribosomes 137, 139
transposons 153, 426-7, 430, 435, 438
transvestism in butterflies 415
triplicate codon table 89
triplicate DNA codes 83-6, 91-2, 126
triploidy 145
trisomy 511, 524
Triticum aestivum 332-3
Triticum monococcum 332
tRNA *(see transfer RNA)*

'true' standard deviation 590-92
trypsin/giemsa staining 518-9, 524
tryptophan 57-68, 78, 80, 94, 211
tryptophan synthesis enzymes 101
Tschermak, E. 22
Tsingita, A. 86
t-tables 590-1, 594-5, 608
tubers 209
tubules (seminiferous) 279, 547
tumour inducing plasmid (Ti) 257
tumour location 255
tumours 258, 486-7, 502
Turner's syndrome 522
tunicate ascidian 453
two adjacent base changes 83, 342
two cells from one 275
tylosis 550
types of genes changed by mutations 317-8
tyrosine 95, 536
tyrosine tRNA 119

U

UAA termination codon 89, 119
UAG termination codon 89, 119-20
UGA termination codon 89, 119
ultimate upstream gene loci 533, 548
ultrabithorax homeodomain 457
ultrastructure of nucleoli 132-3
ultraviolet illumination of chromosomes 518-9
ultraviolet radiation 319-20, 342, 374, 449, 481, 498
unbalanced hybrid chromosomes 333
unbanded allele in snails 202
unbanded snail shells 204
unblocked DNA strand 108
uncertainties of genetic engineering 243
uncertainties of systematic classification 383, 402, 404
uncoiling of the nucleolus organiser 138
uncontrolled orientation of DNA 246
unequal distribution of chemicals at cell division 311
unequal distribution of chemicals in eggs 453-4, 459-61
unequal migration of broken chromosomes 321
unequal plasmid distribution at mitosis 257
unequal reciprocal translocations 326-7
uneven number of chromosome sets 144-6, 168, 274, 277-8, 286-7, 299, 301, 310, 331
unfertilised eggs 453
unfolded haemoglobin 99
unimodal maximum of butterfly wing spots 409
uninterrupted 1A2 exon 536
uninterrupted pY53.3 exon 543

unique chromosome sets at meiosis 286-7, 31
unique diploid chromosome sets at meiosis 285
unique genetic composition of gametes 287-8, 310
unique mutations 396
unit evolutionary time 357
University of Oregon 86
University of Osaka, Japan 86
unmethylated C = G pairs 537
unpaired DNA bases 107, 268
unpaired RNA bases 104
unpalatable species 414
unsuitable gene combinations 379
unusual environmental factors 32
unwinding of DNA 266
unzipping of paired bases 268
upper limits of mutation rates 316
Uppsala University, Sweden 496
upstream genes 452, 493, 545
upwind Agrostis tenuis plants 419
uracil 81, 86, 103-4, 112
useful properties of viruses 439
u/v radiation 319-20, 342, 374, 449, 481, 498

V

V set of genes for antibody synthesis in mice 465-8
vacated operator 159
vaccination 254, 260-1
vaccines 260, 440, 501
vacuolation 215
valine 94, 342, 506
Valeriana spp. 334, 377
Van Benenden, E. 22
Vandiemenella viatica 382
Vane-Wright, R.I. 413
variability 376, 392
variable amino-acid sites in polypeptides 346
variable genes 374
variation 288, 315, 390-1, 423
variation:-
  creation of 156, Chapter 10
  in diploid number of chromosomes 329
  in DNA 51
  in embryonic development 463
  in genotype 277-88, Chapter 10
  in plasmid inheritance 218
  in population numbers 394
  multifactorial 404
variegated plants 223
varieties 51, 374
variety 322, 369
vas deferens 547
vectors (of microorganisms) 378, 386
vegetal pole 454
vegetative cell divisions 256

vegetative cells 225-6
vermillion eye colour 56
vertical transmission of viruses 494
veterinary uses of genetic engineering 231
viability:-
  of chromosomes after centromeres change positions 326-8
  of chromosomes after structural change 322-330
  of hybrids 374
  of inversions 323-5
  of new mutations 394
  of phenotypes 354
  of positional changes of centromeres 326-8
viable:-
  alleles 319
  chromosomes 326
  deletions 322
  genes damaged by mutations 312
  hybrids 377, 384, 386-7
  mutations 314, 317, 320, 370-1, 391-3, 396, Chapters 10, 11, and 12
  sex chromosome crossings-over 540
  structural changes to chromosomes 322-30
  super-alleles 319
viral:-
  assembly 435
  base sequences 492
  DNA 150, 152, 252, 435, 481, 499
  DNA that causes cancer 488-9, 492
  genetic engineering 261
  genome 252, 259
  oncongenes 488-9, 492
  plasmid equivalents 252
  polypeptides 258
  RNA 435, 499
  RNA that causes cancer 489
virus:-
  /antibody complex 259
  B-type 494
  C-type 494
  φ X 174 149
  genetic engineering 440
  hybrid particles 503-6
  SV 40 493
viruses 20, 40, 86-7, 92, 126, 151, 209-10, 258-9, 261-2, 379, 426, 429-30, 435, 499, 501, 562
viruses and cancer 489-497
viruses and genetic engineering 235
visual approach to genetic crosses 166
vitamin deficiency 56
vitamins 56, 481
Vogt, R. 501

W

Wallace, A. 21, 23, 31, 33, 369

walnut comb shape in poultry 189-90
warning (aposematic) colours 415
Wasteson, A. 496
water as a polar molecule 98
Waterfield, M.D. 496
watery medium and the natural selection of polypeptides 98
Watson, J.D. 41, 47, 146
Watson-Crick DNA model 40
Watts-Tobin, R. 80
weak chemical bonds 45, 50-5
Weinberg, R. 503
Weissman, A. 22
wetting (and drying) of DNA 320
wheat germ 449
wheat 331-3
wheat/grass crosses 332
whiptail lizards 331
white:-
  blood cells 258-62, 440-1, 483, 486, 498
  blood cells ingesting cancer cells 486-7
  coat colour in mice 191-2
  eye in D. melanogaster 185-8
  feathers in chickens 198
  leghorn chickens 198-200
  mice 192
  snapdragons 196
  spores in Sordaria brevicollis 303
  sweet pea flowers 193
  tobacco flowers 194-5
  Wyandotte chickens 198-200
Whitehead Institute for Biomedical Research 534
Whittle, N. 496
whole cell insertions for genetic engineering 253
widely separated bacterial genes – mapping of 56-66
widely separated chromosomes 310
widespread polypeptides 357
Wigler, M. 503
wild grasses 332-3
wild-type 57-9, 84-5, 147, 171, 252, 302, 355, 549
wild-type:-
  alleles 171, 175, 232, 230, 317-8, 343-9, 356, 516
  apparent reversion to, in sweet peas 192-4
  bacteria 57
  Drosophila embryos 463
  homozygote 548
  point mutations becoming established in cell lines 348-53
  Sordaria brevicollis 302
  symbol 175
Wilkins, M. 41, 47
wing shape in D. melanogaster 171
wing spots in butterfles 404, 406, 409

wobble hypothesis 110, 125-6
Wollman, E. 56
women's age and chromosome aberrations 519
worlds's atmospheric oxygen 212
wound cells in humans 496-7
'wrong' base pairing 50-5
'wrong' base tRNA 125-6
'wrong' bases 125, 318
'wrong' cDNA 242
Wyandotte chickens 198-9

X

X and Y chromosome exon similarities 536-7
X and Y chromosome homologous regions – divergence 541
X chromosome 185, 205, 521, 523-4, 528, 533, 536-7, 539-42, 545-7
X chromosome boundary 542
X chromosome fragments 523
Xenopus laevis 131, 139, 216, 450, 453, 456-7, 460
Xg^a allele 518
XO females 522
X-ray diffraction 41
X-ray plates 242
X-rays 481, 498
XX females with deletions 522
XX humans that develop male gonads 538-9
XX male mice 545-7
XX males 544
XX zygote transformation using Sry 547
XXY males 521
XY humans that fail to develop male gonads 533-4
XYY males 522

Y

Y chromosome 150, 205, 322, 331, 524, 532-3
Y chromosome – human 150, 532-3, 536-42
Yale University 88
yeast 174, 211, 218, 222, 234, 456, 534
yellow snail shell allele 202
yolk concentration of chemicals 454

Z

Zaire basin pygmies 373
Zea mays 316
ZFX gene 533-4, 537
ZFY gene 533-4, 537, 544, 548
ZFY gene – absence of 539, 542-8
ZFY gene and spermatogenesis 537-8
ZFY gene locus 533-4
zinc cations 535
zinc finger polypeptides 160, 452, 533
zinc fingers 160, 452, 533, 535-6, 545
zones of hybridisation 387-8
zygote 168, 213, 226-7, 253, 270, 277-8, 297, 354-5, 378, 392, 441, 444, 472, 546